George Dadd

The American reformed horse book

A treatise on the causes, symptoms, and cure of all the diseases of the horse

George Dadd

The American reformed horse book
A treatise on the causes, symptoms, and cure of all the diseases of the horse

ISBN/EAN: 9783337143060

Printed in Europe, USA, Canada, Australia, Japan

Cover: Foto ©berggeist007 / pixelio.de

More available books at **www.hansebooks.com**

THE

AMERICAN REFORMED

HORSE BOOK.

A TREATISE ON THE

CAUSES, SYMPTOMS, AND CURE

OF ALL THE DISEASES OF THE HORSE, INCLUDING
EVERY DISEASE PECULIAR TO AMERICA.

ALSO EMBRACING FULL INFORMATION ON

BREEDING, REARING, AND MANAGEMENT.

DESIGNED FOR POPULAR USE.

By Prof. GEO. H. DADD, V. S.,

Late Prof. of Anatomy and Physiology in the Veterinary Institute of Chicago,
Author of the " Anatomy and Physiology of the Horse," etc.

ELEGANTLY ILLUSTRATED.

NEW YORK:

ORANGE JUDD COMPANY,
751 BROADWAY.
1889.

PREFACE.

AMERICANS rank high in the estimation of the scientific world for their genius and industry in the great work of intellectual, moral, and scientific improvements. See what has occurred in the arts of printing, electrotyping, etc. The eloquence of the great literary magicians of the New World, the honored graduates of free institutions and free schools, no sooner falls from their lips than the phonographist transfers it to paper, the nimble fingers of the compositor puts it into "form;" next the "battery" deposits on the face of the type a more durable coating of copper, and "Hoe's press" sends off "impressions" with almost magical rapidity. No less surprising are the wonderful feats performed by American mechanics, in the construction of fast yachts and steamers, which often pluck a laurel from the so-called "Mistress of the Sea;" and old "Uncle John" gives us due credit for perfecting a race of fast trotters that can beat the world of horse-flesh.

Improvements in every department of Science and Art are constantly treading on the heels of improvement; yet, as regards the science of Veterinary Medicine and Surgery, America seems almost the last among civilized nations to put her shoulder to the wheel. What can be the reason? Her youth may be offered as a plea for excuses. The daughter of the Old World has not yet, in this department, arrived at years of discretion; she is yet in her teens. Her sons have borne the heat and burden of the day in establishing their liberty, in founding cities, extending commerce, and in taming the wild face of soil, by uprooting the giant oak, and in clearing away the forest, so that the husbandman, with plow and harrow in hand, might follow on, and next impregnate mother earth with germs for fructification, which, in the day of the harvest thereof, might be gathered and garnered for the benefit of man and the inferior orders of creation.

And long ere the woodman's ax ceases to vibrate its clear stroke-notes in the ears of the new settlers, a giant enterprise is in contemplation. Space and distance must be comparatively annihilated; the iron horse, with his lungs of steel and breath of steam, must have his highway and byway, and the result is, the choice spirits of progression have been earnestly engaged in dotting a vast surface of United States Territory with a magnificent net-work of railway and telegraphic wire; and various other enterprises, with which the reader is familiar, have so occupied the American mind, that the Science of Life, as it applies to animals, has been almost at a stand-still.

In the prosecution of these objects, so essential as they have proved in the economy of an infant republic, it is not surprising that our science should fail to receive attention, and that America should be in the rear of the veterinary squadrons of the Old World; yet, notwithstanding this seeming indifference, veterinary science does, in some parts of this country, occupy as high a platform as that which obtained in England, about seventy years ago, when the Apostle of Mercy, St. Bel, first landed on British soil.

The science which we here advocate is as valuable and charitable to nature's menials, for the purpose of ministering to their physical wants, relieving their aches and pains, and of prolonging their lives, as that practiced on the more noble of created life; and, at the present day, testimony can be produced going to show that among us can be found "good Samaritans" who are always ready to minister to the wants of the inferior orders of creation— a work as acceptable to the God of Nature, and as creditable to manhood, as when the welfare of one of us is concerned. And should we estimate the science in exact ratio to the value and usefulness of the legion host of domestic animals that have been reared in this country, and those which, without regard to cost, have been imported, to whose welfare it directly contributes, the reader will readily perceive that it is entitled to the consideration and support of a nation of husbandmen.

It must be borne in mind that a knowledge of the principles of veterinary medicine can not be acquired without diligent study and close application. It is only the property of the industrious and devoted student.

In order to acquire the necessary tact, skill, and experience for the practice of our art, some of the shining lights—the early disciples—have spent the May-day of their existence in solving its problems, and, when near its close, have declared, with a modesty indicative of true genius, that their education was yet incomplete.

Such testimony goes to show that there is no republican nor royal road to veterinary knowledge. In fact, after a man has perseveringly pursued the routine studies of the most popular collegiate institutions, he may still be found " wanting."

Let us contemplate for a moment the superior advantages which human practitioners have over our craftsmen, both as regards their scientific qualifications and the superior means at their command for ascertaining the character, location, and intensity of the various maladies peculiar to the human race. For example, the practitioner of human medicine, if he has been a faithful and industrious student, enters upon the active duties of his profession, having acquired a valuable fund of knowledge of anatomy, physiology, pathology, and other necessary branches of study; and, having disciplined his mind in the regular school, and in that of experience, he is expected to be able to rejoice in the possession of a well-trained mind, which enables him to understand and interpret the physiological laws which preserve health and life in the constitution of organized beings; also to comprehend the why and wherefore of disease; to be able to institute sanitary and other regulations, and to select suitable medicinal preparations, in view of meeting the various indications of each and every form of disease. And when a man becomes the subject of sickness, and the doctor is called upon to exercise his skill, the latter receives valuable aid, in view of making a correct diagnosis, from the patient and his friends. Each can be questioned, and their responses throw considerable light on the history and nature of the malady. Now, as regards the latter advantages, they have no parallel with the men of our craft. Our patients are deprived of the power of speech, and we can only judge of the state of their health, and the nature of their maladies, by signs revealed or elicited through physical exploration. We have other difficulties to encounter of no less magnitude. For example, our patients are often located in situations unfit for a *well* animal to reside in, and they do not

always receive that attention from their owners which the urgency
of the case demands.   In fact, the good nursing, kind attention,
etc., which often proves of so much value in the restoration from
sickness of one of our race, is often denied the inferior creature;
and other matters, too numerous to mention at the present time,
tend to show the advantages which our brethren of a sister science
have over us, and how necessary and important it is that a practitioner of our art should be thorough master of his profession.
It is a lamentable fact that the great majority of our husbandmen have not hitherto realized that the superstructure of veterinary science rests on the same intelligent basis as that of human
medicine; but such is really the case.   In the language of the la-
mented PERCIVALL, we are assured that " whether we prescribe for
a man, horse, or any animal, the laws of the animal economy are
the same; all require the same treatment," that is to say, the
propositions for restoration are all founded on the same intelligent
basis.

The importance, therefore, of educating men for the profession,
and of furnishing reliable works for study, is very evident; and
the necessity is further illustrated in the daily experience of those
persons who, as a matter of necessity, are obliged to do the best
they can for the sick and dying of their flocks and herds, having
no knowledge of those advantages which a thorough course of
training affords.   Such persons must, necessarily, feel that they are
groping in darkness, and when a ray of light does flit across their
medical path, it only serves to make them better acquainted with
their own want of skill.   Persons thus circumstanced, unable to
procure the services of a good veterinary surgeon, in consequence
of a scarcity of the "genuine material," are often compelled to
assume the responsibility and risk of medical attendant; and, no
doubt, they prescribe with good intention and honesty of purpose,
but disease and death has the vantage of them, for they know not
the modes of pathological warfare, nor are they acquainted with
means best calculated to insure a favorable issue; and they will
certainly hail, as a great blessing, any and every effort to improve our veterinary literature, and thus diffuse knowledge so
sadly needed.

It is well known that prior to the introduction of a rational
system of veterinary medicine in the mother country, millions of
domestic animals were annually sacrificed at the shrine of igno-

rance; but science came to the rescue, and now some of the disciples of St. Bel, Coleman, their co-workers and kindred spirits, can, by the aid of their vast materia medica, their anæsthetic agents, scalpel, etc., accomplish unheard of wonders. We are now working to dispel the awful cloud of ignorance and superstition which has too long lingered around the stable and barn. We aim to illuminate the dark spots that have existed for many hundreds of years, and to obliterate the false theories that have been handed down to us by the Egyptians and Arabians, and the modern jugglers of this science.

There never was a period in the history of the United States when the services of educated veterinarians were so much needed as at the present time; for the live stock of this country do not enjoy immunity from those pestiferous epizöötic maladies which have formerly operated as a withering simoom on the enterprise of British husbandmen in the breeding and raising of live stock. For example: the people of the Western States complain that a disease occasionally makes its appearance among cattle, to which they have applied the name of "trembles," or "milk sickness," and it has so scourged both the superior and inferior orders of creation, that the former have often abandoned the old homestead, in view of seeking a location where there seemed to be some prospect of enjoying immunity from the pestilential scourge. A contagious and infectious disease often prevails among swine, carrying them off by hundreds and by thousands, yet many of us are in the dark regarding its cause, nature, and treatment. Typhoid affections, puerperal fever, apoplexy, and dropsy of the brain are just as prevalent here as in England. Miscarriage or abortion is fearfully on the increase. Diseases of climatic origin are more rife in this country than in England. This is, perhaps, owing to the diversity which exists in the climatic temperature of our vast territory, and to our various faulty modes of management. In fact, there is scarcely any disease known to veterinarians of the Old World but that has prevailed in the New.

We must have reliable text-books and educated surgeons, in order to understand the nature and treatment of the diseases incidental to domestic animals. It is not only a matter of national, but of individual, policy and interest; and should we view the matter with the eyes of business men, we shall see that such enterprise must surely pay.

Persons who have paid any attention to the rise and progress of veterinary science in this country, are painfully aware of its great imperfections, and the author's object in preparing this work is to endeavor to lessen and remove them, by giving the reader the benefit of an experience which extends through a period of twenty-five years.

In view of furnishing reliable information, and of showing how little of medicine is required for the treatment of various forms of disease, a number of cases, recorded in the author's note-book, are introduced in various parts of the work. These cases may, perhaps, also give the unprejudiced reader juster views of the relations of nature and art to diseases; for it is a notorious fact that very many well-instructed persons of all classes have hitherto exercised a blind faith in the medical art for the cure of disease, when it is a fact well known to those who practice rational medicine that Nature possesses vastly greater powers than Art in curing diseases. What is here meant by nature, is the conservative power inherent in the living body. For a better understanding of this subject, the reader is referred to the works of Sir JOHN FORBES, OLIVER W. HOLMES, and Professor BIGELOW.

In view of aiding nature in the cure of disease, the author has introduced to the reader's attention a new class of remedies, viz.: Fluid Extracts of a sanative character. They have proved more safe and efficacious in the practice of rational medicine than all the other heroic arms of physic.

These remedies have been carefully tested for several years by the author, and those students who have, from time to time, placed themselves under his instruction, and the result has been very satisfactory; otherwise, they would not be recommended in this work.

Finally, the author feels it due to himself to state that the almost constant occupation of his time, professionally, has given him less opportunity than he desired for the production of this work, yet he entertains a hope that he has not labored in vain; and thus this mite is cast into the common treasury of Veterinary Science.

G. H. D.

CONTENTS.

SKELETON OF THE HORSE.

FROM A CELEBRATED ANATOMICAL MODEL.

(Engraved expressly for this work.)

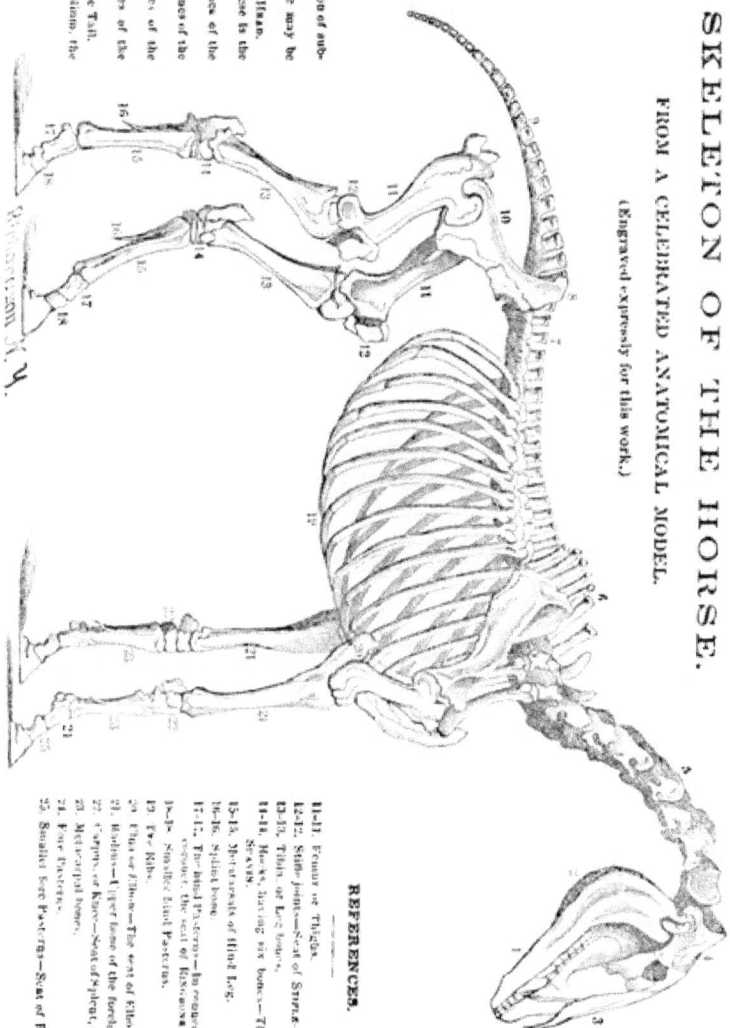

REFERENCES.

1. Lower part of Jaw-bone.—Location of sub-maxillary gland.
2. Upper part of Jaw-bone.—Where may be found the Pulse.
3. Skull and Face.—The seat of Big Head.
4. Parietal bones.—Just back of the seat of Poll-Evil.
5. Seven Cervical Vertebræ, or bones of the neck.
6. Eighteen Dorsal Vertebræ, or bones of the back.—The seat of Fistula.
7. Six Lumbar Vertebræ, or bones of the Loins.
8. Five Sacral Vertebræ, or bones of the Haunch.
9. Caudal Vertebræ, or bones of the Tail.
10. The Haunch, consisting of the Ilium, the Ischium, and Pubis.

REFERENCES.

11-11. Femur or Thigh.
12-12. Stifle joint.—Seat of Stifle-Out.
13-13. Tibia, or Leg bone.
14-14. Hock, having six bones.—The seat of Spavin.
15-15. Metatarsals of Hind Leg.
16-16. Splint bone.
17-17. The hind Pastern.—In connection with 16, the seat of Ringworm.
18-18. Small or Hind Pastern.
19. Toe Bone.
20. Elbow or Elbow.—The seat of Elbow Tumor.
21. Radius.—Upper bone of the Forelet.
22. Carpus, or Knee.—Seat of Splint, or Splint.
23. Metacarpal bone.
24. Fore Pastern.
25. Smaller Fore Pastern.—Seat of Ringbone.

THEORY AND PRACTICE

OF

VETERINARY MEDICINE AND SURGERY

SECTION I.

ON THE CAUSES OF DISEASE.

Remarks on Veterinary Science, and the Relation it bears to Social Science—Veterinary Science; how to Inaugurate it—What Benefits have we derived from its Study—Inflammation; its Nature and Treatment.

To explain the cause or causes of disease occurring among domestic animals is an herculean undertaking, when expounded by the theory and faulty logic as applied to the ancient method of reasoning on "false facts;" but, thanks to the genial spirit of progression and improvement, we are enabled to throw some light on the subject. We have been taught to believe that many of the causes of disease are invisible and mysterious; but this is not the fact. We are in possession of evidence going to show, very conclusively, that the causes of disease may be classified very distinctly and intelligently under two heads, viz.: Predisposing and Exciting; yet this classification is not offered by the author as absolute, but is a more convenient arrangement than any other to study the multifarious conditions which people believe are remotely or immediately concerned in the production of disease. For example, many interesting and valuable facts are on record, going to show that the peculiar properties, characters, and qualities of the parent—whether good or bad, healthy or diseased, external or internal—are transmitted to the offspring; or, in common phraseology, are predisposing and hereditary.

"Among horses and cattle, we find, as in the human subject, ample illustration of the hereditary tendency of external form, disposition, habit, and disease. The parent transfers to its offspring size, shape, and general conformation similar to its own; and the aphorism, 'like produces like,' is as applicable to faulty and disproportioned as to beautiful and symmetrical form, to diseased and debilitated as to healthy and vigorous constitution, to gentle and tractable as to fiery and indomitable disposition. The size, weight, general appearance, expression of countenance, fleetness, and temper of the horse are all hereditary. Many illustrations might be given of particular families being remarkable, during several generations, for good or bad *points*, as for well or ill-formed head; for high and well-developed or for low and weak withers; for fine, strong, and well-turned, or for coarse, weak, and ill-formed limbs. Peculiarities of color often extend through many generations, and are so constant in their transmission as sometimes to form one of the distinctive characteristics of a race. Indeed, most breeds of horses have a prevailing color, to which there are few exceptions. The heavy horses of Lincolnshire, for example, are generally of black; the Cleveland, bay; and the wild horses of the plains of Eastern Siberia, dun. Particular markings, also—as white spots on various parts of the body, stars and blazes on the face, one or more white feet or legs—often continue for many generations, and are peculiar to certain families.

"There are some maladies in which it is comparatively easy to trace the connection between conformation and disease. In the horse, certain sorts of limbs notoriously predispose to certain diseases. Thus, bone spavins are most usually seen where there is a disproportion in the size of the limb above and below the hock; curbs, where the os calcis is small and the hock straight; strains of the tendons of the fore-leg, where the limb is round and the tendons and ligaments confined at the knee; and navicular disease, where the chest is narrow and the toes turned out. Among horses so formed, these diseases are unusually common, and are developed by causes which would be quite inadequate to produce them in animals of more perfect conformation. But it appears to us that internal and constitutional hereditary diseases also depend upon the altered conformation or texture of the parts specially affected, or of some disturbance of the relation which should subsist between the different constituents of these parts. This abnormal state of the internal parts is seldom within the limits of our means of observation or investigation; but its existing in animals having an hereditary predisposition to disease can not, we think, be doubted, as we shall now endeavor to show. The ground of our reasoning rests chiefly on the analogy which subsists, in all respects, between external and internal parts. The same law which regulates the hereditary transmission of form, texture, and relation of external and visible parts, also operates with equal force in regard to the form, texture, and relations betwixt the component parts of internal, and, it may be, inaccessible to ordinary powers of investigation. Then if, as we have shown, external hereditary diseases, such as lamenesses, are traceable to external hereditary peculiarities of conformation, we do not think it pushing our analogy too far in asserting that, in like manner, internal hereditary diseases must, in great part at least, depend upon some inherent hereditary peculiarity of the internal parts affected."

The common disease known as bone spavin is inherent, or dependent on predisposition, as related above; for the author never knew a horse to be the subject of this malady unless he had faulty hocks; or, in other words, in so far as the conformation of the parts were concerned, predisposition was evident. The same remarks apply to ringbone. Either the pasterns are too upright, thereby causing jar and concussion, or they slant unnaturally, and the consequence is, strain of the ligamentary structures, ending often in ringbone and other osseous diseases.

Ample evidence can be adduced to show that various diseases of the eye are hereditary. The tendency to ophthalmia shows itself in a peculiar conformation of the eye, and parts in the vicinity of it. Periodical ophthalmia is notoriously hereditary, and usually ends in cataract.

Colic, also, can be traced to inherent tendencies; for animals subject to flatulent colic are known to have a morbid appetite; they have, also, a large, roomy abdomen, which gives them an ungainly appearance. Yet it is true that diseases may occur accidentally, without the intervention of predisposition; but predisposition exists in almost all cases, and it only requires some anatomical and physiological knowledge on the part of husbandmen to detect it.

F. Dun, a lecturer in the Edinburgh Veterinary College, tells us that diarrhea and colic are, to a certain extent, hereditary, "inasmuch as they are very prone to attack horses of particular form and constitution, as those with narrow loins, large flat sides, and of what is generally termed a *washy* appearance. If such animals be overworked, (especially soon after being fed,) if their food be suddenly changed, or if they be allowed an unusual quantity of fluid, they are almost certain to be attacked either by purging or colic. The tendency to these diseases appears, in such cases, to depend on a want of adjustment among the different organs of the body ; a want of balance among the different functions of digestion, circulation, and respiration.

"Many farm-horses, as well as others without much breeding, are remarkable for consuming large quantities of food, for soft and flabby muscular systems, and for round limbs containing an unusual proportion of cellular tissue. These characters are notoriously hereditary, of which indubitable evidence is afforded by their existence in many different individuals of the same stock, and their long continuance, even under the best management and most efficient systems of breeding. Such characters indicate proclivity to certain diseases, as swelled legs, weed, and grease. If horses of this description stand long, the circulation of the blood through the limbs is retarded; for, as the contraction of the muscles which materially aid circulation are wanting, the blood in the veins rises with difficulty against its own gravity, while the soft and lax condition of the venous coats, and of the muscles in contact with them, permits the passage of the fluid parts of the blood, giving rise to a serous effusion, which is soft, and pits on pressure. This anasarcous condition, although troublesome, and frequently recurring, is easily removed by friction, exercise, or a little physic, and does not unfit the animal for ordinary work.

"But the same conformation and constitution which induce simple swelled legs also give rise to the more serious affection known as *weed*, or a shot of grease. This consists in a disturbance of the balance which naturally subsists between the waste of the system and the supply of new material to repair that waste. Food is assimilated in larger quantity than the wants of the system require; the chyle so formed accumulates in the absorbent vessels and glands, which become, in consequence, irritated and inflamed. That part of the absorbent system situated in the hinder extremities is usually the principal seat of the disease. The animal suddenly becomes lame; the inguinal and other glands in the groin become enlarged and very painful, and the swelling and pain gradually extend downward along the course of the absorbents, while the limb becomes a great deal larger than its natural size. There is, at the same time, a good deal of constitutional fever, with a full and bounding pulse. The swelling of the leg is, in the first instance, inflammatory, being hot and tender, and the skin, over the part affected, hard and tense. Such swellings may, by judicious treatment, be removed; but, in cases of a chronic character, or where the same limb has been previously affected, lymph is effused, forming hard and nodulous, and even diffuse swellings, which often cause lameness, by interfering with the motions of the joints or tendons. These indurated swellings must be carefully distinguished from the serous effusions above noticed, which, although giving the animal an unsightly appearance, do not materially impair his usefulness.

" *Grease* consists in a morbid condition of the sebaceous glands of the horse's heels and fetlocks. It occurs in various degrees of intensity; sometimes as a mere scurfy itchiness of the skin about the fetlocks, more commonly of the hind extremities; sometimes attended with much inflammation, causing great heat, pain, and swelling, and an ichorous fetid discharge; sometimes causing falling off of the hair about the heels, and the formation of deep cracks and fissures; and sometimes becoming so violent and inveterate as to cause eversion of the sebaceous glands, formation of granulations, and secretion of pus, constituting the loathsome complaint termed the *grapes*. There are few diseases better deserving the epithet of hereditary than grease, and few in which the hereditary nature can be more easily discovered and traced. Almost every practitioner can bring to his recollection cases showing the tendency of this disease to descend from parent to offspring. A friend of mine, some years ago, purchased a valuable four-year old entire horse, adapted for agricultural purposes. When bought, he appeared perfectly sound, and his limbs were nearly black, well-formed, and fine; within a short time, however, they became thick and greasy. And, although the mares to which he was put were perfectly free from such faults, the progeny have shown, in every case where they can be traced, unmistakable evidence of their inheriting the greasy diathesis of their are. They have all been found liable to swelled legs when they stand idle for a few days; most of them have been the subjects of repeated attacks of weed; all are affected, particularly in spring, with scurfiness of the skin of the hind extremities and excessive itchiness, and lose, at a very early age, their flatness and smooth-ness of limb. The faults occur, to a greater or less degree, in all the stock of this horse, by many different mares, and are dis-tinctly traceable to the third generation. But, although grease is undoubtedly hereditary, and is, therefore, readily induced by comparatively simple causes, still it is frequently caused, and is always aggravated, by neglect of cleanliness; and of this there is ample evidence in the fact that it is most common in foul and badly-managed stables, and where no pains are taken to keep the horses' feet and legs clean and dry."

The scrofulous predisposition is very marked in certain breeds of horses; it occasions rickets, softening, deformity, and various forms of disease in the bones, as, for example, big head, big jaw, etc. The same author, just quoted, says in reference to scrofula:

" From their weak and unsound constitution, horses of a scrof-ulous diathesis are unusually prone to *glanders and farcy* —two forms of a disease peculiar (at least as an original disease) to the equine species. As has been already remarked, it is characterized by a specific unhealthy inflammation, identical in all important characteristics with the syphilitic inflammation in man. From the dire and loathsome nature of glanders, and the terror in which it is held, animals affected by it are never used for breeding, so that we have little opportunity of judging of its hereditary nature. There is no evidence (so far as I know) which proves it to be di-rectly hereditary,[*] but there is no doubt that the progeny of a glanderous horse would exhibit an unusually strong tendency to the disease. Its ordinary predisposing causes are, many of them, hereditary; it is very prone to attack animals of a weak or vitiated constitution. It is emphatically *the* disease which cuts off all horses that have had their vital energies reduced below the healthy standard, either by inherent or acquired causes. Glan-

---

[*] "Though I am not aware of any facts proving glanders to be congenital, yet I think there is every probabili y that such is the case; for it is notorious that syphilis, the analagous disease in the human subject, is congenital, and often appears at birth in the children of women affected by that disease."

ders is, also, sometimes caused by inoculation; is frequently produced in healthy subjects by mismanagement, as by insufficient food, want of shelter, and overwork; and often supervenes on bad attacks of influenza, strangles, diabetes, and other diseases which debilitate the system, or impair the integrity of any of its more important parts. These causes appear to possess the power of engendering in the constitution of a horse a peculiar poison, which, as it reproduces itself, and spreads to all parts of the body, gives rise to the characteristic symptoms of glanders, causing, sooner or later, a breaking up of the system, and a fatal prostration of the vital powers. This poison produces in the blood abnormal changes, which vitiate that fluid, and unfit it for healthy nutrition.* From the irritant action of the morbid fluids passing through them, the lymphatic glands and vessels become inflamed, and lymph is deposited. This, however, being of an unhealthy nature, soon runs on to softening, which extends to the skin overlying the part, and ulcerating farcy-buds are formed. On the surface of the more vascular mucous membranes effusions of tubercular matter are also poured out; these take on an unhealthy inflammation, and degenerate into chancrous ulcers, which may generally be seen on the mucous membrane of the nostrils in most bad cases of glanders.

"These are the most common scrofulous diseases of horses; but an animal of the scrofulous diathesis, besides being specially subject to these, is little able to withstand ordinary morbific causes, and hence is also unusually liable to many ordinary diseases; in such a subject, too, disease is very apt to be severe and complicated, and to be acted on tardily and imperfectly by all remedies."

Having now furnished the reader with a few valuable facts bearing on hereditary predisposition, we shall briefly discourse

---

* "A comparison of the two subjoined analyses will show the great difference in composition between the blood of healthy and of glanderous horses—a difference consisting chiefly in a diminution of the red corpuscules, and a proportional increase of the fibrine and albumen.

	Blood of Healthy Horse.	Blood of Glanderous Horse.
		A.	a.
Water	804.75	842	859
Fibrine	2.41	8.07	8.7
Blood corpuscules	117.13	58.30	44.29
Fat	1.13 }	76.30	82.27
Albumen	67.85 }		
Soluble salts	6.82	5.04	5.38

in reference to some of the exciting causes which in general produce those diseases to which the horse, by virtue of his constitution, is liable. The following quotation, brief and pointed, will be found pertinent :

"By far the easiest and best mode of curing diseases of animals is to keep the diseases and the cattle separate. With cattle, as with ourselves, disease is caused, in most instances, by *an excess* of carefulness or of carelessness. Too much and too rich food is as injurious as too little and too poor. Perhaps fewer diseases, in proportion to the numbers treated in each way, spring from privation than from surfeit. Too little as well as too great exposure to extremes of heat and cold has a tendency to create disease ; in the one case it causes disease, in the other it renders man or animal more liable to it. Let a wealthy and a warm-hearted man own and take charge of a favorite animal, and the chances are that it will be ' killed with kindness ;' if it is tended, in part, by a hireling, the equilibrium may be restored. In our experience—no very limited one, by the way—*care* is more needed among cattle than cures ; and all the diseases which came under our treatment sprang from the want of judicious care. Has a cow a cough, has she the hoven, has she the red-water, or the puerperal fever? it is from *overcare*—too much warmth or too much feed. Is she mangy, lousy, hide-bound, or affected with horn-ail? the chances are that she has been neglected. Of course there are very many exceptions to this and to every other general rule. But this is our experience. Let *few* farmers, however, feed their animals less than they now do, or give them more airy lodgings than they now do, because the greater number of the cattle in this country suffer from privation and exposure to the inclemencies of the weather rather than from pampering. But animals, as well as men, will, with the best care we can bestow on them, become sick at times. Our judgment is often at fault when our intentions are the best. Sometimes our back is turned, or our eyes are not opened sufficiently wide. At all events, our stock *will* become sick, some of them ; and recourse must be had to medical remedies."

It is usual to classify causes ; yet, for all practical purposes, it is only necessary for us to know that exciting causes are those from which disease most directly springs. For example, suppose that in the act of shoeing a horse a nail is driven in a wrong direction, euters the sensitive tissues of the foot, and lameness occurs; the nail, in this event, may be classed as a special or mechanical exciting cause; and, under the same head, we class poisons and injuries of all sorts.

When horses become the subjects of glanders, or farcy, by being exposed to the maladies, they are then said to have taken it either by infection or contagion, and the cause is said to be *specific*. The terms infection and contagion are now used as synonymous, and, undoubtedly, if reference be had to the etymology of the words, the arrangement is correct. Coming now to plain matter of fact, the exciting cause of disease is that which, when taken into the body, or applied outwardly, is followed by disease or derangement.

The exciting causes of disease may, therefore, be summed up thus: Debility, induced by bad management, exposure, unwholesome food; plethora, brought on by over-feeding and want of exercise; overdosing animals with improper medicines and poisons; overtaxing the muscular powers of the animal, and the breathing of an impure atmosphere.

A great proportion of the diseases occurring among horses and cattle in the Western States is due to the fact that they are not properly housed at night, and the food that they often gather in their exposed situations lacks the nitrogenous and nutritious matter for the manufacture of good, healthy blood and strong muscles.

Disease sometimes, however, occurs when no exciting nor any other cause appears; yet, with the exception of those maladies supposed to be produced by miasmatic influences, it is probable that the cause, in such cases, has not been discovered. Finally, it may be said that the causes of disease among animals are within their bodies as well as around and about them.

Veterinary Science, and the relation it bears to Social Science.— The necessity which exists for the services of educated Veterinary Surgeons throughout the United States, and in cavalry regiments of the United States army, is evident to every thinking man; yet, prior to the advent of our late national troubles, it was found almost impossible to interest the right kind of men, so that they might use their influence for the purpose of directing the attention of the General Government to the value and importance of veterinary science; but the probability is that Congress will soon legislate on the subject, and before long we may expect to hear of the appointment of a Veterinary Surgeon and Assistant Surgeon to each cavalry regiment.

There exists no earthly reason why our profession should not rank as high in America as in England and France. The great losses which are continually occuring among army horses is a monstrous and growing evil, and, in view of preventing unnecessary diseases and premature deaths, the Government must organize a veterinary corps.

It is very important that our science should attract the notice of legislators; for it bears a very intimate relation with social science, inasmuch as men and animals are subject to similiar diseases, which are communicable one to the other, and which often arise spontaneously, and are transmitted from first to second. For example, there is a disease which often occurs among cattle, known as "Anthrax." It is a malignant disease, and many persons have lost their lives by absorbing the virulent "anthrax poisons" from the carcasses of dead animals.

A cutaneous disease often occurs among the bovine species, known as "ring-worm," which affects the hands and arms of some persons who attend such animals.

It is asserted, also, that the vesicular murrain, which at times prevails among cows, gives rise to the development of a virus which is often squeezed into the milk-pail, and produces very bad effects on children when fed on the same.

It has been discovered that the little vesicles found in measled pork are tape-worms, in a certain stage of development, and if not destroyed by thoroughly cooking the infected meat, may develop the dangerous parasite in the human intestines.

That awful disease known as virulent "glanders" (spontaneous in some horses) has destroyed many of our race. A small portion of the glandered matter coming in contact with an abraded or absorbing spot on any part of a man's body, will surely cause him to die the most horrible of all deaths; and the same remarks apply to the disease known as "malignant farcy."

We might enumerate many other forms of disease which are communicable from the superior orders of animality to the inferior, and vice versâ, but the above must suffice, in view of attracting attention to the value and importance of the subject.

in regard to the origin of diseases, it is known to the profession that the diseases of men and animals are often due to similar causes; that the evils of domestication, which operate to develop disease in animals, are as notorious as the evils of civilization, which induces unnecessary diseases and premature death among members of the human family; therfore, our science does bear an intimate relation to social science, and it appears to us that no scientific mind can fail to appreciate the advantages to be derived from a more extended knowledge of the fundamental principles of veterinary science.

Let the husbandmen and stock-raisers of this country put their shoulders to the wheel, in view of establishing veterinary schools and colleges throughout the length and breadth of our immense agricultural domain, and soon we can boast of having a class of veterinary surgeons, educated on American soil, conversant with the diseases incidental to the live stock of America. In that event, we can do our own doctoring, without foreign interference.

We are not disposed to find fault, nor scold about the itinerant veterinary practice which prevails throughout the United States, for we are aware that all try to do the best they can for the relief of the inferior orders of creation. The fault is not with the practitioners, but with the people, who have failed to furnish the means of education.

Resume.—Since writing the above, the " *Trichina* " disease has made its appearance in Berlin, creating a terrible consternation among the consumers of pork. The disease in its symptoms somewhat resembles cholera, but on making post-mortem examinations, there was soon discovered in the flesh of some persons an immense number of small microscopic parasites; these were traced back to the flesh of the hog, where they were discovered in the form of *chrysali* imbedded in a little shell of lime; these, on being subjected to a slight degree of heat, speedily brought forth myriads of small parasites, of various forms and sizes, a dozen of which were frequently found in a piece of tissue as large as a pin's head. The process of incubation was discovered to have been produced in the intestines of those who had eaten the pork, by the animal warmth of the stomach. Thus freed from their shells, the creatures make their way through the walls of the intestines in every direction, choosing the hard and fleshy tissues especially, and none of them have yet been found in the blood. Most remarkable to relate, the "Trichina" is said to have come from the "*field-mouse*," which the hog, by eating, introduces into his system, from whence it is introduced into the system of man. A few years ago, a number of persons in London were very seriously affected with vomiting and dysentery after partaking of sausages, the meat of which was found to be in a putrid state. Liebig, the great chemist, informs us that "several hundred cases are known in which death has occurred from the use of bad sausages in a state of putrefaction." This kind of testimony might be multi-plied to an indefinite length, but what has been written must suffice. The reader can not fail to perceive that, in the study of comparative anatomy and physiology, a wide field for investiga-tion and research is before us, which will amply repay the devoted student, and be the means of accomplishing much good.

Veterinary Science—How to Inaugurate it—What Benefits have we derived from its Study?—Shortly after the commencement of our late national struggle, the author was consulted in reference to the most feasible plan of inaugurating Veterinary Science, so that it might be made available for the restoration of sick, lame, and otherwise disabled horses in the Government service, which drew forth the following reply: Up to the present period there exists no law authorizing the employment of regularly-educated veterinary surgeons; therefore any rational plan calculated to ameliorate the condition of army horses, or to prevent the many unnecessary cases of disease and premature deaths which are now constantly occurring, or to guard against the great pecuniary losses which the Government and people of this country have been compelled to submit to in the condemnation and forced sales of deteriorated horses, can not, as I understand the matter, be considered by the powers that be until Congress shall legislate upon the subject. I now propose to inform the reader how to render the knowledge we possess available.

In the first place, I should follow the systems of instruction and the organizations of the veterinary schools of London, Ed-inburgh, Alfort, and Saumur, and prepare suitable text-books adapted to the wants of a nation of people who have never given the subject that attention which its importance demands.

In the selection of pupils, I favor the plan of General McClellan, which is as follows: "The pupils for the veterinary school might be selected from among the best recruits; indeed, it is not improbable that the advantages of such an institution would induce excellent men to enlist for the purpose of availing themselves of its benefits. Should such be found to be the results, it would be well to require them to enlist for longer than the usual time, as a compensation for the time spent at school." Cavalry officers, farriers, or blacksmiths should be permitted to attend lectures on anatomy, physiology, and the obvious diseases and lameness of horses.

In cases of emergency, a competent veterinarian should be employed in each regiment, whose duty it should be to select an intelligent farrier from each company, and instruct such person, in a brief and practical manner, on the management of cavalry horses in camps, the prevention of diseases, and the most rational method of treating diseases incidental to camp life.

Should it be found impossible to secure the services of a sufficient number of competent veterinarians for the above purpose, let those who have shown themselves qualified, by an examination before a medical board, march from camp to camp, give instructions, see that the sick and disabled are separated from healthy animals, and that the former be placed in a covered hospital, and then issue special regulations for the use of mounted troops in garrison and in the field, for the purpose of guarding against the consequences of ignorance and abuse of the animal machine.

It would be advisable to establish a veterinary professorship at West Point, and there locate the National School. The Government has there a very fine French model of the horse, a skeleton, and various preparations of morbid anatomy, etc., which, together with other material available, will furnish all that is necessary for the purposes of veterinary tuition. It would also be advisable for the Chief Veterinary Surgeon to issue to his subordinates such orders as shall secure proper sanitary regulations in camp. The horses should be as well cared for as the troops, for in cavalry and artillery service, the success of a campaign, or attack and defense, depends much on the health and efficiency of horses. It should be the business of some members of the veterinary corps to station themselves at accessible points, and there erect temporary hospitals for the reception of sick and lame horses, where the latter shall receive the benefits of a rational system of medication and nursing, whereby, in their restoration to usefulness, many millions of dollars might be saved to the Government in a short time. I contend that the condemnation and ruinous sale of sick and lame horses is a wanton waste of property, and it would be just as rational, if rational at all, to sacrifice, neglect, or abandon sick or disabled soldiers simply because they are not in fit condition for present duty.

Importance of Veterinary Education.—The importance of educating men for the scientific practice of Veterinary Medicine and Surgery is self-evident, and is further illustrated by the daily experience of those persons whom, as a matter of necessity, have been obliged to prescribe for the sick and dying of their flocks and herds, at the same time, having no knowledge of those advantages which a thorough course of training affords, must necessarily feel that they are groping in comparative darkness, and when a ray of medical light does flit across their path, it often serves to make them better acquainted with their own want of skill. Persons thus circumstanced, unable to procure the services of veterinary surgeons, in consequence of a scarcity of the "genuine material," are often called upon to assume the responsibilities of medical practitioners; and they, no doubt, prescribe with good intention and honesty of purpose, but disease and death have the vantage of them, for the disease is not often rightly diagnosed, nor are they acquainted with the means of securing a favorable issue; and they will hail, as a great blessing, every attempt made for the purpose of diffusing a knowledge of Veterinary Science.

We all know that there is greater need for scientific qualification on the part of those who propose to practice on brutes than in the case of a practitioner of human medicine; and if a special course of study and qualification, obtained under the guidance of competent teachers at college or elsewhere, be essentially necessary in the one case, surely they can not be dispensed with in the other. Now, it is well known that prior to the introduction of a rational system of veterinary medicine in the mother country, millions of domestic animals were annually sacrificed at the shrine of ignorance; but science came to the rescue, and now the disciples of ST. BEL, COLEMAN, their co-workers (the founders of the science in England), and kindred spirits, can, by aid of their vast *materia medica*—their anesthetic agents, their scalpel, and other appliances—accomplish unheard of wonders.

The Samaritans of our craft have solved the problems on which the fundamental principles of physiology and pathology are founded; they have, to a certain extent, dispelled the awful cloud of ignorance and empiricism which had reigned in the British Isles. They have succeeded in illuminating the dense and dark spots that have existed in horse and cattle practice since the advent of the ancient Egyptians, up to that period which introduced a new era in our art. They have done more than this : our craftsmen have often robbed death of some of its victims, and have succeeded in blunting the keen edge of many an epizoötic sword; and thus they have succeeded in stealing a march on the grim monarch of epizoötic destiny. Their weapons were often such as were wielded by JENNER, the benefactor of our race. Our efforts may not always be successful, but as our veterinary fathers have accomplished much, we ought not to despair.

The field of enterprise in the United States is composed of virgin soil, and the " bridegroom of promise" hath not yet put on the complete mantle of fertility; so that there is a splendid chance for the teachers and graduates of American schools to distinguish themselves and earn the laurel of fame.

The husbandmen of this country have also suffered immense losses by the death of valuable animals, and since the landing of the Pilgrims on Plymouth Rock, up to this very moment, death hath run riot in the camping-ground of horses and cattle, and people have scarcely been informed of the " why or wherefore." The legion host of " itinerant " practitioners which necessity and circumstances have created, remain remarkably silent and non-committal ; and probably this is the best course they can pursue, for, if " ignorance is bliss," and that pays well, it would be decidedly unbusiness-like in them to assume wisdom which might not pay.

There never was a period in the history of this country when the services of educated veterinary surgeons were so much needed as at the present time. The live stock of " Young America " do not enjoy immunity from those pestiferous epizoötic maladies which have in former years operated as a " withering simoom " on British husbandry. For example, a dreadful malady known as pleuro-pneumonia has made sad havoc among live stock in Massachusetts, and the terrible alarm there created in consequence, is keenly felt, not only in that State, but in many others. In some of the Western States, a malady known as 'milk-sickness," or trembles, afflicting both man and beast, sometimes prevails, attended with such disastrous results that many farmers have often been compelled to abandon the old homestead in view of seeking a location where there was some prospect of enjoying immunity from the pest.

Hog cholera, or pleuro-pneumonia, (among swine) has often slain its millions and tens of millions. Contagious typhus prevails, and puerperal fever has carried off hundreds of thousands of our best stock. Apoplexy, softening of the brain, dropsy of the brain, are just as prevalent here as in England. Abortion, or miscarriage, is fearfully on the increase. Diseases of climate and spontaneous origin are constantly occurring. In fact, to shorten a lengthy story, there is scarcely any disease known to the veterinarians of the Old World that has not prevailed in the New. The mere mention of these matters should at once arouse stock-owners to a sense of duty to themselves and their domestic animals.

We must have educated veterinary surgeons for the treatment of diseases peculiar to the live stock of this country. It is a matter of national policy, a work of benevolence; and should we take a business view of the affair, it is a cent and dollar enterprise that will surely pay.

We want schools and colleges in this country, where the young men of America can acquire a *republican* veterinary education, so that in the era before us, the people in this country can employ the graduates of their own schools, endowed and fostered by their own enterprise and industry. Some persons have an idea that veterinary science can only be acquired under the banner of royalty, but this is a great mistake. I am aware that the people of this country have been so absorbed in promoting its growth and developing its resources, that there has been but little chance for great scientific achievement in our calling, yet within the past few years some of the most learned of our profession, educated under the auspices of royalty and aristocracy, have been amazed at the contributions to science furnished by a few devoted men in this country that never saw the inside of an European college.

As a national affair, therefore, it should be the pride of every American citizen to put his shoulder to the wheel for the purpose just indicated, for the enterprise will surely redound to the honor and glory of the nation. The science which we are laboring to render popular is vast and boundless as the ocean ; and although it has been explored by many distinguished scholars, it is capable of unfolding new truths, and when the united intellect of intelligent Americans shall be brought to bear upon it, will advance with a rapidity truly astonishing. What the American people have accomplished, during the past few years, in rendering other sciences popular, is proof positive of what shall occur when the current of investigation is directed to the science of veterinary medicine and physiology.

In the absence of educated veterinary surgeons throughout the country, it is of vast importance that stock-owners generally should study, as far as they can, the diseases to which their property is subject, and acquire a knowledge of, at least, the ordinary remedies. By doing so, thousands of dollars' worth of valuable horses, cattle, and swine may be annually saved to the country, and the intense misery which these poor brutes suffer frequently be relieved. It is our object in this work to simplify the science so that it may be acquired, to a very useful extent, by any intelligent man. If we shall succeed in enabling stock-owners to practice the healing art on their dumb but faithful companions, we shall feel that we have done a good work.

Inflammation ; Its Nature and Treatment.—Inflammation is said to consist of pain, heat, and redness. It is usually considered as an enemy to the patient ; and we have been taught to subdue it by means of bleeding and purging, at all hazards ; and some practitioners, when they undertake the job, generally succeed, yet they lose the patient for their pains.

Some writers have made inflammation cut up all manner of pranks. It is now sanative in healing a wound or injury, in repairing damages which occur in various parts of the animal economy ; and in the same paragraph the writers declare that inflammation is the cause of nearly all the deaths that occur. This is sheer nonsense, and will not stand the test of logic, nor the daily experience of that practitioner who deals in fact, and does his own thinking, independently of the dictum of some of the schools, and the ancient method of reasoning on false facts. A distinguished medical reformer, writing on this subject, says, " Numerous hypotheses or opinions respecting the true nature of inflammation have for ages been advanced, and, for a time, sustained; but even at the present day the various doctrines appear to be considered altogether problematical."

We often hear, in common parlance, that, "if inflammation sets in, the horse will die." Now, the very reverse of this is the case; for, while inflammation lasts, there is strength, vitality, and augmented action, though there may be much pain. This should be interpreted as the warning voice of Nature, to apprise us of a loss of equilibrium in the circulation of the blood, or some deranged condition of the system; and we should heed the warning, and array ourselves on the side of Nature, in view of promoting a full and free circulation of the blood over the system, so that there shall be no excess anywhere; and the derangement, wherever it exists, must next command our attention.

While inflammation and pain exists, there is hope. Take, for example, inflammation of the intestines. While that stage lasts, the animal is not in danger; but so soon as inflammation and pain ceases, (often accomplished through the means of bleeding and narcotism,) we then open up a new page of pathology. In the course of a few hours the animal dies. Of what—inflammation? No. He dies of mortification, or gangrene of the bowels. Therefore death was not caused by inflammation.

So in regard to phrenitis (inflammation of the brain). A certain amount of pain and activity in the parts are observable; the inflammation may now be attacked after the true Sangrado fashion; the inflammation is subdued, but still the animal is no better. He dies. Of what does he die? Not of inflammation, for the autopsy reveals, very often, softening of the brain, and fluid is found within the lateral ventricles. Softening of the brain is equivalent to gangrene (death of a part). Then, again, horses can not be said to die of inflammation when the practitioner declares that he heroically subdued it some time previous to the death of the animal.

There is no proof that bleeding has any directly beneficial influence over the course of inflammations, either external or internal. Surgeons never bleed now in external inflammation; and educated and liberal physicians have given up all arguments in favor of the practice for internal inflammation. GOOD, in his "Study of Medicine," seems to have given us some definite information regarding the effect of blood-letting in inflammation,

which should agree with the sentiments of all men desiring to see
veterinary science joined in the march of improvement which is
now progressing in almost every other department of science and
industry. "The immediate effect of profuse and repeated bleeding is exhaustion. While this exhaustion continues there is a
diminution of action of every kind, and hence an imposing appearance of relief to the symptoms of disease; but it no sooner
takes place than an instinctive effort is made by the *vis medica-
trix natura* to remedy the evil hereby produced, and to restore the
system to its former balance of power. This balance is called a
rallying, or reaction of the living principle. The arteries contract to adapt themselves to the measure of blood that remains;
the sensorial organ is roused to the secretion of a large proportion
of nervous power to supply the inordinate drain that takes place.
During the general commotion, all is in a state of temporary
hurry and urgency, and, for the most part, irregularity of action,
while the instinctive effort is proceeding. And hence, no sooner
is the immediate effect of prostration and exhaustion overcome
than the heart palpitates, the pulse beats forcibly with a jerking
bound, the head throbs, and the eyes flash fire. Now, it often happens that these concurrent signs are mistaken for proofs of latent
or increased vigor, instead of being proofs of accelerated action;
and action, too, that adds as largely to the exhaustion as the depletion that produced it; and the unhappy patient is bled a
second, a third, and even a fourth time,* till no reaction follows,
at which time it is strangely supposed that the plethora or inflammatory diatheses is subdued and lulled into a calm, because the
patient has been so far and fatally drained of the living principle
that there is no longer any rallying or reactive power remaining,
and gives up the ghost, in a few hours, to the *treatment* instead of
the disease." This is the case with thousands and tens of thousands of valuable animals that annually die in this country, in

---

* My attention was recently called to a valuable stallion, which had lately
been brought from Nova Scotia. He was attacked with what the attending
would-be physician called "founder," but which was a pure case of acute rheumatism. In three different bleedings, forty-two pounds of blood had been abstracted; and I found the patient so weak and prostrated that he was scarcely
able to stand. It seems to me that every man having the least spark of charity
for so noble an animal as a horse, should discountenance such outrageous treatment as this.

consequence of our misguided notions of inflammation, and our want of knowledge of a suitable remedy to treat it. But a brighter day is dawning, and the antiquated notions of disease and its treatment are fast giving way to a more rational and successful system; and the day is not far distant when blood-letting for the cure of inflammation will be entirely abandoned.

It is very gratifying to the author to be able to record that many of the progressive medical writers of the present day are on the right side of Nature, teaching us that "she is ever busy, by the silent operation of her own forces, in the cure of disease;" and they are inaugurating a very great and desirable revolution in the theory and practice of human as well as veterinary medicine.

Nature of Inflammation.—The physical characteristics of inflammation are, as I have just written, redness, heat, pain, and sometimes swelling. It is, and always was physiologically, operating for the good of humanity and the inferior orders of creation. Its curative power none can dispute. We see it, in the form of a blush, on the cheek of offended humanity. Friction, injuries, poisons and disease, etc., excite Nature to hoist the symbol of distress—inflammation. She calls loudly for help, but she does not always get it; and instead of acknowledging her autocracy, and furnishing what she wants to use in her own way, viz.: the water, oil, and wine of the ancient Samaritans, we offer fire, knife, and poison.

Inflammation being an exalted condition of local arterial circulation, it can only be excited by some mental emotion, injury, loss of function, or by what is known as disease, in parts adjacent or remote from its seat. Hence, all diseases of an acute character are preceded and accompanied, to a certain stage or period, by inflammation. Hence, also, according to ancient usage and the dictum of *alma mater*, we are constrained to talk and write as though inflammation was the great evil or disease which required our services; and thus we coquette with Nature by means of sharp-edged tools, while the actual disease steals a march on us, and we lose the patient in consequence of our want of knowledge.

Treatment of Inflammation.—Inflammation being more or less active, according to the intensity of the disease of which it is a forerunner, or rather an indication, and being accompanied by pain, it becomes our duty merely to restore the equilibrium of the circulation. This removes the inflammation, and promotes a free and full circulation of the blood all over the system, so that there shall be no excess (inflammation) anywhere. This is what we call subduing the inflammation, with this difference, however, that we resort to natural means and sanative medication instead of using the fleam and cathartics. The means are warmth, moisture, stimulants, and friction; and clothing to the extremities and to parts of the body that are cold; and cold applications to parts that are unnaturally hot. One of the best equalizers of the circulation is the "Roman bath," the vapor bath, and wet packing. Should the inflammation exist internally, then give plenty of warm water, emolient drinks, and emolient enemas, and keep up a physiological action and heat on the surface of the body by means of friction and warm clothing. In this simple way the circulation of the blood is equalized, or, in other words, the inflammation is "cured;" and, at the same time, this treatment may cure the disease as well as the inflammation.

SECTION II.

DISEASES OF THE EYE.

AMAUROSIS ("GLASS EYE"—GUTTA SERENA.)

AMAUROSIS is known to most men as "glass eye." Physi-cians, however, have named the disease gutta serena. The abnormal condition, which consists in dilatation of the pupil, un-influenced by light or darkness, is occasioned by paralysis of the optic nerve and its ultimate expansion.

Causes.—Some horses, of an excitable, nervous temperament, often become the subjects of dilated pupil, without any assignable cause; while that form of amaurosis occurring among plethoric subjects, or those whose digestive organs are deranged or occupied by a large quantity of undigested food, (they, the subjects border-ing on that state known as stomach staggers,) can easily be ex-plained on the well-known law of sympathy.

We may, however, with propriety, assign a cause for its occur-rence in the eyes of nervous, excitable horses; for this very con-dition of the nervous system, which gives rise to the excitability, perhaps goes to show that the brain is actually diseased, either in function or structure. Animals subject to this affection are gen-erally of a "bony," spare, muscular development, and have wiry sinews, thin tapering ears, delicate lips and nostrils, diminution in the quantity of the hair in the mane and tail, but remarkable for compactness of texture. We generally find, under ordinary circumstances, the black color preponderates in nervous horses over the various shades of equine coloring-matter found in the

hair, eyes, etc.   I had a case of this kind, and the owner's statement was as follows: The animal (mare) had been in his possession about five years.   She had four successive attacks of blindness,
which had never been treated by a certified surgeon, and it was
concluded that she must have a periodical attack in spite of all
efforts to prevent it.   Now, this is what we call a case of idiopathic periodical amaurosis, and if it does not owe its origin to
some hereditary idiosyncracy, it is the result of an abnormal condition of the brain.

When our attention shall be called to a case of this character,
we must bear in mind the fact that this animal survived four
periodical attacks of the same malady, and performed the ordinary
duties of a horse, in the intervals of health, with far-seeing eyes,
and without the aid of science ; and this goes to show that this
condition of the nervous structure of the eyes of some horses, as
well as many other diseases incidental to them, are often " self-
limited," and the subjects will recover, if carefully nursed and
left unembarrassed by "poisonous drugs."   This opinion is indorsed by some of the most distinguished surgeons of the present
day.   I have no desire to try to saddle the reader with any peculiar notions of medication emanating from my own fancy or
brains.   I am not offering sensible people a mirror of my own
reflection, but I wish to present the truth in that mirror, so that
they may see it and judge for themselves.   Excessive medication
is a quicksand which we must try to avoid.   Thousands of well-
meaning medical aspirants have actually committed professional
suicide by clinging to the absurd theory of medication as practiced in bygone days.   The lancet, and many of the heroic medicines which our predecessors placed great confidence in, and in
their writings have urged us to use, are now almost discarded.
It is not because an essential change has taken place in the nature
of equine maladies, but because we have greater faith in Nature as
our curative agent, and have become better acquainted with anatomy, pathology, and physiology.

But to resume the subject of amaurosis.   If from the history
of the case, we have reason to believe that it is of a periodical
character, and if it be the sequel to, or an accompaniment of,
some hereditary affection, we may be sure that the malady will
exist for a certain length of time in spite of all we may do.   In
such cases, all that is necessary is to have the animal comfortably

housed, fed on a light, laxative diet, and pay careful attention to
the condition of the secretory and excretory functions of the
body.

That form of amaurosis occasioned by a deranged condition of
the digestive organs, plethora, or over-distension of the same, will
cease so soon as the cause is removed, and may be considered as
a sympathetic affection.

What that cause is we must endeavor to decide on by careful
examination, and then frame our plan of treatment in accordance
with the indications to be fulfilled.

Should we consult authorities on amaurosis, we shall find that
much diversity of opinion exists. YOUATT, who is quoted in
this country by some persons as high authority, has mistaken the
disease, or else has strangely distorted facts. He contends that
" the treatment of amaurosis is quite as difficult as that of cataract." Now, every one knows that cataract occurring among
horses is notoriously incurable ; and PERCIVALL writes, " A man
must be able to perform miracles ere he cures cataract."

Now, I contend that amaurosis, or dilatation of the pupil, is not
of itself a disease, but the symptom of one; hence, between cataract—a disease resulting from altered structure of the eye—and
amaurosis there exists no analogy. Whenever altered structure
exists in the mechanism of the eye, then we have an incurable case.

Mr. SPOONER, the reviser of YOUATT's work on the horse, as-
sures us that amaurosis is often connected with diseased liver,
thus controverting YOUATT's theory of incurability ; for animals
often recover from very severe attacks of liver disease.

I have called the attention of our readers to the opinions of
YOUATT, PERCIVALL, and SPOONER merely to show the importance of exercising our own reasoning faculties. We must not
place too much confidence in mere book authority, no matter how
high the source, unless experience and intelligence indorses the
theory or facts in the case.

Speculative and false theories are only surface deep. They need
not the eye of the philosopher, nor the mental genius of a Webster, in order to discover their flimsy texture and develop the
truth. Those who propose to do their own thinking have the
ability to judge between right and wrong. Let us commit to
memory all matters which appear as facts, and forget the rest.

Amaurosis can be artificially produced by administering a few doses of extract of belladonna, and this amaurotic state of the pupil is also observed during etherization, and disappears so soon as the animal is restored to consciousness.

PERCIVALL informs us that amaurosis may proceed from some abnormal condition of the optic nerve or brain, which may prove, as he says, the last link of a series of marked phenomena, origi nating in some remote part of the body, operating sympathetically on the nervous system, and, through it, extending to the eye. Cases are occasionally met with which may be termed symptomatic amaurosis, occasioned by congestion, tumors, or hydrocephalus.

Symtomatic amaurosis may follow cranial fracture, and, con sequently, depression. In such cases, the skill of the surgeon is called into requisition to trephine the uninjured region of the skull, in order to elevate the depressed and fractured parts. The potent remedy is the removal of the pressure. Amaurosis is oc casionally associated with anæmia. Now, anæmia is a condition of the body which indicates a loss of the circulating fluid, either of quantity or quality. The remedy for this condition is carbo naceous food, mineral and vegetable tonics. Preparations of iron and golden seal are the best remedial agents.

Symptoms.—I shall now turn the reader's attention, for a mo ment, to some of the symptoms attending the amaurotic condition. The head of the patient is elevated, and he is known, in popular language, as a "star-gazer." His ears are kept in constant mo tion, one turned backward and the other forward, in quick suc cession, to catch the sound of approaching vehicles, so that he may have timely notice of their whereabouts. When started, and while walking, the fore-feet are raised from the ground to an unnecessary height. This is probably done for the purpose of sounding the ground, and, at the same time, of avoiding obstacles which may lie in the road; for the sight is defective, and the animal must depend more on the senses of hearing and feeling.

Should such an animal be attached to a vehicle and left alone in the street, we shall observe that when an omnibus or heavy loaded wagon approaches him he becomes restless, and exhibits signs of fear. Such are the principal symptoms which usually accompany defective sight, or total paralysis of the optic nerve, producing amaurosis.

Treatment.—Sympathetic amaurosis should be treated by means of antispasmodics and alteratives. One ounce of tincture of assafetida may be given every morning, and in the evening, one ounce of the fluid extract of sassafras. So soon as the animal regains his natural vision, the medicine may be discontinued.

FOREIGN BODIES WITHIN THE EYELIDS.

When foreign bodies, such as small particles of hay or dirt, get within the eyelids, they create great pain and uneasiness, and if allowed to remain there, produce a very grave form of disease, often ending in disorganization and total blindness. Should any thing of the kind be discovered, it may be removed by raising the upper or depressing the lower lids; then introduce and explore the eyeball by means of a camel's hair pencil, to which the foreign body will usually adhere, when it is easily brought away. The parts should then be sponged with lukewarm water. After the lapse of a few hours, should the membranes of the eye and lids appear much reddened, use the following:

No. 1. Rose water .... 4 oz.
      Fluid extract of gelseminum........... 2 dr.

Put the patient on a diet of sloppy bran mash, and place him where the rays of sunlight shall not affect the eye.

FILMS OR SPECKS ON THE EYE, KNOWN AS OPACITY OF THE CORNEA.

Opacity of the cornea sometimes results from injuries; at others they are symptomatic, and accompany other diseases. They are of very common occurrence in the various stages of influenza and other catarrhal affections, and sometimes a distinct speck will remain long after the disease which gave rise to it has disappeared. The author remembers many cases in which the sight has been completely destroyed by using substances of a corrosive or stimulating character. The proper mode of treatment is to bathe the eye occasionally with a portion of the following lotion:

No. 2 Fluid extract of bloodroot.............. 1 oz.
      Water .... 8 oz.
      Mix.

Give the patient one ounce, morning and evening, of fluid extract of phytolacca decandra (poke-root). This acts as an alterative, and will expedite the cure.

CATARACT.

Cataract in the horse is usually the result of previous attacks of disease. It consists of opacity of the crystalline lens, or the capsule of the same. The disease is incurable, because if an operation was performed for the breaking-up or depression of the cataract, such as is practiced in human medicine, vision would still be as imperfect as ever without the aid of magnifying glasses, which would be a great inconvenience to the horse, and scarcely of any practical service.

WORM IN THE EYE.

Small parasites are occasionally observed in the eyes of horses. They float about in the watery humor, often for a great length of time, without appearing to do any mischief; yet, being a foreign body, they are, to say the least, injurious. The author has never seen but one case of worm in the eye, and that occurred in a horse, the property of a resident of Chicago. The animal was kept on exhibition for the inspection of the curious, and was the source of considerable profit to the owner. It is possible that such worm might be extracted by a surgical operation. A writer in the "Veterinary Review" thus alludes to this affection :

" It may also be here noticed that there is a disease which frequently affects the eyes of horses in India, but which, as far as I know, has never been observed in Europe. A *worm*, which, from the description I have received from different people, may be compared to the common *ascaris*, is generated in the anterior chamber, and can at times be distinguished swimming about, with apparently great vigor, in the aqueous humor. It produces a great deal of irritation and inflammation, the effects of which ultimately destroy the organ. The natives of India cure the disease by making an incision through the cornea, and extracting the worm. Though I have never had an opportunity of examining an eye affected with this curious disease, the circumstantial accounts, from several accurate observers, leave no doubt in my mind of its existence; and the fact accords with what is known to take place regarding the formation of worms, not only in the human body, but more particularly in the liver, lungs, brain, and other organs of the inferior animals."

OPHTHALMIA, OR INFLAMMATION OF THE EYE,
(SOMETIMES CALLED "HOOKS.")

This is a very common affection, yet a most painful one, an can not well be mistaken. The animal keeps the eyelids closed. They are more or less swollen, and the internal surface of the same appears very red. There is an unnatural flow of tears running down the cheeks. The *haw*, or nictitating membrane, is swollen and very vascular. This condition of the haw is known to some persons as the *hooks*, and serves as an excuse for the performance of a barbarous operation, not sanctioned by science or common sense, viz.: amputation of the haw, which only aggravates the disease, and deprives the horse of a useful appendage to the organs of vision.

In the early stage of this disease, and when scalding tears run copiously over the cheeks, we denominate it catarrhal ophthalmia. The disease is not of a very serious nature, but often becomes so in consequence of the outrageous treatment practiced, such as bleeding, purging, and the local application of lunar caustic. The proper mode of treatment is to keep the patient on a light diet, and occasionally bathe the region of the eye with an infusion of poppies or hops; sometimes with tepid or cold water, as the case may seem to indicate, according to the rational judgment of the owner of the afflicted animal. It is the very best application that can be made, and is much more scientific, according to the dictum of the God of Nature, than many of the agents heretofore used. The fact is, many indications of disease designated as inflamma-tion are curable only through the intervention of time and a rational expectancy. Don't do any violence to nature in the use of destructive or poisonous agents, and should the case be curable, it will terminate favorably under this treatment. Ac-cording to the dictum of the schools of veterinary medicine, inflammation of the eye, as well as inflammatory action of other parts of the body, has to be treated heroically, which system of treatment kills more than it ever cures. As the subject of inflam-mation is a very important one, worthy the attention and con-sideration of American husbandmen, I here introduce an article from the pen of A. S. COPEMAN, who was formerly associated with me in conducting the "Veterinary Institute of Boston," Massachusetts:

Treatment of Inflammation.

" It must be admitted by all who contemplate the actual state
of medical practice at the present day, that the use of blood-letting and of other antiphlogistic remedies, has, within a recent
period, greatly declined. According to Youatt and Percivall,
such remedies, and more especially blood-letting, were formerly
highly successful in arresting diseases, in the treatment of which
we now know they not only fail, but are even highly injurious.
The inference has been drawn from these supposed facts, that inflammation itself is no longer the same; that its type has altered
from an inflammatory to a typhoid character. In short, it seems
to have been the opinion of certain writers that an advanced
knowledge of physiology and pathology has had little influence
in producing this great revolution in our treatment, but that the
constitutions of animals are fundamentally altered, and that medical men were as right in bleeding thirty years ago as they are
correct in now abstaining from it. In opposition to these views,
it will be my endeavor to show, 1st, That little reliance can be
placed on the experience of those who, like Blain, Percivall,
and Youatt, were unacquainted with both histology and organic
chemistry, and, per consequence, the nature of inflammations;
2d, That inflammation is the same now as it has ever been; 3d,
That the principles on which blood-letting and antiphlogistic
remedies have hitherto been practiced are fallacious, and opposed
to pathology; 4th, That an inflammation once established can
not be cut short, and that the object of judicious medical practice
is to conduct it to a favorable termination; 5th, That all positive
knowledge of the experience of the past, as well as the more exact observations of the present day, alike establish the truth of
the preceding propositions as guides for the future.

1st Proposition.—*That little reliance can be placed on the experience of those who, like Blain, Percivall, and Youatt, were
unacquainted with histology, and, per consequence, the nature of
inflammations.*

Inflammation, for many years, was generally recognized, especially in external parts, by the existence of pain, heat, redness,
and swelling, and in internal parts by fever, accompanied by
pain and impeded function of the organ affected. In fact, groups
of symptoms, in accordance with the nosological systems of the day,

consti , ed inflammation. But the school of morbid anatomy, by showing that inflammation was a diseased condition of a part, entirely overthrew the errors and confusion inherent in all such nosological systems; while more recent histological research, by exhibiting to us that inflammation is, in truth, a disease of *nutrition*, governed by the same laws that determine growth and waste of the tissues, has united physiology and pathology into our science, and has removed our present knowledge still further from the traditional errors of the past. Now, if it could be shown that the group of symptoms formerly called inflammation always induced the same morbid lesions, former experience might still be useful to us. But we contend that this is what clinical observation proves to be impossible. Such are the contradictory statements and the confusion resulting from the unacquaintance of the past race of practitioners with a correct diagnosis and pathology, that no confidence whatever can be placed in their impressions as to what cases were benefited by bleeding. Medicine is not a scientific art, which is dependent for its principles on the study of a commentary on the older writers. What they thought and what they said are not, and ought not, in a question of this kind, to be our guide as to what was or is. On the contrary, it is the book of Nature, which is open to all, that we ought to study; and why should we read it through the eyes of past sages, when the light of science was comparatively feeble and imperfect, instead of bringing all our improved modern appliances and advanced knowledge to elucidate her meaning?

2D PROP.—*That inflammation is the same now as it has ever been.*

The essential nature of inflammation has been already alluded to, viz.: a series of changes in the function of a part, terminating in exudation or effusion of lymph. Now, what proof is there that any of these necessary changes have, of late years, undergone any modification? If a healthy animal receives a blow, or any other injury, are the resulting phenomena, in these days, in any way different from those which took place in the days of YOUATT and PERCIVALL? Were the effects which followed wounds in 1830 different from those which resulted from similar injuries in 1860? This has not yet been shown. Again: if a healthy horse, nowadays, be exposed to wet and cold, and be seized with an inflammation of the lung or pleura, is not the lung hepatized in the one case? and does not effusion follow in the other, in exactly the same way as formerly? But what should this asserted change in the nature and character of inflammation lead us to infer? It is said that inflammation and its results are entirely changed within the last thirty years. It is, then, argued that horses, in all parts of the world, since the days of BLAIN, YOUATT, and PER-CIVALL, have become so debilitated and deteriorated; that their constitutions have been so altered for the worse; that, attacked by the same lesion, and to the same extent, there is no longer the same reaction. If so, where is the evidence of this?

For my own part, I have earnestly sought for but can not dis-cover a shadow of evidence for such a belief. Moreover, I have a most lively remembrance of all the facts and circumstances con-nected with the bleeding of hundreds of patients, thirty years ago, when I first commenced the study of veterinary medicine, and my impression is, that not the slightest difference exists between the character of inflammation now and what it was then.

3D PROP.—*That the principles on which blood-letting and anti-phlogistic remedies have hitherto been practiced are fallacious and opposed to pathology.*

Large and early bleeding have been practiced, under the idea that, by diminishing the amount of circulating fluid, 1st, the *materus morbi* in the blood would be diminished; 2d, less blood would flow to the inflamed parts; 3d, the increased quantity of blood in the part would be lessened; 4th, the character of the pulse was the proper index to the amount of blood that ought to be drawn. Let us examine a few of these principles of practice. The increased throbbing and circulation of blood in an inflamed part may be shown not to be the cause of inflammation, but the result of it, and that the idea of so-called determination of blood to inflamed parts is fallacious. Now, if we attend to what takes place in the finger from a thorn entering the skin and remain-ing unextracted, we find the irritating body first acts upon the cellular constituents—the nerves and blood-vessels of the part; then comes on the congestion and exudation, and, lastly, follows the throbbing, which is the evidence of so-called determination, and result of the inflammation, and not a cause of it. The blood, in this case, instead of being sent by a *vis a tergo*, is, in fact, drawn by a *vis a fronte*, and, as we shall endeavor to show, for the most important purposes. But why should Nature, in cases

4

of inflammation, draw an increased amount of blood toward the part? She does so, it seems to us, in obedience to one of her wisest laws, but one which has been too much ignored by medical practitioners. It must be obvious, however, that an inflammation having occurred, the great work now to be accomplished is the removal of the exudation—to eliminate from the injured part either directly by discharge externally, or by passage into the blood, to be finally excreted through the emunctories. In all such cases, the blood is not sent or determined, but drawn to the part, in consequence of the increased actions going on in them; in short, it is absolutely imperative that the part in which these changes go on should receive more blood than in health. But, hitherto, medical practitioners have supposed that this phenomenon is injurious, and ought to be checked by blood-letting and antiphlogistics. The rapid flow of blood, which is so necessary, they have sought to diminish, and the increased amount in the neighborhood of the part which is so essential for the restoration to health, it has been their object to destroy. In doing so, we argue they act in opposition to sound theory, and, as we shall afterward attempt to show, good practice, also.

The inconsistency of the theraupeutical rules on this head will become more manifest when we remember that it is necessary, in the opinion of many medical practitioners, not only to weaken the pulse when it is strong, but to strengthen it when it has been made weak. Now, although it is obviously good practice to support the strength when the calls upon the nutritive functions have exhausted the economy, it is injurious to diminish, by blood-letting, the nutritive processes themselves, when they are busily engaged in operating on the exudation and eliminating the morbid products. In short, the phenomena of fever and excitability accompanying inflammation have been wrongly interpreted, and danger is to be apprehended from them, not directly, but from the subsequent exhaustion which all great exertions of the animal economy produce. In themselves, these exertions are *sanative*, and indicate the struggle which the economy is engaged in when attempting to get rid of the diseased processes; and whenever we lessen the vital powers at such a critical juncture, we diminish the chances of that struggle terminating favorably. This proposition seems to be universally admitted in the case of essential fevers, and its truth ought to be accepted equally in inflammation.

It has been argued, however, that the immediate effect of blood-letting justifies the practice. But, assuming it is granted that in some cases the pain is for a time relieved by bleeding, and that in pneumonia the respiration temporarily becomes more free, at what a cost are those advantages obtained, should the patient be so weakened as to be unable to rally! Even if he does rally, a large bleeding almost always prolongs the disease. Of course, we are now speaking of a true pneumonia, and not of that combination of symptoms commonly *called* a chill, or lung fever.

4TH PROP.—*That an inflammation once established can not be cut short, and that the object of judicious medical practice is to conduct it to a favorable termination.*

There was a time when it was supposed that the progress of influenza, distemper, and many other specific fevers which are now generally allowed to run their natural course, could be arrested by medical interference. But with regard to them has been established the principles, first, of prevention, and, second, when this fails, of simply conducting them to a favorable termination.

It appears to me that the same rule ought to hold with regard to internal inflammations, and that this will be admitted when it is made apparent, not only that every inflammation, once formed, runs through a definite course, but what that course is. This I now proceed shortly to consider. If a violent blow or injury has been received, a greater or less amount of exudation is infiltrated among the contused and torn tissues, which undergo disintegration, is absorbed and excreted from the economy; if not, it collects, in the form of a fluid, and constitutes an *abscess.* The series of changes here referred to have always been found to be best perfected in animals of vigorous constitutions, while in the weak, poor, and broken-down they proceed slowly, or not at all. Surely, it can not be maintained that in cases similar to the above, we can favor the reparative processess by blood-letting and lowering the strength of the economy?

But in internal inflammations, say of the lungs or intestines, are the processes different? Certainly not. But because the processes by which Nature relieves herself have been hid from view, physicians have supposed that instead of treating the inflamed parts, as the surgeon does, he ought to attack the general symptoms which result from the lesion. In cases of fracture and contusion there are also febrile symptoms, increased pulse, and so on;

but does the surgeon imagine that callous will form better, or an abscess be resolved or reach maturity sooner, by general blood-letting and antiphlogistics? Experience teaches him otherwise; and in the same manner it may be most reasonably argued that such treatment can not favor the natural termination of internal inflammations.

5TH PROP.—*That all positive knowledge of the experience of the past, as well as the more exact observation of the present day, alike establish the truth of the preceding propositions as guides for the future.*

Before it is possible, however, to determine with exactitude the value of any practice, it is essential to ascertain the natural dura-tion of the disease we propose to treat. Fortunately, we have now some data which will enable us to arrive at this information with regard to many diseases. We have seen many severe cases of pneumonia submitted to homeopathic remedies—that no rea-sonable medical man can suppose to be any thing else than inert—yet most of these cases got well, and, I think, may be considered as excellent studies of the disease left entirely to Nature. Many years' experience and close observation have convinced me that uncomplicated pneumonia, especially in young and vigorous con-stitutions, almost always gets well, if, instead of being lowered, the vital powers are supported, and the excretion of effete pro-ducts assisted. It is in exactly these cases, however, that we were formerly enjoined to bleed most copiously, and that our systematic works even now direct us to draw blood largely, in consequence of the supposed imminent danger of suppuration destroying the texture of the lung. Such danger is altogether illusory, and the destruction to lung tissues, so far from being prevented, is far more likely to be produced by the practice. In fact, the only cures in which it occurs are in the aged or enfeebled constitutions, in which nutrients, and not antiphlogistics, are the remedies indi-cated. We can, however, readily understand now blood-letting, practiced early, and in young and vigorous constitutions, does less harm, or, to use a common expression, is "borne better," than when the disease is advanced, or the patient weak, and this be cause then the vital powers are less affected by it. But that it cures the greater number of animals attacked, or shortens the duration of the disease, is disproved by every fact with which we are acquainted. Before closing we have a few words to offer on

'mercurials.' The confident belief in their power of causing absorption of lymph, by operating on the blood, is not only opposed
to sound theory, but, like blood-letting, is not supported by ex
perience, which has been so confidently appealed to in their favor.
I can not, therefore, resist the conclusion that the principles which
led to an antiphlogistic practice in inflammation were erroneous,
and are no longer in harmony with the existing state of pathology.
I think it has been further shown, that in recent times our success
in treatment has been great just in proportion as we have abandoned 'heroic remedies,' and directed our attention to furthering
the natural progress of the disease.

Internal inflammations are cured, not by bleeding and drugs,
but by a natural process as distinct and definite as the process of
normal nutrition. What we may do by our interference, may
either aid, promote, and even accelerate, this natural tendency to
get well, or it may very seriously impair and retard, and even
altogether stop, that salutary process. If, then, this view of the
nature of the means by which inflammation is resolved in internal
organs be correct, it is not unreasonable to assume that a very depressed state of vital power is unfavorable to the healing process.

Indeed, if you watch those cases in which nothing at all has
been done, or in which nothing has been done to lower the vital
powers, you will find that the mere inflammatory process itself,
especially in an organ so important as the lung, depresses the
strength of the patient each day more and more. You will perceive, then, that, according to these views, there are strong *à priori*
reasons in favor of the policy of upholding our patients, even in
the earliest stages of acute diseases, by such food as may be best
suited to their digestive organs, such as is most readily assimilated, and calls for the least effort, the smallest expenditure of
vital force for its primary digestion—nutritive matters, tea
sweetened milk, etc., and also alcohol, which is directly absorbed
and tends to keep up the heat of the body.

If, then, it has been satisfactorily shown, in consequence of our
advanced knowledge of diagnosis and pathology, that an antiphlogistic practice is opposed to the cure of diseases, it follows that
many of the principles which have hitherto guided us in their
treatment must be considerably modified. That medical practice
has undergone a great revolution during the last fifteen years, is
a fact already so well established that it can be no longer denied.

In my discourse on inflammation, reference was made to the views held by the 'neuro-pathologists.' Now, nervous pathology has been in medicine the 'great scape-goat' upon which more professional sins have been heaped than any other. 'Nervous exhaustion,' nervous irritation, etc., are a few of the many terms with which we seek to cloak our ignorance of the real nature of many disorders, the intimate nature of which is beyond our ken. Many accomplished practitioners still maintain that abnormal, vital phenomena may be, and are likely to be, occasioned by dynamic aberrations alone, and that such phenomena are correctly designated as functional disease. We can not concur in this opinion. What is called force of every description is connected with, if not dependent on, changes in the atoms of matter Force is the hypothetic agent which underlies the phenomena of material change; and to affirm that dynamic modifications of vital function may exist without alteration of material organization, is to ignore the fundamental principles of philosophic physiology. All diseases, therefore, in our opinion, is organic, even mental and nervous diseases of every kind and form. Not a thrill of sensation can occur, not a flashing thought or a passing feeling can take place, without changes in the living organism; much less can diseased sensation, thought, or feeling occur without such changes—changes which we are not able to detect, and which we may never be able to demonstrate, but which we are, nevertheless, certain of. For, whether we adopt the theory that the states and things which we call heat, electricity, vitality, etc., are distinct entities of what is called 'imponderable' matter, or the far more probable theory that they are only phenomena belonging to ordinary ponderable matter, an atom or a cell, charged with electricity or heat, or in a state of chemical activity, is essentially in a different condition to a cell or an atom in chemical or electrical equilibrium with surrounding substances. Organic actions can not exist without corresponding changes in material condition. The only force capable of explaining any of the phenomena of life is the chemical one, and this only in a state of constant activity and interminable change. In disease, the chemical composition of the cells, or general matter, is altered from the standard of health, and this alteration of chemical composition is the real groundwork of organic disease. Those abnormal states which depend upon an altered condition of the blood, are not less st strictly organic than all other diseases; for not only can no change take place in the composition of the blood without in some degree affecting all parts which are nourished by it, but that this fluid is, strictly speaking, itself a living organism, and every change which *takes* place therein is organic. Life can not be arrested without material change in the organs necessary to life. Every autopsy in which we fail to discover these changes, proves merely that our knowledge of life and death is defective, not that the changes do not exist."

PURULENT OPHTHALMIA.

Purulent ophthalmia takes its name from the profuse discharge of pus, or matter, which escapes from the eyelids. In regard to the human subject, the authorities contend that there are several varieties of this malady prevalent in man, one of which, the Egyptian ophthalmia, is contagious; but no such disease was ever known to occur among horses.

Causes.—The various maladies which affect the eyes sometimes owe their origin to accidents; at other times they occur in badly ventilated stables, where a large number of horses are congregated together, and then they are supposed to be contagious. But it appears that there is no necessity, in such case, to advance the theory of contagion by way of explanation of the outbreak; for the same cause which produced the malady in the first horse would be more or less operative in all the rest. A hot, foul atmosphere has a very bad effect on the eyes of both men and horses. The membrane lining the eyelids, which also covers the eyeball, is very vascular and sensitive, and is extremely susceptible to irritation from the action of the ammoniacal gases which prevail in foul stables; hence, in order to prevent the disease, proper attention must be paid to ventilation and cleanliness.

Treatment.—For the treatment of purulent ophthalmia the author recommends the following lotion:

No. 3. Powdered chlorate of potass............½ oz.
      Fluid extract of matico.................1 oz.
      Water................................8 oz.
      Mix.

A portion of this lotion should be applied to the eyes, by means of a soft sponge, two or three times a day; the eyes as well as the lids must, however, be previously cleansed with lukewarm water. The patient should have a few doses of fluid extract of poke-root. About half an ounce of the extract, night and morning, will suffice, which will act as an alterative.

SPECIFIC OPHTHALMIA.

This form of disease affects animals periodically, and is dependent on some peculiar predisposition; hence it may be termed hereditary.

Symptoms.—There usually comes on very suddenly, perhaps in a single night, a great tenderness in one eye, commonly marked by the eyelids being shut, a copious secretion of tears, the white of the eye appearing slightly red, and the whole anterior chamber of the eye dim and clouded, there being no distinct speck on the cornea, as takes place in the common inflammation of the eye. The redness of the eyeball is never very remarkable, even though the disease assumes its most aggravated form; but the dimness of the anterior chamber increases rapidly, and in two or three days, or even a shorter period, a yellow spot appears at the bottom of that cavity, arising from the formation of pus. Sometimes the quantity of pus is very considerable, and I have seen it fill at least two-thirds of the anterior chamber. After lasting one, two, or three weeks, the inflammation and watering usually begin gradually to subside. The pus, though in very large quantity, is sometimes almost entirely absorbed, so that scarcely any vestige is to be seen; and in other instances thin webs of opaque matter remain, which destroy the transparency and luster of the eye, and which, by their adhesion to the edges of the pupil, interfere with its motions and destroy its form.

It is astonishing how acute dealers in horses are in discovering an eye which has had an attack of this kind.

Sooner or later, while the horse appears in a state of perfect health, the eye is again attacked, the disease being accompanied by the same symptoms, making a similar progress, and having the same termination, while each new attack is accompanied with the deposition of more and more opaque matter. These attacks succeed each other at very different, and sometimes at very distant, intervals, until the whole pupil is filled with an opaque white matter, and the sight of the eye completely destroyed.

During this progress, the disease is often confined to one eye; at least one eye is usually much more severely affected than the other. In some cases the two eyes are simultaneously affected, and, finally, by a succession of attacks, the horse becomes completely blind. The probability is, that after a horse has suffered several times from periodical attacks he will finally become blind.

Treatment.—The same treatment as recommended for common inflammation of the eye is applicable to this complaint.

SECTION III.

DISEASES OF THE BRAIN AND NERVOUS SYSTEM

ABSCESS WITHIN THE SUBSTANCE OF THE BRAIN—HYDROCEPHALUS, OR DROPSY OF THE BRAIN—STOMACH STAGGERS—APOPLEXY, OR CEREBRAL HEMORRHAGE—APOPLEXY OF SPINE, OR SPINAL HEMORRHAGE, PRODUCING PARAPHLEGIA—EPILEPSY, OR FITS—TETANUS—VERTIGO—ACUTE OR GENERAL DISEASE OF THE BRAIN, KNOWN AS ENCEPHALITIS—SLEEPY STAGGERS—CEREBRAL MENINGITIS—CHOREA, OR STRINGHALT.

ABSCESS WITHIN THE SUBSTANCE OF THE BRAIN.

THE author, having some doubts about the curability of abscess within the brain, can not offer the reader much encouragement as regards the cure; yet, for the purposes of research, and in view of prosecuting our Samaritan-like calling on a noble and valuable animal, it is proper that the reader be put in possession of the facts in the case. The symptoms and morbid conditions connected with various forms of brain disease are, at the present state of our knowledge, very obscure. Even in human practice, the authorities contend that the subject of symptoms and conditions is full of uncertainty and apparent irregularity. Doubtless there is some constant and uniform connection of cause and effect between the altered physical states of the brain and the altered manifestations of its functions, but we have not yet been successful in our search after those settled relations, nor have we but partial and imperfect glimpses of them.

Causes.—It is probable that the same causes which produce abscess in other parts of the body, not clearly traceable to local injuries, are operative in regard to the brain. Abscess is, as a matter of course, always preceded by an active stage, known as inflammatory, and when not called into existence by local injury, must depend on both predisposing and exciting causes. The predisposition lurks in breed, and the exciting causes may be among

A HORSE BENT ON MISCHIEF—THE SUBJECT OF PHRENITIS.

those which produce disorder in the digestive function or organs. It is well known, however, that this, like some other diseases of the brain, is constantly occurring among members of the human family as well as the equine species, they being the subjects of constitutional defect in the form of scrofula. Derangement of the digestive organs almost always affect the brain, in consequence of sympathetic relations existing between the two. Hence, in view of preventing disease of the brain, we must keep the stomach in good working order, by means of an intelligent system of dietetics, and the exhibition of sanative medicines when they seem to be needed.

Symptoms.—The observable symptoms of abscess within the brain do not differ materially from those which are present in dropsy of the brain. In the early stage, the animal appears lethargic, sleepy, and, when urged to move, reels and comes near falling. The head is usually somewhat depressed, yet it is often inclined to one side: the pupil of the eye is dilated, and the membranes of the lids are congested and reddened. As the disease advances, a state of torpor sets in. Blindness, from pressure on the brain, ensues; the animal gets upon the floor, soon abrades the skin from the regions of the hips and shoulders, until, as a matter of charity, the owner puts an end to the sufferings of the patient.

The lethargic condition, therefore—dilated pupils, torpor, and reeling of the animal—indicate this serious difficulty in the substance of the brain or its cavities.

Treatment.—Should the surgeon or the proprietor of the affected animal desire to attempt a cure, the author recommends the following:

No. 4 Fluid extract of stillingia. (queen's-root,) 4 oz.
   Fluid extract of bloodroot............. 3 oz.
   Powdered iodide of potass............. 1 oz.
   Water................................. 6 oz.

Dissolve the iodide in the water, and then add the stillingia and bloodroot.

Dose, two ounces, twice daily.   Local treatment in this malady does not amount to much, if any thing at all, for the only way in which the matter within the brain can be got rid of, if got rid of at all, is by exciting the absorbents to action, so that the matter may be absorbed.   At the same time, alteratives are to be used; and, for this purpose, the above prescription is recommended.

HYDROCEPHALUS (DROPSY OF THE BRAIN).

Hydrocephalus is the termination of some disorder in the brain itself, or the membranes surrounding it; yet some medical writers contend that dropsy is rather a symptom of disease than disease in itself.   It may be so in some cases, which constitutes the exception to a general rule; but the author regards dropsy of the brain as the last stage of organic disease of that organ.   On the dissection of the brain of some horses that have died of this malady, scrofulous tubercles are often found in the substance of the brain, and tubercular deposits also manifest themselves on the membranes covering the brain.   In fact, the pathological appearance revealed on dissecting the brain of a horse which has died of hydrocephalus does not differ from that found in the human subject.   CONDIE says:

" Tubercles, varying in size from that of a pin's head to that of a pea, are very generally found scattered irregularly over the surface of the pia mater, following it between the convolutions.   Occasionally, however, they occur in distinct patches of an inch or more in extent.   They are commonly hard and semi-transparent; sometimes opaque, and of a whitish, grayish, or yellowish color.   They are found upon all parts of the surface, the convex and lateral portions as well as the base, in the infractuosities of the convolutions, and in the fissures. According to Rilliet and Barthez, they are more frequent upon the convex surface of the hemispheres than at the base. Dr. Hamernjh found them more frequently at or near the base of the brain. They are much more abundant upon the brain than upon the cerebellum. They are met with, also, imbedded in the gray matter of the brain, and are here often surrounded by a halo of redness, usually connected with an enlarged vessel, ramifying from the pia mater. More rarely, tubercles are detected in the medullary portion of the brain, where they are often overlooked, in consequence of their pale, semi-transparent, yellowish tint. The plexus choroides is, also, often covered with tubercles. They are very commonly met with, likewise, on the serous membranes of the thorax and abdomen, in the lungs, and occasionally in the substance of the liver. In *twenty-seven* out of *thirty-three* cases of hydrocephalus, Barthez and Rilliet found tubercles or granulations associated with inflammation of the pia mater; in *four* cases the meningitis was unattended by any trace of tubercular deposition in the encephalon; and in *two* cases the granulations, or meningial tubercles, were unattended with any traces of inflammation. In all the thirty-three cases the symptoms were nearly identical."

Causes.—Among the causes of this disease, therefore, we may consider the scrofulous diathesis as being the most potent; yet we can not always prove the fact until after the death of our patient. Yet an autopsy held on such a case is really valuable to the inquiring mind; for, in discovering the true pathology of the affection, we are not only enabled to comprehend its character, but also the *modus operandi* of treatment.

The exciting causes of the malady are not always so apparent. The author has known it to follow castration, puncture of the foot, staggers, and acute disease of the stomach.

In some cases, faulty nutrition is the exciting cause; in short, the symptoms sometimes supervene on the subsidence of some other malady.

Symptoms.—As regards the symptoms of hydrocephalus, they do not differ materially from those alluded to in the preceding article. At first, torpor, unconsciousness, unsteadiness on moving; dilation of the pupil; the animal gets upon the floor, in many cases never to rise again. Stertorous (grunting) breathing takes place; the animal tosses his head about in an unnatural manner, often throwing it backward---a very noticeable feature of this affection—occasionally raising it from the ground. Finally, convulsions set in, and the subject dies in that state, perfectly unconscious.

When hydrocephalus occurs after phrensy, or what is known as "mad staggers," it always proves fatal.

Treatment. — For the treatment of hydrocephalus give the fol lowing :

        No. 5. Fluid extract of buchu................ 4 oz.
               Water.............................. 6 oz.
               Iodide of potass.................... 2 oz.
               Mix.

Dose, two ounces, morning and evening.

The patient should have injections of soap-suds, once or twice daily ; and should the disease have made its appearance very suddenly, or, in other words, be of an acute character, so that the parts in the region of the brain feel hot, they should be then sponged very frequently with cold water; then give two drachms of fluid extract of gelseminum twice daily, until the pulse feels soft, or until the acute stage subsides.

STOMACH STAGGERS.

The cerebral disease usually denominated stomach staggers prevails among horses which are overfed, whereby the function of digestion becomes deranged ; and thus the food given accumulates, and finally gorges the stomach, producing cerebral derangement, which makes the horse reel and stagger like a drunken man. The horse may be said to be drunk from the effect of food. Sometimes the cause is accidental. A horse gets loose in the stable, and, finding a lot of meal or oats incautiously exposed, he devours voraciously a large quantity, and very soon after becomes the subject of stomach staggers.

Symptoms.—A stomach surcharged with food, without any accompanying distension, does not appear to occasion any local pain, but operates with that kind of influence upon the brain which gives rise to symptoms, not stomachic, but cerebral; hence the analogy between this disease and staggers, and the appellation for it of "stomach staggers." The unnaturally-filled stomach produces, for the first time, a sense of satiety ; the horse grows heavy and drowsy, reposes his head upon the manger, falls asleep, and makes a stertorous noise. All at once he rouses from his lethargy, and violently thrusts his head against the rack or wall of the stable, or any thing, in fact, that happens to oppose him, and in this posture paws with his fore-feet, or performs the same action with them as he would were he trotting, evidently all the while unconscious of what he is about. His eye, which at first was full of drowsiness, has now acquired a wild, unmeaning stare, or has already become dilated and insensible to light. The respiration is tardy and oppressed; the pulse slow and sluggish; the excretions commonly diminished.

The late Professor COLEMAN used to relate a circumstance, in his lectures, connected with this disease, which throws considerable light on its origin. The artillery horses stationed in London during the winter of 1817 suffered very considerably from stomach staggers; so much so that it was considered to be endemical, and of an infectious character. With his usual penetration, he soon discovered the cause, and found that, from some new regulations about that time, the stablemen were not allowed any candles, and during the winter the horses were bedded up at five o'clock in the evening, and not fed again until eight o'clock on the following morning, when they consumed their breakfast voraciously, gorging their stomach, not to the degree likely to produce acute indigestion, but sufficiently distending them as to oppress the blood-vessels and the circulation through them. This practice, continued day after day, caused a *specific* inflammation of the stomach—an inflammation of a peculiar character, differing from gastritis or inflammation of the part. The symptoms produced were regarded as resulting from the sympathetic connection between the stomach and the brain, united to the effects that would arise from the daily distension, throwing a vast quantity of blood on the brain. An order was obtained for candles for the use of the stablemen, which enabled the horses to be fed at a later hour in the evening, and an earlier one in the morning, when the disease disappeared.

A common error still prevails, in many districts, that staggers is a contagious disease; but should the horses on a farm be attacked occasionally with slight fits of this kind, the farmer may rest assured that there is mismanagement somewhere in the feeding department.

From such evidence as this, it will be inferred that there exists no doubt regarding the cause of stomach staggers.

Treatment.—We now propose to show how this disease ought to be treated. The proposition of cure is, that the digestive function shall be aroused, and the only way to accomplish that is by administering bitter tonics and stimulants. In this view, the following prescription is offered:

No. 6. Fluid extract of black pepper.......... 4 oz.
      Fluid extract of ginger................ 6 oz.
      Hyposulphite of soda.................. 2 oz.
      Water............................. 4 oz.

Dissolve the hyposulphite in the water, then add the pepper and ginger. Give the animal a wine-glassful every four hours. A stimulating injection may be thrown into the rectum occasiona.y, composed of a handful of fine salt to about four quarts of water.

The animal should be allowed to stand quietly in the stall, and the medicine must be given with care, for the least excitement may augment the cerebral difficulty. So soon as the medicine arouses the digestive functiын, and the food gradually passes the pylorus into the intestines, the animal will obtain relief. Both food and water should be withheld until there is some marked improvement; the patient has had enough of food for some time, and water only retards digestion.

Apoplexy (Cerebral Hemorrhage).

' As regards the cause of apoplexy, the author has nothing to offer, except he has noticed that the subjects of this affection generally have short, thick necks, and, as the saying is, " chunky " heads. From this he infers that, in so far as conformation is concerned, there lurks in the system of such animals a peculiar predisposition.

Symptoms.—An animal may be on the road, trotting along as usual, without any apparent impairment of health, when suddenly he falls down; the pupils of the eyes become dilated; stertorous breathing sets in; a deprivation of the sense of feeling and of motion immediately occurs; a tremulous motion of various parts of the body is observed; the pulse beats with unnatural force, yet the animal appears to be in a deep, snoring sleep. It may be said that the functions of animal life are suspended, excepting those of respiration and pulsation. The animal is unable to swallow, and if fluids be put into the mouth, they appear to choke him, or they run out again at the corners of his mouth. The prognosis of apoplexy is very uncertain. Some horses die in a few hours, while others live for several days. This depends on the amount ʼblood extravasated on the surface of the brain; but, in most cases of cerebral hemorrhage, the horse falls never to rise again. He may be conveyed home on some vehicle constructed for the purpose, but the finger of Death is on him; his days are numbered, and the owner charitably puts him out of existence.

The reader must bear in mind that in apoplexy the horse falls, and is suddenly deprived of all voluntary motion; is insensible to the prick of a pin, and the breathing, so soon as he is on the ground, becomes stertorous. This disease, therefore, is not to be confounded with others of the brain and spinal marrow. For example, a horse may have an attack of simple apoplexy without hemorrhage, fall down, and, by judicious treatment, recover; or he may be the subject of epilepsy, which occasions a temporary suspension of consciousness, with spasms recurring at intervals. Hence, in case of doubt or mistake, it will be advisable to treat the case in the following manner:

Treatment.—Procure a few ounces of spirits of ammonia, with which saturate a sponge, then apply it to the nostrils. In the mean time, sponge the head with cold water, and rub the body and limbs briskly with a brush or whisp of straw. If he revive under this treatment, there may be some hopes of recovery; and should it appear that the act of swallowing can be performed, give a drench composed of

No. 7. Powdered chlorate of potass............ 2 oz.
   Boiling water...................... ½ pint.

When cool, administer. The action of chlorate of potass on the blood is to oxygenize it, and thus liberate carbonic acid gas. With the same object in view, we apply ammonia to the nostrils, viz.: to decarbonize unpurified blood.

Blood-letting is inadmissible, for it can not accomplish any good, neither will it act as a purificator of the vital current which the lungs have fai'ed to arterialize.

APOPLEXY OF THE SPINE, (OR SPINAL HEMORRHAGE, PRODUCING PARAPLEGIA.)

Paraplegia signifies paralysis of the posterior half of the body. Spinal apoplexy may be classed in the same order of disease as cerebral apoplexy—that is to say, in so far as the pathology of the two forms is concerned—and is usually just as fatal.

5

Causes.—These are very obscure, except in cases of fracture of the spine, or injury to the same, by falling. Then the disease is accounted for. It always ends in paraplegia—palsy of hind limbs.

Symptoms.—The disease is usually sudden in its attack. A profuse perspiration ushers it in; next, the hind limbs fail to support the body, and the animal makes desperate efforts to support himself, and gradually crouches and falls to the ground, unable, perhaps, ever to get up again.

It has been noticed that paraplegia may also be occasioned by effusion of serum within the coverings of the spinal cord; yet we can not demonstrate the fact only by autopsy; and if we could, I fear that medicine would be of little value. However, if it is the owner's wish that the case should be treated, then the same course as recommended for cerebral apoplexy must be pursued.

EPILEPSY—FITS.

Epilepsy consists of a temporary suspension of consciousness. This disease, if properly managed, is not often fatal. It varies in duration. Sometimes a horse will suddenly fall, lose all sensibility and consciousness, exhibit spasmodic contraction of the voluntary muscles, go into convulsions, recover, and get up again in the course of ten minutes; or he may lie on the ground and have a succession of paroxysms, which may last for half an hour or more. If protracted beyond an hour or so, the patient is very apt to die. According to the best authority, "the functions that are affected in this disease are functions of the brain. Sensation, thought, and motion, regulated by the will, are the natural functions of that organ. The temporary abeyance of sensibility, thought, and volition,* and violent and irregular action of the muscles, which are thus withdrawn from the government of the will, constitute a paroxysm of epilepsy."

The fit is generally brought on by a derangement in the relation between the arterial and venous circulation within the head and a temporary pressure on the brain; in other words, a determination

---

* Horses think and reason just as man does. Their manifestations of mind do not differ from ours in kind, but only in degree. "The noble and daring war-horse, when he sniffs the distant field of blood, neighing for joy, instils a desperate courage into the veteran trooper's quailing heart, gives evidence of a soul, the proper attribute of man."

of blood to the head.   After the horse has fallen, by his struggles and herculean efforts to battle with the malady, although unconciously, he soon breaks out into a profuse perspiration.   This has the effect of relaxing the capillaries so that the blood circulates more freely and uniformly.   An equilibrium of the circulation takes place, and this is the end of epilepsy for the time being.   But a horse once having had a fit of this kind must be looked upon with suspicion; for he is liable, when under excitement from wanton punishment, or from exercising great feats of strength in drawing heavy loads, to have a re-attack.

Cause.—The predisposing cause of epilepsy has an hereditary origin.   Horses subject to it have a misshapen head.   It is not symmetrical—does not correspond with the conformation of the neck and body.   In the language of the turf, "the head is too coarse."   It has been found, also, among members of the human family, that epileptics have heads of an unnatural shape.   WATSON says: "There is no doubt that a tendency to epileptic disease is frequently hereditary.   It may be bequeathed from parent to child, or it may skip over a generation or two, and appear in the grandchild or great-grandchild; or it may be traceable only in the collateral branches of the ancestry."   Epilepsy, however, may not always arise as an hereditary affection; for a mere passive congestion of the brain, owing to a loss of equilibrium in the circulation of the blood, may produce it.   In regard to the horse, it is very difficult for us to decide on the universal hereditariness of the malady, because we have no reliable history of the ancestry and idiosyncrasies of our equine patients; yet if we study carefully the external conformation of well-formed horses, and make ourselves conversant with anatomy, we shall be better able to judge whether or no such an animal carries about with him the inherent tendencies to particular diseases; and this knowledge will make up, to a certain extent, for the advantage which human medicine has over the veterinary in this department of knowledge.   The principal symptoms of epilepsy are as follows:

Symptoms.—Suppose the horse attached to a vehicle, and traveling along at any given pace.   He gives a sudden, snorting, loud noise, and falls to the ground instantly, as if felled by some unknown power.   Here he lies, to all appearance, totally unconscious, violently convulsed in every limb, his eyes staring as though they would burst out of their sockets; the mouth foams

with saliva, and violent convulsions will sometimes affect the
whole frame.   Such are the principal symptoms attending this
formidable malady.

It will be quite a novelty to some of our readers to be .n-
formed that the horse is actually subject to the same diseases
which afflict his master; and in view of showing that there is a
reality in the author's views on the subject now under consideration, the following quotation from the "Veterinarian" is offered.
It will show, by way of comparison, the features of the disease
as it occurs in the horse and in man :

"In the first place, as to the exciting cause or causes.   It is
well known that among the exciting causes of epilepsy in the
human subject, mental or moral emotions have long been considered as holding a foremost rank.   Without going into detail
on this subject, or offering illustrations of this statement, it will
suffice to remark that the experience of all adequately acquainted
with medical literature, or tolerably familiar with medical practice, can not fail to supply them with numerous instances wherein
this class of causes has been in operation as concerned in the production of the disease in question.   Over and over again I have
myself witnessed cases of epilepsy, either during or after the paroxysm or fit, in which I have had good reason for entertaining
the opinion that certain mental or moral emotions had largely,
if not entirely, contributed to this result.   Now, without denying that, in some instances, fright may so far affect the horse as
to prove an exciting cause of epilepsy in that animal, still, I
think that, almost as a general rule, the class of causes now
under consideration may be excluded from further notice as tending toward the production of this disease in the horse.

Assuming this position for the sake of argument, we thus
eliminate at once, so far as concerns this animal, most of such
cases of epilepsy as, if speaking in reference to the human subject, would be regarded as cases of epilepsy of *centric* origin.
Without altogether denying that, in some instances, the presence
of tumors, or morbid growths, or excrescences of any kind, or
of spiculæ of bone in certain parts of the brain, or spinal cord,
or their membranes, may occasionally operate as causes of epilepsy, yet I am by no means prepared to admit that their presence in such places, by irritation of these structures, is so frequent
a cause of this disease as is affirmed by some medical authorities.

Yet, whatever view of the question, whether affirmative or negative, may be taken, the same reasoning will apply equally to man and the horse in regard to the disease under consideration. We are told that, in some instances, such foreign bodies have been found in these situations on examination after death. This I do not deny; but, at the same time, the evidence that these had much, if any thing, to do in the production of epilepsy, still less that the disease took its origin from them, is, in my judgment, far from complete, and, in some cases that have been recorded, unsatisfactory to the last degree. Again, the circulation of blood in an unhealthy state, in its accustomed channels through the substance of the delicately-constructed brain or spinal cord, appears to me not unlikely to be a frequent cause of this disease in man. The comparatively recent physiological researches of Dr. Brown-Séquard have thrown much light upon the heretofore obscure pathology of epilepsy in the human subject; and I conceive that his arguments, and the conclusions deduced therefrom, are, for the most part, perfectly applicable to an animal so high in the zoölogical series as the horse is well known to be."

Treatment.—So soon as the horse falls, some hay or straw should be placed under his head and around him. Bathe the region of the cranium with cold water, and carefully wash the foam from his mouth, taking care not to let any water, hay, or dirt enter the nostrils. Officious persons are very apt to attempt, by force of strength, to raise the horse on his legs; but this ought not to be done. Let him rest quietly until consciousness returns; then, should he attempt to get up, help from the bystanders may then be of some service. When on his legs, deal gently with him. Let the external surface of the body be rubbed until the skin is dry; then administer two ounces of fluid extract of valerian, and let the patient be provided with comfortable quarters. For a few days the patient should be excused from work, and be fed lightly. The only way to prevent a re-attack is to keep him at light work, and treat him in the most gentle manner, both in the stable and out of it.

In cases of this character, as well as many others, men are apt to place too much confidence in medicine, and dose the animal accordingly; but, as we understand the practice of medicine, it is the province of the good physician to know when to do nothing.

TETANUS, OR LOCKED JAW.

We recognize tetanus by the unnatural condition of the muscles of various parts of the body, more particularly those of the face and neck. But the fault lies not in the muscles, but in the nervous system. This disease constitutes one of the most terrible and fatal to which the horse is subject. It is not only frightful to look upon, but is productive of the most distressing agony to the animal.

THE TEST FOR TETANUS OR LOCKED-JAW.

EXPLANATION.—On placing a hand under the angle of the jaw, and pushing the head in an upward direction, the nictitating membrane (haw) will be suddenly thrown partly over the pupil, as seen in the above picture. This is the diagnostic symptom of tetanus.

There is no difficulty in recognizing this disease. It is characterized by involuntary and persistent cramps of the voluntary muscles. The muscles which seem first to be affected are those of the jaws, neck, and throat; and soon all the other muscles are involved. One of the principal tests for the discovery of the disease is to place a hand under the jaw and raise the head. Immediately the haw, or nictitating membrane, shoots over the eyeball. (See cut.) There is no other disease with which tetanus can be confounded; therefore, a minute description of the symptoms seems superfluous at this point of our argument.

When the disease affects the muscles of the jaw only, it is called trismus; but it is only a variety of the same disease, and, therefore, requires the same treatment. In former years the author met with but little success in the treatment of this malady, yet recently, by using more mild and sanative agents, he has been very fortunate. Bleeding, purging, and blistering has had its day, but that day has now past. We have learned that to do violence to the animal system is not to do good; and our aim now is to "pair off" with Nature, endeavor to sustain the vital powers, or, rather, by sanative medication and nutritious fluids, keep the animal alive, while the disease runs its course. Physicians have no power to arrest the disease, and those who think so only deceive themselves and their employers; and those who attempt the feat of cutting the disease short by heroic medicines, are arrayed in hostility to Nature, and an unnecessary death is often the consequence. Any of our readers who happen to have great faith in drugs will probably feel little comfort in the perusal of the following quotation, uttered by one of the most distinguished physicians of France. It was intended for the benefit of mankind, but it also applies to veterinary medicine, simply from the fact that the diseases of animals are to be treated on the same general principles which apply to man :

" The sick-room no longer resembles the sample department of of a drug warehouse. Our physicians have consciences and common sense. They recognize Nature as the great antagonist of disease, and endeavor to assist her in her struggle to expel it, instead of negatively helping disease by prostrating the physical strength of its victims with drastic cathartics, cantharides, and the lancet. In ailments for which gallons of medicine were given half a century ago, as many ounces are not administered at the present day, and bleeding and blistering have almost fallen into disuse. Not long before his death, the great French surgeon, physician, and medical chemist, Majendie, told his pupils, in the college of France, that the old hospital practice was mere humbug; that he himself had prescribed the drugs of the dispensary at the *Hotel Dieu*, in Paris, without having the least idea why or wherefore, and that on administering bread pills and colored water to one division of his patients, and the preparations of the pharmacopœia to another, he found that the mortality was least among those who took no medicine! 'You tell me,' said this extraordinary man, in one of the lectures of his final course, 'that doctors cure people. I grant you people *are* cured. But how? Gentlemen, Nature does a good deal ; imagination does a good deal. Doctors do very little, wher they don't do harm. You ask, then, what is the use of attending medical lectures? I'll tell you. We come here (to the college of France) to study Nature, to learn to reinforce and aid Nature, not to spin fine theories. I would not give a *centime* for all the theories in the world. Give me stubborn facts.'

" Such was the bold and candid language of one of the greatest anatomists and physicians of the age but a short time before his death. Our medical men do not take quite such strong ground as Majendie took, but they show their lack of faith in what was formerly called ' vigorous treatment' by sparing medication, and a still more limited use of the lancet and other mechanical means of depletion. Nature, after having been professionally misused for centuries, seems at last to have found a friend and ally in the medical faculty."

We now come more directly to the *treatment* of tetanus; and, in view of showing what unwarrantable outrages are perpetrated on the poor uncomplaining brute, the following article is here introduced:

" Wounds of tendinous and ligamentous parts are the common cause of tetanus, more generally known as locked-jaw, so-called because the first symptoms of the violent spasmodic affection are detected in the jaw. Castration, nicking, docking, lacerations, and punctures, particularly of the feet, are the kinds of wounds that end in tetanus. Even slight contusions will bring it on. It is a dreadful and, too often, a fatal disease. The animal is nearly paralyzed by the constant spasm of all the voluntary muscles. The symptoms are unmistakable: First, a certain stiffness about the throat, and difficulty in swallowing or turning the head. This gradually extends to the jaws, contracting the mouth. The horse possesses a desire to masticate, but, in the earliest stage of the disease, the work is imperfectly performed, with great pain. The eyes become vivid in appearance, and present a retroverted aspect. The disease continues to spread, and when it affects the voluntary muscles of the trunk and the extremities, the animal becomes a living picture of agony and distress. The cocked yet quivering tail, the distended legs, the contraction of the abdominal muscles, the tremulous and irregular pulse, the hurried respiration, the sweat-marks, and the fixed and sunken eyes, all speak unmistakably of the violent and painful nature of the disease. The animal sometimes suffers for six or seven days, when it falls, completely exhausted by continued irritation and inanimation; or it expires in terrible convulsions. Tetanus, indisputably, is a nervous affection. An injury to a small fiber of some nerve extends to the origin of the nerve, when the brain becomes affected, and a diseased action of the body ensues. When locked-jaw results fatally, it will be before the expiration of nine days, as the horse dies, without a remission of the spasms, from sheer exhaustion. The free use of the lancet is regarded as one of the surest means of curing tetanic cases, as by a flow of blood we open the bowels and tranquilize the system. Venesection is the most powerful sedative for muscular spasms. The nervous influence passes off with the blood; therefore the flow should be most copious. We may bleed the horse until he falls, before the triumph is complete. The permanent strength of the animal is not hazarded by the free use of the lancet, but we simply make an attack upon the seat of the disease. Physic, also, is a matter of much importance. Profuse bleeding will cause the muscles of the jaws to relax, when the dose may be introduced into the mouth. Clysters assist the action of the purgatives; therefore are useful. The application to the spine of green sheep-skins, warm from the slaughtered animal, will sometimes diminish the sufferings of the patient. By gently rubbing the spine with the hand, and then by using an opiate liniment, a relief may be afforded. From eight to ten drachms of aloes should be administered as a physic. The farina of croton-nut is also a powerful purgative. A solution of Epsom salts constitutes a safe injection as a clyster. Opium, when given in doses, varying from one half to two drachms, is a valuable medicine in cases of lockjaw. The disease is a violent one, and to check it prompt measures must be resorted to."

Contrast this with the following case: While in the city of St. Louis, a short time ago, I was requested to visit a bay gelding, aged eight years, the property of Captain SILVA. The messenger informed me, not knowing that it was a case of tetanus, that the horse was "all stiffened up." On an examination, the following symptoms were presented: The muscles in the region of the neck and back, rigid; the eyes had a sort of squinting appearance; the nose protruded; the ears were erect and stationary; the nostrils were expanded to their utmost capacity; the head, neck, and trunk seemed to be immovable, so that it was impossible to make him turn in any direction, or describe the least segment of a circle.

The abdomen appeared " tucked up," as the saying is.   This arose
from the rigid state of muscles of the abdomen.   The hind limbs
presented a straddling appearance, and the fore ones were unnaturally advanced far beyond the axis of the shoulder-blade; the
bowels were constipated ; the pulse was wiry, and the respirations
were accelerated and laborious.   The case was diagnosed as tetanus
from puncture of the off hind-foot.   It appeared that the animal
had picked up a nail, five days previous, which was withdrawn by
a blacksmith, and the foot was dressed in the usual way.   The
treatment was as follows :

The rigid muscles were rubbed, twice daily, with a portion of
the following antispasmodic liniment:

No. 8. Cod-liver oil........................ 12 oz.
       Oil of cedar......................... 4 oz.
       Sulphuric ether...................... 3 oz.

The application of the liniment was continued during a period of
ten days, in which time four drachms of Indian hemp were daily
placed on the tongue.   The jaws now relaxed, so that the patient
could eat bran mashes.   At the expiration of two weeks all signs
of tetanus had disappeared.

VERTIGO (DIZZINESS OR GIDDINESS).

The term vertigo signifies giddiness or dizziness.   It is usually
the result of some latent disease within the brain ; and when that
is present, any exciting cause, which we shall presently refer to,
may bring on a fit of vertigo.   The following article, by surgeon
HAYCOCK, will, perhaps, prove interesting to the reader :

" By vertigo is meant a chronic disease of the horse, chiefly in
dicated by a disturbance of the sensitive faculties, occasioning
derangement in the ordinary functions of life.   Much that is incorrect has been written regarding the seat, properly so-called, of the
evil.   At present, most veterinary surgeons are agreed in seeking
the proximate cause, not as formerly, on the brain, but in the abdominal organs, and on considering the cerebral affection as purely
secondary."

Vertigo often succeeds acute disease of the brain.

Causes.—Its chief exciting causes are confinement in hot and
badly-aired stables, cold, extreme fatigue, blows and injuries on
the head, indigestion, unwholesome or too much food in proportion

to the exercise taken. The fear of punishment, especially of the whip, occasionally gives rise to it, in sensitive and irritable animals. Some horses have an hereditary predisposition to it, and mares are considered more subject to it than stallions. Further, it is scarcely ever observed except in hot weather; and as it is generally at the beginning of summer that it commences to appear, it goes away always in autumn, at least with respect to its chief symptoms.

Symptoms.—The horse having, previous to the attack, been lively and active, begins, all of a sudden, to appear heavy and indolent. He is dejected, and prefers to keep himself in the darkest corner of the stable. Eyes, dull; look, fixed and stupid; eyelids, half shut; inattention to every thing, forgetting even himself, and, as it were, asleep, his head hanging down, or resting on the manger. His gait is heavy, slow, and unsteady; he raises his feet very high, and puts the entire sole to the ground, raising and letting down the limbs in a manner purely mechanical, and, as it were, unconsciously. He exhibits much awkwardness in turning, and can not be pulled back except by depressing the head very much, and pushing it latterly. He also leans to one side in walking. To maintain his equilibrium the better, he places the fore-legs beneath the belly, and moves his ears backward in a peculiar manner. According as the disease progresses, he becomes less and less sensible to external impressions. Mastication is performed slowly. He takes, from time to time, a mouthful of food, masticates it, swallows a portion of it, but keeps the remainder in his mouth. He prefers taking his food off the ground rather than in any other way, and when drinking, he plunges his head into the water, even above his nostrils. During and after some rather violent movements, his symptoms become much aggravated, and the signs of complete insensibility become more and more marked. The animal runs on quite blind till some obstacle stops him, or turns round, or remains tranquil, with his head depressed, and the legs crowded beneath the body, without being able to change this unusual attitude, unless assisted to do so. There is never any fever. The pulse is often from ten to twelve pulsations slower than in the normal state.

In the same way, also, the respiration is constantly slow, deep, and frequently of a sighing character. In almost all cases, the tongue is foul, and the mouth dry and clammy. With respect to treatment, the remedies which have succeeded best with me are, camomile (some doses), then sulphur and nux vomica. In a particular case, where, independently of the symptoms peculiar to vertigo, the conjunctiva, tongue, and mouth were more yellow, the horse frequently flexed his fore-legs, seldom lay down, the fæces were hard, and he passed but little urine.

The reader will perceive that many of the symptoms above alluded to are present in other diseases of the brain; therefore, some difficulty may be encountered in diagnosing the case. It is best for us, however, to diagnose the case on the symptoms of dizziness or giddiness; then a mistake is not likely to occur. Yet, after all, a mistake in the true nature of the malady will not prove disastrous, provided the patient be treated on the general principles laid down in this work, by means of sanative medicines and little good common sense.

Treatment.—For example, should a horse have an attack of vertigo on the road, the driver must immediately stop and loosen the throat-latch and check-rein. Then let the animal stand in quietude for a few minutes, during which time he may possibly recover, and soon be able to resume the journey. If not, he must be taken out of the harness, and carefully led to the nearest stable, or where he shall be free from annoyances of every kind. After the excitement is over, the animal may be led home, and put into a roomy stall, where he must be dieted according to his condition. If fat and plethoric, bran mashes are indicated, into which should be stirred a drachm or two of hyposulphite of soda. Should the patient be poor in flesh, a few good oats are indicated, to which add a small quantity of powdered ginger and balmony. The surface of the body being cold, it must be warmed, after the usual fashion, by means of clothing. But should the surface of the body feel hot, more particularly in the region of the head, then tonics, stimulants, and food are to be withheld. A bucket of cold water is then to be placed before him, into which about four drachms of nitrate of potass may be stirred.

INFLAMMATION OF THE BRAIN (BLIND, SLEEPY OR MAD STAGGERS).

The contents of the cranium are called, collectively, the encephalon; hence the term encephalitis, which signifies inflammation of any, or all portions of the contents of the cranial cavity. Various terms are applied to disease of this character, such as "sleepy staggers," "coma," "phrenitis," "cerebritis," and "cerebral meningitis," the latter being formerly recognized as blind or sleepy staggers. These several terms merely apply to the various stages of the acute disease as it gradually invades the membranes covering the brain, or the substance of the brain itself. It sometimes appears to invade at once the whole of the parts within the skull, or, beginning in one part, it extends rapidly to all the rest, so that the term encephalitis seems to be more applicable than those just enumerated. It is a matter of impossibility for us to tell precisely what are the pathological conditions of the parts affected. Nor are the symptoms always the same. They may range from a state of phrenzy to one of coma. Still, in our treatment, we shall not be led astray; for, being an acute affection, (or affections, as some persons may term it,) we have to treat it on the same general principles which obtain in many or all diseases of an acute character, viz.: by means of sedatives, laxatives, cold water, spare diet, and rest.

Should the patient die during the acute stage of disease of the brain, an autopsy will reveal great vascularity and softening of the cerebral mass, and thickening of its membranes; but should the disease run on unchecked to a fatal termination, pus and fluid may be found within the lateral ventricles of the brain. This enables us to explain the difference between the symptoms which prevail in the early and latter stages of the malady; for, at the commencement of the acute stage, the loss of equilibrium in the circulation sends the red arterial blood, in undue quantities, to the brain—the part, perhaps, most predisposed to diseased action, or, it may be, at the time actually in a pathological condition; hence the loss of equilibrium in the circulation—which, in consequence of accelerated respiration, becomes highly charged with oxygen, acts as a potent stimulus, not only to the nervous system, but to the muscular system also, producing those active and phrenitic symptoms which have led us to infer that the patient is going or is actually mad; hence the name which some persons have applied, "mad staggers." This activity can not last long; for it is potent to exhaust the vital forces. Organs and parts of the body become overworked; then comes organic changes—softening of the brain, effusion, formation of pus, which.

ends in coma, which is the latter stage just referred to, devoid of
any mad or phrenitic symptoms, but marked by lethargy, or a
prolonged comatose condition, from which no stimulus will
arouse the patient. He is not only the subject of altered structure within and around the brain, but the latter is compressed by
the serum, or pus, as the case may be, and thus sensibility is
more or less destroyed. The degree of coma, however, will be
directly proportioned to the degree of pressure. Now, the reader
will perceive that the proposition of treatment, just referred to,
for the acute stage will not apply to the comatose condition; for
here we have to stimulate and give tone to the system, so as to
sustain the failing vital powers.

In regard to this subject, Professor COPEMAN observes: "In
inflammation of the brain, phrenitis, mad staggers, I have told
you that it may be said to be impossible to distinguish, with any
thing like certainty, in individual cases, acute inflammation of
the substance of the brain from those affecting the membranes
or ventricles. The reason will be obvious when we reflect that
the phenomena, in every instance, are, in fact, attributable to
pressure on the entire contents of the skull, encephalon, viz:
cerebrum, cerebellum, medulla, oblongata, and membranes; and
if this be rapid and general, it can matter little whether it originates from the membranes or the *brain*. Acute inflammation
does, however, sometimes appear to invade at once the whole of
the parts that are lodged within the skull; or, beginning in one
part, it extends rapidly to all the rest. The symptoms which
usually mark an attack of *cerebritis* are the following: The horse
becomes sleepy, heavy, or more or less comatose, accompanied by
general fever. After a time a convulsive attack supervenes. He
becomes morose, and shows delirium; perhaps rears both fore-
legs into the manger, from which position he may reel around
and fall, kicking and tearing every thing about him. During
the fit his respiration is much excited, and he sweats profusely
If he gets upon his legs, he makes sudden and violent efforts,
dashing against rack, manger, or stall, totally disregarding any
thing that may be said or done to him. Thus he is continually
struggling, panting, and perspiring, perhaps foaming at the
mouth, leading the ordinary observer to believe he is not only
delirious but actually 'mad.' But all these symptoms vary much
in different cases. Coma occurs frequently, but often only tem-

porarily. Great prostration and muscular debility are generally observed.

The seat of 'sleepy staggers,' or cerebral meningitis, is the so-called subarachnoid cavity, in which is a quantity of loose areola tissue, richly furnished with blood-vessels. It generally results that the exudation poured into this cavity is quickly accompanied or followed by pus. Hence, we find that what is generally called a recent layer of coagulable lymph, covering the convolutions in meningitis, is, in point of fact, a layer of pus, generally presenting a molecular character. As to the diagnosis, notwithstanding the efforts which have been made to distinguish meningitis of the convolutions from that of the base, or either of these from a simple effusion into the ventricles, I have in vain sought for any precise symptoms, which could be relied on, as indicative of the situation of the disease. Drowsiness and coma, causing slow and subsequently rapid pulse, succeeded by restlessness. The horse is excited; he flings himself about, frequently jerking his head up and down, sometimes rearing, perhaps, into the manger; tension of the limbs, thrusting the head into the *rack*. The faculties of the organs of sense are lost, for the horse neither hears nor sees. The state of excitement may terminate, more or less quickly, in convulsions and death, or the patient may relapse into a state of coma, and ultimately result in partial or complete recovery.

The gradual mode of invasion, and the succession of the symptoms to one another, are also characteristic, and differ markedly in degree from those which attend sudden attacks of apoplexy caused by hemorrhage. They are both the result of general pressure on the brain, and hence the reason why mere effusion can not be distinguished from hemorrhage."

Treatment.—The treatment of the preceding forms of acute disease of the brain was formerly, and is at present, to some extent, conducted on the absurd antiphlogistic plan, by blood-letting, purging, and blistering, which practice kills more than it ever cures. The plan now adopted by the author is to give drachm doses of gelseminum every four hours, until relief is apparent. The cranial region is kept constantly bathed with cold water; the rectum is kept free from excrement by means of injections of soapsuds; the bowels are kept in working order, by mixing Glauber salts with thin bran mashes. Half a pound of salts, dissolved in about four quarts of mash, will generally prove laxative. Should it fail to have this effect, after a lapse of about six hours, the dose may be repeated. This plan of treatment is more rational, and has proved more successful, than that just alluded to.

Should the disease progress so that the animal manifests symptoms of coma, or lethargy, then chlorate of potass is the best agent. It should be given in half-ounce doses, every four or six hours, in the form of drench, or it may be dissolved in the water which the animal is allowed to drink. A few doses of the following preparation must also be given:

No. 9. Fluid extract of golden seal...... } each 4 oz.
        Fluid extract of juniper.......... }
        Mix.

Dose, two ounces every morning.

Keep the rectum empty by injections, and, if the case be curable, such treatment as this, followed up by careful nursing, will accomplish the object. COPEMAN, who is authority in this disease, fully indorses this treatment, and says:

" Hitherto the treatment of meningitis (sleepy staggers), whether real or supposed, has been antiphlogistic, but it is impossible to say that any benefit has ever been effected by the practice. The early stages of the disease are probably generally overlooked. So long as the horse retains his appetite and his consciousness, no suspicion of disease arises. It is only when exudation or effusion has been poured out in such quantity as to cause drowsiness and stupor that our suspicions are awakened, and thus it is very difficult to understand how blood-letting or purging could facilitate its absorption. Besides, we have seen that the tendency of such effusion is to pass into the circulation. Hence, the treatment which favors the reabsorption of the exudation, as I have previously explained, must be most effectual. For this purpose time is required, and the vital strength, instead of being lowered, should be supported. In short, the duty of the practitioner is to support the economy as much as possible, to give nutrients with moderate stimulants, to unload the bowels, from time to time, artificially, by injections, etc., and in this way to gain time, which will enable the effused matters to pass through their natural transformations, to be absorbed and ultimately excreted. It has appeared to me that the collection of serous fluid, whether in the ventricles or over the surface of the brain, either with or without exudation, is consecutive on obstruc-

tion of the vessels, and is, therefore, more allied to the dropsies
than to inflammations. It is the collection of serum which does
the mischief—presses on the brain, and causes the coma and stupor. If so, the occurrence of those symptoms should be regarded
as secondary, instead of as primary, and as analogous to ascites,
dropsy, or anasarca, following disease of the kidneys. These
pathological considerations are, it appears to me, wholly opposed
to the idea of blood-letting and antiphlogistics being beneficial
after effusion has occurred."

EQUINE CHOREA, OR STRINGHALT.

Chorea, or stringhalt, consists of an irregular and involuntary
spasmodic action of some of the muscles of the hind extremities.
Equine chorea differs somewhat from human chorea. In the latter case, it usually begins with slight twitches in the muscles of
the face, or in the upper extremities, and various parts of the body
twitch and contort in such a singular and unnatural manner,
that some persons have denominated the disease "insanity of the
muscles."

In the case of horses, no such "insanity of muscles" has been
observed. It is mainly confined to the posterior limbs. Mr.
FERON, a distinguished V. S., contends that stringhalt bears
some affinity to what is known in human medicine as *chorea*, or
"St. Vitus' dance." He does not, however, wish to convey the
idea that they are essentially the same disease, only they are both
of a convulsive or spasmodic character, wherein the mind, will,
or instinct has lost more or less of its control over the voluntary
muscles of the hind extremities, and the peculiar feat is thus accomplished. When the animal has lifted his hind leg from the
ground, which is always done with a convulsive twitch, the fetlock nearly approaches the belly, and, by some other remarkable
irregularities in its action, before the foot can be replaced on the
ground, displays such unnatural movements as to convince us that
volition is impaired. Hence, we may infer that, in a majority of
cases, stringhalt is the result of some abnormal condition of the
nervous system. What occasions that condition is a matter of
fact and argument, which remains as an open question for some
future pathologist to decide. So far as the author's experience
goes, he is satisfied from actual dissection of the parts, that some

stringhalt horses are the subjects of ulcerative disease of the hock-joint, and they finally become sprained.

In such cases, it is possible that stringhalt may be the result of hock disease, which irritates the nerves in the vicinity of the part, and thus affects that portion of the nervous system which controls the muscles concerned. ANDRAL tells us that "chorea, like epilepsy, may be excited by irritation." Hence, in the commencement of spavin, or so soon as osseous incrustations are thrown out, they may, during the active motion of the joint, produce local irritation, of nervous filaments, in the vicinity, which irritation may be communicated to larger and more important nerves, and thus induce stringhalt.

We all know that local irritation produced by the prick of a nail, or sometimes docking and pricking, is often productive of locked-jaw, simply because some nervous or tendinous structure is injured; and so the local irritation in the region of the hock may be productive of the malady now under consideration. The author has paid considerable attention to this subject, and is convinced that almost all horses of the nervous or excitable temperament, the subjects of inter-articular spavin, manifest more or less catching-up of the limb (stringhalt) at some period during the progress of spavin, and that when the bones of the hock are anchylosed (united), and the motion and irritation of the parts have ceased, the stringhalt becomes modified. The reader, however, must not receive this opinion as absolute; for at times, in consequence of some peculiar idiosyncrasy, the stringhalt gradually grows worse. Among some members of the human family, of the peculiar temperament to receive the impression, almost any thing which makes a forcible impression upon the nervous system may act as an exciting cause of chorea. The subject is a very important one, and most veterinary writers of the English school have rather mystified the subject, and at last have come to the conclusion that the disease is incurable. This was formerly the author's opinion, but a change has taken place in his sentiments, and he is unwilling to deprive the afflicted animal of the benefits of progressive science. More light on the subject may enable us to effect many cases of cure; at least it will enable us better to understand the pathology of the case, which is an important step in the right direction. In view, therefore, of furnishing the reader with some useful information, whether he be a veterinary ractitioner or not, we shall borrow from analogy, in the introduction of the following article from "Watson's Practice":

" In certain of M. Magendie's experiments on animals, the following curious facts were ascertained : When a vertical section of the cerebellum of a rabbit was made, leaving one-fourth of the whole adhering to the crest of the right side of the cranium, and three-fourths to that of the left, the animal rolled over and over incessantly, turning itself toward the injured side. The same phenomenon occurred upon the division of the crus cerebelli. The animal lived for eight days, and continued, during the whole of that time, to revolve upon its long axis, unless stopped by coming in contact with some obstacle. How like is this to the symptoms exhibited at one period in the girl whose case is related by Dr. Watts. Nor is Dr. Watts's case a singular one. M. Serres has described another much resembling it. A shoemaker, sixty-eight years old, of intemperate habits, after one of his debauches, exhibited a kind of drunkenness which surprised his friends. Instead of seeing objects turning around him, as a drunken person is apt to do, he thought he was himself turning, and soon began to revolve, and this lasted till he died ; and when his head was examined, extensive mischief was found in one of the pedicles of his cerebellum. Again : M. Magendie noticed that when the upper part of the cerebrum is gently removed in birds and mammalia, they become blind ; but no affection of the locomotive powers is produced. No further result is occasioned by the removal of a portion of the gray matter of the corpora striata ; but when the striated part is cut away, the animal immediately darts forward with rapidity, and continues to advance as if impelled by some irresistible force, until stopped by an obstacle ; and, even then, it retains the attitude of one advancing. The experiment was tried, with the same results upon various species of animals—dogs, cats, hedgehogs, rabbits, Guinea-pigs, and squirrels. It seems that there are horses that can not back, although they make good progress enough in a straightforward direction. Now, Magendie says that he has opened the heads of such horses, and has always found in the lateral ventricles of their brains a collection of water, which must have compressed and even disorganized the corpora striata. It has been further ascertained, by the same experimenter, and by others, that certain injuries of the cerebellum cause animals to move backward contrarily to their will. If the tail of

the animal so mutilated be pinched, he still persists in his retrograde course.   Injuries of the medulla oblongata had the same
effect.   Pigeons, into which he forced a pin through that part,
constantly receded for more than a month, and even flew backward.   A section of the medulla oblongata, where it approaches
the anterior pyramid, gives rise to a movement in a circle like
that of a horse in a mill, the animal, in its walk or its flight,
bearing round continually to the injured side.   Surely we have,
in these facts, supplied by experiments on living animals, and by
observation of the phenomena of disease in the living human
body, some of the materials for a more exact knowledge, both of
the physiology and of the pathology of the nervous system, than
we have reached.   M. Magendie supposes that different portions
of the encephalon are endowed with energies which tend to cause
motion in various directions; that in the healthy state these balance each other, and that a preponderating impulse can be given
to any one of these forces by the will; but that when the equilibrium is destroyed by disease, the will is not sufficient to counteract the tendencies which are then brought into play.   Mr Mayo
offers a different explanation of the phenomena.   He supposes
that the injuries inflicted on the nervous matter produce a sensation analogous to vertigo, and that the animal conceives itself
either to be hurried forward, and makes an exertion to repel im-
aginary force, or to be moving backward, or turning round in
one direction, and endeavors to correct this by moving the corresponding muscles."

   Treatment.—In stringhalt, it is nearly always safe for us to
conclude that it must be treated on the same general principles
which obtain in the management of other nervous disorders of a
chronic character, viz.: in the use of tonics and anti-spasmodics
Take, for example:

   No. 10. Fluid extract of valerian.......... } equal parts.
            Fluid extract of poplar bark....... }

   Dose, one ounce, morning and evening; to be placed on the
tongue.   The spine and affected limb or limbs should be rubbed
every night, for a couple of weeks, with a portion of the following

   No. 11. Fluid extract of poppies............. 6 oz.
            Proof spirit........................ 1 pint.
            Mix.

Should the animal prove to be spavined, the following liniment is recommended:

No 12. Cod-liver oil .... } equal parts.
       Kerosene oil ....
       Mix.

A small quantity of this liniment should be rubbed on the seat of spavin, inside of the hock, once or twice daily, until counter-irritation is accomplished, which shall be known in consequence of the hair falling off; then discontinue the liniment, and lubricate the affected parts with olive oil, and wait patiently, so as to see what Nature will do for the case. "Patient waiters are no losers." The popular, or rather the ancient, method of treatment does more harm than good.

SECTION IV.

DISEASES OF THE RESPIRATORY PASSAGES AND ORGANS.

IMPORTANCE OF VENTILATING STABLES—BRIEF EXPOSITION AND DESCRIPTION OF THE FUNCTION OF THE LUNGS—SPASM OF THE MUSCLES OF THE GLOTTIS AND EPIGLOTTIS—LARYNGITIS, (SUPPURATIVE AND INFLAMMATORY)—CROUP—CHRONIC COUGH—ROARING—POLYPUS—BRONCHOCELE—INFLUENZA, OR EPIZOOTIC CATARRH—PNEUMONIA, OR INFLAMMATION OF THE LUNGS, INCLUDING TYPHOID AFFECTIONS, PLEURISY, AND DROPSY OF THE CHEST.

IMPORTANCE OF VENTILATING STABLES, IN VIEW OF PREVENTING DISEASE OF THE LUNGS.

IT was the intention of the Creator that all animals, so long as they were permitted to exercise their natural instincts, and thus comply with the requirements of physiology—the science of life—should enjoy health and long life. Hence a great amount of disease and death results from the evils of domestication.

One of the conditions which physiology imposes, in order that a horse shall enjoy health, is, that the atmosphere, at all times, and under all circumstances, shall be uncontaminated, so that the blood shall be decarbonized and purified of the defiling elements acquired in the course of circulation.

Let the reader understand that the lungs are something like a sponge, elastic, composed of a myriad of cells. In the former, however, these cells have a vast internal surface, communicating with each other up to their common origin, the bronchial tubes and windpipe. On their internal surface we find a delicate yet highly important membrane permeable to atmosphere. In extent, it is supposed to occupy a square surface equal to that of the external body. In contact with this membrane comes the atmosphere. If pure—zephyr-like—it fans into healthful blaze the flame of life, upheaving from the living Vesuvius arid lava, in the form of carbonic acid gas, almost as destructive to animality as that issuing from its great prototype proves to vegetation.

The stable atmosphere being pure, and the lungs in working order, the blood is well arterialized, capable of supplying the waste of the animal machine and renovating its tissues. On the other hand, should the atmosphere be impure, it fails to vitalize the blood. The latter is unfit for the purpose of nutrition, and may be considered a non-supporter of vitality. Hence the need of pure air, the breath of life.

But are horses always furnished with pure air? Let the owners of unventilated, crowded, filthy, down-cellar and low-roofed stables answer. Let those who have stables in the region of swamp, sewer, and stagnant pools of water answer. In such locations disease and death run riot, and the noble companion of man, instead of being within the ramparts of the science of life, is on the margin of death's domain. He may exist for several days without food and water, yet the consequent result is nothing when compared to that occasioned by breathing an atmosphere highly charged with emanations arising from his own body excrements and decomposing bedding.

A horse is said to consume in the lungs, in the course of twenfour hours, ninety-seven ounces of carbon, furnished by venous blood. In order to perform this feat, he requires 190 cubic feet of oxygen. Now, suppose there are ten horses occupying the stable. They require, in the same time, 1,900 cubic feet of oxygen, and consume 970 ounces of carbon. They are supposed, also, to give out from the lungs a volume of carbonic acid gas equal to that of the oxygen inspired; and supposing the atmosphere to be saturated with only five per centum of the former, it is a non-supporter of life. Hence, a horse shut up in an unventilated stable must, sooner or later, become the subject of disease. The evil may be postponed, but the day of reckoning is sure and certain.

Diseases, such as horse-ail, influenza, catarrh, strangles, and glanders, often originate and prevail to an alarming extent in the unventilated stable and pest spot; while in other locations, favorable to the free and full play of vital operations, the favored ones seem to enjoy a remarkable immunity from the prevailing disease, or epizoötic.

Stablemen and husbandmen are often led to remark, that when

they keep but few animals, disease and death, except in cases of accident or old age, are quite rare, but so soon as they crowded the same, sickness and death were the consequences. In view of supporting this theory, we may be permitted to remark that ship and jail fevers may be manufactured *ad libitum*, at any time when a large number of persons are congregated together in a given space, no provision having been made for the admission of pure air. The unfortunate prisoners in the Black Hole of Calcutta are an example, and the mortality occurring on board our emigrant ships furnishes another illustration.

A number of horses were once shipped from England to Spain, and on the passage, a violent gale arising, it became necessary to batten down the hatchway. The consequence was that most of them ultimately died of either glanders or farcy. We contend, therefore, that the active or morbid germ of disease enters the living citadel through the pulmonary tissue in an insidious manner, and, therefore, much oftener than the generality of men would be likely to realize. Therefore, it is a matter of vital importance that attention be paid to the ventilation of our stables. If proper sanitary regulations were established, and fully carried out in all our stables, glanders and other infectious diseases would be exceedingly rare. They are so among horses free from the control of man, whose stalls are broad as from ocean to ocean, their height ranging from earth to regions above, the space pervaded by a pure atmosphere concocted by the Great Chemist, pure as the pearly drops and refreshing as the morning zephyr. In such locations death has no terrors nor disease any victims.

BRIEF EXPOSITION OF THE FUNCTION OF THE LUNGS.

The principal function of the lungs is to arterialize or decarbonize the blood; that is, purify it. This arterialization of the blood, which goes the rounds of the circulation, is more essential to life than either food or water; for men and animals can exist for several days, perhaps for two weeks, without food, yet the same can not live over a few seconds unless supplied with a sufficiency of atmospheric air. Hence, in a popular sense, pure air is the "breath of life."

The functional acts of respiration are necessarily divided into two parts; and in cattle the number of respirations are about twelve per minute, varying, however, according to the temperament of the animal and the condition he may be in at the time of making the observation; while in horses the respirations are more frequent, varying in health, and when at rest, from fourteen to twenty-five; yet, under excitement and disease, they sometimes number over one hundred.

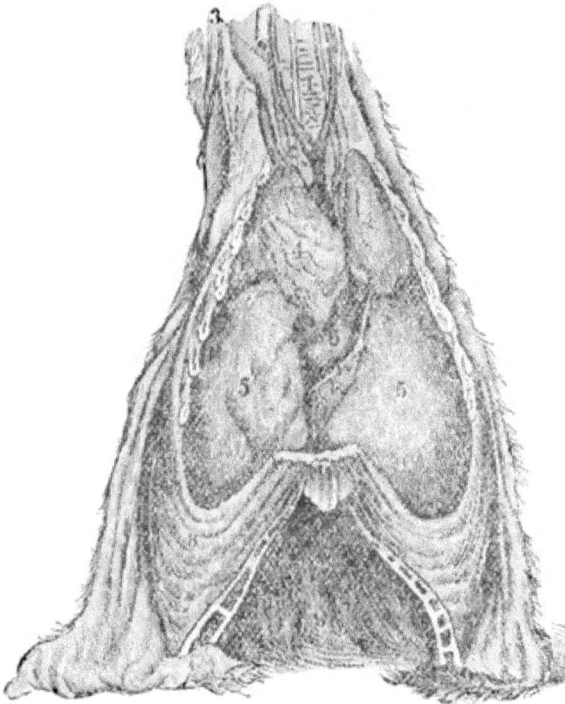

THE CONTENTS OF THE THORAX.

EXPLANATION.—Fig. 1, Trachea ; 2, Bifurcation of the carotid artery ; 3, Internal carotid artery ; 4 4, Anterior lobes of the lungs: 5 5 Posterior lobes of the lungs; 6, The heart; 7, Coronary artery; 8, Cartilages of the false ribs; 9, The diaphragm.

When an animal is located in a pure atmosphere, and the lungs are in good working condition, all the impurities contained in venous blood are brought into the presence of oxygen through the lining membrane of the air-cells, and thus a change in the color and character of the blood is immediately effected. In the first place, the venous blood, as it appeared before having been submitted to the action of the atmosphere, was of a dark purple color   By union with the oxygen of the atmosphere, it has changed its color to one of scarlet. Next, the union of oxygen with the carbon of the blood liberates carbonic acid gas and vapor. The blood is now fit for circulation and for the renovation of the tissues. It appears, therefore, that as carbon exists in the venous blood, the lungs must be the pulmonary furnaces. The air-cells are flues or safety-valves, the membrane of the cells being permeable to oxygen, yet suffers not the blood to escape. Pulmonary combustion, therefore, bears some analogy to the combustion of carbon or charcoal in a stove; for, in that case, the oxygen causes the generation or evolution of carbonic acid gas.

BRIEF DESCRIPTION OF THE ANATOMY OF THE LUNGS.

The lungs, in common parlance, are known as the lights. They occupy the thoracic cavity, or chest. They are divided into right and left lobes, with a septum or partition between, which makes a double organ. This partition is a duplicature of the pleura, or membrane which completely lines the chest. When the lungs are healthy and properly inflated, they occupy the whole cavity of the chest; but when an action of expiration is effected, they are in a comparative state of collapse, occupying but a very small portion of the thoracic cavity. The lungs are composed of arteries, veins, absorbents, bronchial tubes, air-cells, and also what is known as their parenchyma, or substance. A healthy lung, when thrown into water, will float on the surface; while, on the other hand, a diseased lung, in a state of hepatization or condensation, sinks like a stone. In the fetal state, and when the lungs have never been inflated, they also sink when thrown into water.

SPASM OF THE MUSCLES OF THE GLOTTIS AND EPIGLOTTIS.

The aperture leading into the larynx and windpipe is termed the glottis. Spasm of the muscles of the glottis is one of the most terrible accidents that can possibly occur in either man or horse I view it in the light of an accident, because it usually occurs without warning or any other premonitory symptoms, selecting its victims in the very prime of life, and carrying them off, usually, in the course of a very few seconds. The terms cramp and spasm mean the same thing. When a man, while bathing in the river or sea, is suddenly seized with cramp, it is nothing more nor less than spasm of the flexor muscles of his limbs. Sometimes, how ever, the extensor muscles are affected. In either case, unless assistance be at hand, the person is apt to find a watery grave. Cramp or spasm of the intestines is the same form of affection, only it is confined to the muscular fibers of the intestines; and whenever it occurs in the limbs or intestines, it is always accompanied by excruciating pain and torment.

Treatment.—As regards spasm of the muscles of the glottis, it is very apt to prove fatal, either in consequence of lack of knowledge of the proper mode of treatment, or in failing to apply the remedy which the urgency of the case demands. I allude to the operation of tracheotomy, which consists of making an incision into the windpipe and inserting a tube into the same. A tube may not always be at hand, but this must not deter us from operating; for, by some means or other, air must be admitted, even if it be necessary to dissect out a piece of the trachea, which I always do in the case of a horse, whether I have a tube by me or not. Very little pain attends the operation, and that only occurs when cutting through the skin; for the windpipe, being composed of cartilage, is comparatively insensible. It may be policy, when the subject is not in immediate peril of his life, to resort to some counter-irritant and antispasmodic liniment (equal parts of spirits of camphor and tincture of lobelia); but when the danger is imminent, and the finger of Death is plainly on the patient, we only waste precious moments in the use of outward applications.

The following case, reported by J. B. Dobson, V. S., may possibly prove both interesting and instructive to some of our readers:

"At night a messenger came, saying the horse was very ill. Upon entering the stable, the animal presented the following symptoms: He was stretched out his full length in the stable, apparently in the agonies of suffocation; and such was the difficulty attending respiration that he positively screamed, in performing the act, so as to be heard at some considerable distance. I had not been with him many seconds, however, before he was slightly relieved, and in about ten minutes the spasm passed off, leaving him, with the exception, of course, of great exhaustion, otherwise as well as ever. Viewing the case as one of spasm of the muscles of the glottis, I applied stimulants, and ordered constant fomentations to the larynx, and left him with directions to be closely watched, in case the spasm might return. In the middle of th night I was again called, in haste; and this time the breathing seemed, if possible, worse than before, and it was evident that, if no relief were afforded, the animal must soon be suffocated. I accordingly, with some difficulty, owing to his struggles, performed tracheotomy. The relief was instantaneous, and he was soon on his legs, and anxious to feed. From this time the tracheotomy tube was kept in his throat for a week, when, considering that the glottis might have resumed its normal functions, I placed a cork in the orifice of the tube and sent him to work, as an experiment, ordering the cork to be removed if any symptom of suffocation should come on. We, however, found that the slightest effort at exertion brought on a fit of bad breathing, and it became evident that he would not work without the tube. The horse was accordingly sent to his regular work with the tube in his trachea, and, for three or four months, he did his accustomed labor with it in. At the end of that time, however, by an accident, the tube fell out, when he was some miles from home, and the horse coming home tolerably well without it, the horse-keeper neglected to inform me of it for a day or two, and when I saw my patient, the aperture in the muscles of the neck had closed. As the horse now seemed to work well without the tracheotomy tube, it was not reinserted, and from that time to this (more than two years) he had not another attack."

Should a case of this kind occur, and the consulted party have no tube at hand, he must pass a ligature through each side of the orifice, including a portion of skin and muscles. Each ligature is then to be passed over opposite sides of the neck, and tied at the upper part, just tight enough to keep the lips of the wound apart. In this way air is admitted into the windpipe, and thus the animal is out of danger.

LARYNGITIS (SUPPURATIVE AND INFLAMMATORY).

Laryngitis signifies inflammation of the lining membrane of the larynx. On applying the ear to the region of the throat, the locality of the affection is very evident, from the fact that no such embarrassed sound can be elicited in the lower part of the windpipe. Hence it must be inferred that the disease is located in the upper, or superior, passage of respiration

TRACHEOTOMY TUBE INSERTED IN THE WINDPIPE.

The inflammatory condition exists during a limited period, varying from one to five days. Then commences the suppurative stage, which is accompanied by a copious discharge, from both nostrils, of a secretion somewhat resembling pus, or matter.

Causes.—As regards the special cause of this affection very little is known. It may be supposed, however, to originate in a common cold, or catarrh; yet it is well known that horses are sometimes attacked with it that are not exposed to the ordinary influences which induce a catarrhal affection. Therefore we may infer that such disease will occasionally occur, either spontaneously or accidentally, as the case may be, in spite of our best efforts to prevent it. As the old saying is, "Nature is ever busy in maintaining the integrity of the vital forces of the animal economy. Any deviation from the prescribed laws which physiology imposes is a sure and certain cause of disease."

I now propose to introduce a case which, at the time of its occurrence, attracted much attention and curiosity, both on account

of the great value of the animal, and the desperate state of the
case at the time it fell into my hands. It will also serve to indorse the old aphorism that "where there is life there is hope."

Record of the Case.—The patient was a Patchen colt, aged four
years, a very handsome and promising animal. He was purchased by Mr. McPHERSON, of this city, for the sum of two
thousand dollars, and, at the time of his sickness, was considered
worth three thousand. But disease neither respects man nor
horse. Each have, once in awhile, to suffer and groan, swallow
drugs and get well, or die, as the case may be. And as regards
the horse, the more costly shall he be, the more likely is he to
get sick, from the fact that valuable horses are almost always
overfed and petted, and receive too much care and attention. On
arriving at the stable where the animal was located, I found him
in a dangerous condition. He appeared to be gasping for breath.
A loud stertorous noise, which could be heard at some distance,
indicated the nature of the difficulty as depending on obstruction within the larynx. The pulse at the angle of the jaw was
very indistinct. Both pupils of the eyes were dilated, or in a
state of amaurosis. The extremities and external surface of the
body were deathly cold. The tongue and visible mucous surfaces
were livid, indicating speedy death. Once in awhile the animal
would be seized with a convulsive or spasmodic fit of coughing,
which, every time, seemed to threaten his life. In those fits of
coughing he passed from the nostrils a sort of cheesy matter,
which appeared to be mixed up with a diphtherial exudation—
yellow secretion and froth. Under the above circumstances, any
attempt to administer medicine was deemed unsafe. I therefore decided to perform the operation known as tracheotomy,
which was done, in the following manner: Having secured the
services of a couple of assistants, the horse was led to a convenient spot in the center of the stable. An incision was then made
through the skin, to expose the trachea, about midway between
the jaws and breast-bone. The trachea was now punctured by
means of a pointed scalpel, and a probe-pointed bistoury was used
to dissect out a piece of the windpipe, corresponding to the size of
the tracheotomy tube. This tube, after being inserted into the
windpipe, was secured in place by means of elastic tape, which was
passed around the neck and tied. On the introduction of the
tube the alarming symptoms immediately subsided, and the act

of breathing was performed through the tube. I then applied a counter-irritant to the throat, composed of cod-liver oil and spirits of camphor, equal parts; ordered a warm bran mash, into which was sprinkled one ounce of powdered chlorate of potass.

For several days a very copious discharge from both nostrils occurred; also from the orifice made in the windpipe; so that the tracheotomy tube had to be removed and cleansed several times, both during the day and night. Four days after the operation, at night, the tube accidentally slipped out of the windpipe, and the mishap was not discovered until morning, when I found that air from the lungs had escaped into the cellular tissue, and inflated it so that the soft parts in the region of the neck and head appeared much swollen. I gave the patient, twice daily, one ounce of hyposulphite of soda, and had the swollen parts well rubbed with liniment. On the fifth day I removed the tube from the windpipe, and allowed the animal to breathe through the artificial opening, he still being unable to respire through the nostrils. At the expiration of fifteen days from the commencement of the animal's sickness, the wound in the windpipe was closed, natural respiration having been restored. During the treatment of the patient, he was allowed, occasionally, flaxseed tea, with about one ounce of chlorate of potass per day. He had good attention, or, rather, good nursing, and finally was again restored to usefulness.

In the early stage of laryngitis, or, rather, the inflammatory stage, cold water bandages should be applied to the throat, and a few doses of fluid extract of gelseminum, at the rate of one fluid drachm per dose.

CROUP (CYNANCHE TRACHEALIS).

Croup (cynanche trachealis), or, as it is sometimes termed, tracheitis, manifests itself in the form of a violent and sudden disease affecting the mucous membrane of that portion of the air-passages which lies between the laryngeal cartilages and the primary passages of the trachea, or windpipe. CONDIE contends that, in the majority of cases, the inflammation in croup commences in the mucous membrane of the larynx, and from thence extends into the trachea. The disease is, therefore, strictly speaking, a laryngeo-tracheitis. In many instances, however, we have reason to believe that the inflammation commences in the bronchi, and from thence extends to the trachea; but there are, we suspect, very few cases indeed, if any, in which the disease is confined to the trachea. In cases of croup that have terminated rapidly in death, the inflammation and pseudo-membranous exudation which form so striking a feature in this disease are found only in the larynx and upper portion of the trachea. When death occurs at a later period, the exudation is often present in the trachea alone, or in the trachea and bronchi. It is never found to exist in the latter tubes alone. It has been attempted to be shown, by JURIN and others, that in the ordinary form of croup the disease is, in its first stages at least, simply a tracheitis, and that in the more violent and rapid form (suffocating croup) the inflammation is confined to the larynx. Although this is not strictly true, as is proved by the result of numerous dissections, yet our observations have shown us that in cases marked by symptoms of great violence, which are sudden in their onset and rapid in their progress, the indications of inflammation are to a much greater extent, and the pseudo-membranous exudation more copious about the larynx, glottis, and upper portion of the trachea, than in cases in which the disease succeeds to bronchitis, runs a more protracted course, and is attended by symptoms of less violence.

The disease generally makes its appearance during the latter part of winter and in early spring, and most frequently occurs among horses just introduced to city life. The subjects such as the author has had occasion to treat were young, most of them of the lymphatic temperament, having short, thick necks, which abounded in a profuse development of adipose and cellular tissues, which results in a rotund bodily conformation. This, perhaps, goes to show that some hereditary idiosyncrasy plays a part in the creation of the disease.

Causes.—The indirect causes of the disorder are embraced among the evils of domestication, which include errors in diet, impure air, and bad management. It usually dates its origin from what are generally considered as the causes of common cold. Several days previous to the attack, the animal appears "dumpish," drowsy, and inactive, having little desire for artificial food, and less for water. The eyes appear somewhat tumefied, and the respirations are laborious, having a peculiar, croupy sound, accompanied by a distressing cough. The respirations can be heard at a distance (as the disease progresses); and on making pressure in the region of the larynx, the animal evinces signs of uneasiness, suffocation, and pain. Soon the mucous, schneiderian, and conjunctivial membranes become reddened, the eyes protrude, the tongue tumefies, and external tumefaction in the region of the throat may be observed. Febrile symptoms are also present, manifested by coldness of the exterior and increased temperature of the interior parts. The nose appears dry, and the mouth is full of frothy saliva (sometimes lymphy mucus), and some difficulty is encountered in opening the latter.

History of the Disease.—The following case, from the author's note-book, will probably prove interesting to the reader: The patient, a roan mare, aged seven. Temperament, lymphatic. Just arrived, in company with several other horses, from Ohio. The property of Mr. Banford, Merrimac stable, Boston. Our patient had been "ailing" some three or four days before our services were secured, at which time she was laboring under "alarming symptoms" bordering on suffocation. The pulse was indistinct, small, thready, and difficult to number, yet probably exceeded sixty. The breathing was croupy, or stertorous, and could be heard at a distance of several yards. The nostrils were dilated, and from them issued a lymphy discharge, mixed with traces of blood. The submaxillary space was somewhat tumefied, hard, unyielding, and seemed to constrict the larynx and its associate muscles. The eyes were prominent, their membranes congested, and streaks of tears escaped over the inner canthus. The flanks were bedewed with a cold, clammy perspiration. The tips of the ears were quite cold, and the limbs the same. Auscultation, in the region of the larynx, revealed a very laborious and crepitating mucous rale. The trachea itself seemed to be free from obstruction. Thoracic auscultation elicited nothing tending to show that the lungs were involved.

The patient appeared to be in a deplorable, in fact dying, condition, making a sort of gasping effort to inflate the lungs. There was little time to spare; and being satisfied that the larynx was the seat of obstruction, we immediately cut down upon the trachea, about five inches below the thyroid cartilage, and amputated, from between two rings of the trachea, a piece, of an oval shape, corresponding in size to the caliber of Arnold's tracheotomy tube, which was the instrument used on this occasion. So soon as the opening into the trachea was effected, the patient experi-

7

enced instantaneous relief from suffocation, and all danger from the same seemed to have immediately passed away. The tube was now introduced and secured around the neck in the usual manner, after which the patient received an aloetic enema, and had a strong counter-irritant applied to the submaxillary space and throat, after which the fauces were swabbed with a weak solution of alum, by means of sponge secured to a piece of whale bone. On examining the "swab," a sort of lymphy or albuminous concretion adhered to it; the mucous membrane of the larynx was either injected or œdematous. The pharynx was not involved, as the animal, shortly after the operation, drank two quarts of water, containing two drachms of nitrate potassa.

The tube remained within the trachea for a period of five days, during which time very little occurred worth recording, except that one night pneumatosis (distension of the cellular membrane with air) appeared, which yielded to a dose of hyposulphite of soda and an outward application of liquor ammonia acetatis; and, also, on the fifth day, a submaxillary tumor was punctured, which discharged freely. At the end of the above period, it was ascertained, by holding a lighted lamp to the nostrils, at the same time stopping up the orifice in the instrument, that the animal breathed through the usual channel. Then the instrument was removed, the parts cleansed, and secured together by suture. The wounds healed by the usual process, and, at the end of three weeks, the animal was disposed of, and went to work.

The operation of tracheotomy, formidable as it may appear to some, is unattended with danger; yet, like every other process of surgery or medicine, is only calculated to relieve certain states peculiarly adapted to the remedy. For example, the operation is admissible in cases of nasal obstruction, from the presence of nasal polypus or other causes; spasm of the larynx, threatening suffocation; suppurative laryngitis, when the animal is in great distress for "breath;" and in case of any foreign body occupying the larynx, which precludes the possibility of continuing the process of respiration. On the other hand, the operation is worse than useless in cases of lung difficulty, bronchial obstruction, or when obstruction occurs within the trachea posterior to the usual point selected for the operation.

COUGH IN ITS CHRONIC AND ACUTE STAGE.

According to veterinary jurisprudence, a cough, so long as it lasts, renders the animal unsound; hence it is very important that we should know something about its cause, nature, and treatment. Chronic cough is often the result of indiscretion in the treatment of influenza, distemper, and disease of the respiratory apparatus. It usually depends on a morbid and irritable condition of the membrane found on the interior of the respiratory passages. The cough is generally aggravated by over-exertion, especially when the roads are dusty. Food of an inferior quality, and that of a musty character, has the same effect.

It is generally supposed, by those persons who have not had the benefit of a medical education, that the presence of acute or chronic cough indicates diseased lungs, or disease in some parts of the organs of respiration. But this is not always the case; for, whenever the liver becomes diseased, the subject is very apt to be tormented with a harrassing cough, which lasts as long as that organ shall be the seat of disease. However, there are several symptoms to be observed in cases of functional or organic disease of the liver which are not present in lung disease; hence there is no difficulty in the way of making a correct diagnosis. And for the benefit of the non-professional, the author would inform them that, in all cases of liver disease, a marked yellow tinge will be observed on the visible surfaces within the mouth; the tongue slightly coated; the dung unusually dark colored, and voided in hardened lumps, and the urine also of a dark yellow color.

Treatment of Chronic Cough.—Give the patient daily one ounce of the fluid extract of Indian hemp (foreign), and offer him occasionally some flaxseed tea; or sprinkle in his food, every night, a handful of unground flaxseed. Should this fail to effect a cure, prepare the following:

> No. 13. Fluid extract of bloodroot............ 4 oz.
> Fluid extract of pleurisy-root... ...... 6 oz.
> Common syrup...................... ½ pint
> Mix.

Dose, two ounces per day.

The acute cough may be treated in the same manner as above, by substituting for the syrup half a pint of syrup of squills.

In the case of liver disease, accompanied by cough, give the animal one ounce of fluid extract of mandrake every six hours, until it operates on the bowels, or the membranes of the mouth lose their yellow tinge.

ROARING.

Roaring is usually the result of structural alterations within the larynx, or upper part of the windpipe bordering on the trachea. In mild cases of roaring, we usually find a thickened state of the membrane lining the upper portion of the respiratory passage; and when roaring is occasioned by thickening of this membrane, its degree depends on the ratio of decrease in the caliber of the tube breathed through. Roaring is a very aristocratic disease. Many of the very best and fastest horses in England were, and are now, notorious roarers. "Flying Childers," as fast a horse as ever wore horseshoes, was one of the worst roarers ever known. The story runs that when "Childers" was at full speed, his roaring resembled juvenile thunder ! He could be heard when distant half a mile!

The worst form of this disease is whistling. This is the sharp shrill note not only occasioned by the thickening of the lining membrane of the primary passages of respiration, but by alterations in the form and structure of the larynx, the larynx being, in popular language, known as the " voice-box."

Roaring is more prevalent among stallions than mares and geldings, and the kind of horse most subject to it is the one having a thick, chunky neck, and having the angles of the jaws in very close proximity with the neck. Roaring scarcely, if ever, admits of a radical cure; and when of a hereditary or congenital origin, a cure is impossible. A roarer should never be incumbered with a check-rein; for it has the effect of causing undue pressure on the larynx, and thus augments the difficulty. Roaring can, however, be relieved by an operation known as tracheotomy, which is performed at a point a few inches below the larynx.

At a meeting of the Imperial and Central Society of Veterinary Medicine, M. LEBLANC read a communication on tracheotomy which was performed on a carriage horse. The operation had been performed because the horse was a severe roarer; and he wore the tube eighteen years and a half, doing fast work all the time. The animal was destroyed at twenty-three years of age the owner not desiring to make further use of him nor to sell

him.   Since the operation, LEBLANC had not observed any
change in the horse, except a depression of the bones of the face.
After death, the larynx was found very narrow, the mucous membrane and submucous cellular tissues were thickened, the epiglottis deformed, very obtuse, and everted at its free margin.   The
changes in the larynx were the original cause of roaring.   The
depression of the bones of the face was connected with constriction of the nasal chambers, and was evidently secondary to the
change in the course of the air in the process of respiration.   The
parts of the trachea in contact with the tube had undergone a
transformation into very hard tissue.   It filled the trachea above
the point where the tube had been introduced, and intermixed
witl this firm fibrous deposit was a cartilaginous and osseous tissue,
which offered great resistance to the scalpel.   Roaring, thick wind,
whistling, etc., are often the sequel of strangles, influenza, laryngitis, and other affections of the respiratory passages, and hence
have an accidental origin.   In such cases, we may entertain a hope
of doing some good by means of medicinal agents and counter-
irritation.

Treatment.—The medicines which have proved most successful
in my practice are as follows:

    No. 14. Iodide of potass...................... 4 oz.
             Fluid extract of stillingia............ 12 oz.
             Water .... 4 oz.
             Mix.

Give the patient two ounces daily, by means of a small vial;
and rub the region of the throat every night with a portion of
the following:

    No. 15. Cod-liver oil.................. } equal parts.
             Spirits of camphor............. }

POLYPUS WITHIN THE NOSTRILS.

Polypus is the name given to an excrescence, or tumor, which
has its origin from the surface of the membrane lining the nasal
cavities.   It is designated as growing from a mucous membrane,
having a narrow neck and body, resembling, in shape, a pear.
The usual symptoms attending the presence of polypus in the nasal
cavities are as follows: Difficulty of breathing, evidently occasioned
by obstruction in the air-passages.

POLYPUS
WITHIN THE
NOSTRIL.

PERCIVALL informs us that the true polypus is attached to mucous membranes, and is usually found in the nasal cavities. He tells us that attending the difficulty of breathing is a mucous discharge from one or both nostrils, sometimes attended by a discharge of blood. Occasionally, however, pure blood runs continuously from the nose. Inspection in a full light discloses, higher or lower in the nostril, the rounded base of a polypus.

Treatment.—The services of a veterinary surgeon are, as a matter of course, here needed. The patient must be cast and the head fixed in a position so as to take advantage of the light. The operator then passes into the nostril and around the tumor an instrument called an ecraiseur, which will remove the tumor without loss of blood. If the instrument is not at hand, the surgeon will pass a ligature around the base of the tumor, and in the course of a couple of days it will be detached. PERCIVALL recommends that, in bringing down the tumor for operation, we must not use any great force. The pedicle being but a duplicature of the skin, and not a portion of the polypus itself, may be divided anywhere. In some cases, the polypus is so high up within the nostril that, in order to get at its base, it becomes necessary to make an incision through the wall of the nostril.

CHABERT, in his " Veterinary Instructions," relates the following: "A horse in a cavalry regiment had been gradually losing flesh, and was quickly and painfully blown at every little exertion. Fetid matter began to run from his off nostril, and the gland correspondent enlarged. The horse was supposed by the sergeant-farrier to be glandered, (there being no veterinary surgeons then in the French service,) and was treated accordingly. After a time, to the confusion and astonishment of the man, a fleshy substance began to appear in the nostril, and which rapidly increased in size. At length a great mass protruded, and the farrier cut it off. No benefit followed; the nostril was still stopped, the breathing laborious, and the horse daily became thinner and weaker. After the lapse of a twelvemonth, the case attracted the attention of M. Tears, the surgeon of the regiment. He cast the horse and slit up the nostril, when he not only found it completely filled with polypus, and the septum narium bulging into the other division of the cavity, but, from long-continued inflammation and pressure, it adhered to the membrane of the nose in so many points,. and so extensively, that it was impossible to get round it or move it. He contrived, at length, to pass a crucial bandage around it, and it was torn out by main force. Four considerable portions of the turbinated bones were brought away with it. The hemorrhage was excessive. He however filled the nostril completely with tow, and brought the divided edges of the false nostril together by sutures. In three days they were all torn out by the incessant attempts of the animal to get rid of the obstruction; but the horse eventually did well. The polypus weighed two pounds seven ounces."

GOHIER relates a case of a horse who had in his left nostril a polypus as large as a turkey's egg, of a grayish color and glossy surface, too high up to be reached with the finger, which prevented his breathing on that side, and gave rise to offensive effluvium, to enlargement of the lymphatic glands, but not to roaring. GO-HIER slit up the nostril, and, with an iron rod with a notch upon its end, contrived to inclose its neck in the slip-knot of a ligature. In drawing this tight, however—which was, of necessity, done in an oblique direction—the pedicle was cut through. Little hemorrhage succeeded, although the tumor weighed twenty-four ounces. The slit nostril was sewn up, and cold water injected into its cavity. A copious discharge from both nostrils followed, with swelling of the lymphatic glands. This was met by proper treatment, and in fifteen days the patient was sent out of the hospital.

BRONCHOCELE (GOITRE, OR BIG NECK).

Bronchocele, or enlargement of the thyroid glands, is a disease which very frequently makes its appearance among certain breeds of horses and sheep, and is supposed to owe its origin to hereditary predispositions and influences. It is also very prevalent among members of the human family. It seems to acquire a home in certain localities in Derbyshire, England, where it is said to be a very common disorder; but its occurrence in other parts of that country is by no means frequent. Among the inhabitants of the Alps, and other mountainous countries bordering thereon, it is a disease very often met with.

The uncastrated animals seem to suffer most. After they get

to be about eight years of age, the tumors acquire such magnitude that they press on the vocal organs, so as to decrease the caliber of the larynx, and thus the animal becomes a "roarer." Judging from what we know of the disease in the human subject, the glands, while undergoing enlargement, do not occasion much pain. The danger arises from mechanical causes, and the death of the subject, if it occur, is due to asphyxia, or suffocation. Among horses there are very few fatal cases on record. The reverse is the case as regards sheep. When these glands are much enlarged, and the animal is near or past the adult age, it is very unsafe to attempt their removal by means of the knife; for at this stage they are highly vascular, and the arteries which run into them are much enlarged. The operation has been suc-

SHOWING THE TUMOR OF BRONCHOCELE IN THE REGION OF THE THROAT.

cessfully performed on lambs, but it must be done when they are quite young, and the artery must be secured before the gland is extirpated, or the animal will bleed to death in a few seconds. It is well known among the members of the profession that the disease is incurable; and the same remarks apply to all hereditary diseases, yet the growth of the glands may be retarded by means of local and constitutional treatment.

The thyroid glands are two ovoid bodies, varying in size from a filbert to an egg, located in the region of the thyroid cartilage (throat), one on each side of the trachea (windpipe). Their attachments are cellular. When cut into, they exhibit a porus texture, highly vascular, well supplied with blood-vessels. Very little is known of their physiology. They are called vascular glands without ducts or outlets, and thus they correspond with the spleen, thymus gland, and supra-renal capsules, all of which, when in active operation, are largely supplied with blood.   As regards the function of each, they may be supposed to separate certain materials from the blood, and only differ from ordinary glands in not having a direct outlet.   Consequently, not knowing the precise part which the thyroid glands play in the animal economy, it would be very unwise to remove them, when their enlargement, in some cases, amounts to little else than an "eyesore."   Their enlargement is generally the consequence, and not the cause, of disease.   Men, horses, dogs, and cattle of the scrofulous diathesis, are known to have what may be termed chronic enlargement of these glands, and this peculiarity would seem to indicate that the thyroid glands are somewhat associated with the lymphatic system.   Then, again, we find these glands enlarged in cases of throat and lung difficulties—in catarrh, influenza, distemper, etc.—so that their extirpation would not remove the original difficulty.   In such cases they decrease in size as soon as the original malady ceases.   The enlargement is not then of a permanent character.

Treatment.—Should the enlargement appear to exist independent of febrile symptoms, twenty-five grains of the iodide of potassium may be given daily, in water (which the patient will not refuse to drink), and a small portion of the ointment of iodide of potassium may be rubbed on the enlargement, with decided advantage:

No. 16. Iodide of potassium................ 1 part.
         Lard............................ 8 parts.
         Mix.

On the Action of Iodine.—Iodine and its compounds are the principal agents used by veterinary surgeons for the treatment of glandular affections, and they supersede, to a certain extent, the preparations of quicksilver, in the form of mercury, which are apt to be absorbed into the system, and thus produce mischief.

MORTON, in his "Manual of Pharmacy," thus alludes to iodine: "The action of iodine and its compounds is markedly seen on glandular structure, and newly-formed and abnormal growths. For the latter they seem to manifest a decided preference, which renders them so valuable as therapeutic agents.   Their influence is that of a stimulant to the absorbents; and by means of these vessels, when those agents have been long and injudiciously given, it is recorded that the mammæ of women and the testes of men have almost disappeared. A case illustrative of the effects of iodine on the glandular system was related by Mr. Wardle, who employs this agent largely, and with considerable success. He had been for some weeks exhibiting the iodide of potassium, and also applying it externally to a bull for an enlarged parotid gland. The reduction of the swelling having been accomplished, his attention was directed to the testicles of his patient, which had become so much diminished in size; and it was also found that the animal had no desire to copulate. Four months generous feed, however, effectually restored the parts to their pristine state.

In chronic enlargements of the submaxillary, parotid, mammary, and other glands; in tumors of long standing; for thickening of the integuments, and indurated swellings about the joints, in all our domestic animals, and for unhealthy ulcerated surfaces, the use of iodine and its compounds is indicated, combining both their internal and external employment. The latter should be accompanied with friction; and so soon as soreness is induced, the application of the compound must be suspended. A desquamation, or scaling off of the cuticle, or skin, usually follows this, and a reduction of the swelling will soon after be perceptible. Unfortunately, it is too often the case that, if the expectations of the practitioner are not at once realized, the agent is discarded as useless. Now, the compounds of iodine rank among those substances whose operation is slow, but which, at the same time, constitute an important class, as their effects are permanent."

INFLUENZA, OR EPIZOOTIC CATARRH.

Influenza is a disease of a catarrhal character, always accompanied by great weakness or debility. It is apt to migrate from one tissue to another, and thus produce various complications. In stables that are badly ventilated, or over-crowded with horses, the malady is apt to make sad havoc, and assume a malignancy ending in typhoid or typhus fever.

In some cases the heart or lungs become congested, and thus their function is embarrassed. The blood accumulates carbon, the heart fails to propel, and the lungs to vitalize the blood. This state demands stimulants of a character that shall excite vital action.

and tend to effect chemical changes in the blood. The most valuable agent, in this view, is carbonate of ammonia. In veterinary practice this agent is considered as one of the most valuable diffusible stimulants. It acts first on the nerves of the stomach, and through them excites the whole nervous system. It may be given in doses from one to three drachms, either in the form of bolus or drench. Its stimulating influence over the heart and nervous system may be increased by the addition of one or more drachms of good Jamaica ginger. This treatment may be continued until congestion subsides, which may be known by the fullness of pulse and heat of the external surface of the body, and by other symptoms which usually attend fever in its inflammatory stage.

The exciting or stimulating plan of treatment must never be entirely abandoned. To husband the powers of the system, and thus guard against subsequent prostration, must be our chief object. Although the case may require a sedative to-day, in the form of two or three drachms of fluid extract of gelseminum, yet, in the course of a very few hours, the prostration becomes so apparent that we are again compelled to exhibit life-sustaining agents. Excitants, or counter-excitants, are also, at times, needed externally. If the patient has a deep-seated cough, strong liniment, composed of oil of hartshorn (and sometimes oil of camphor), must be applied. Often I anoint the parts with a thick paste, composed of mustard and vinegar. The membranes of the fauces (throat) are excessively sore; then the same course has to be pursued. At the same time, a mucilaginous drink, composed of flaxseed sweetened with honey, must be allowed. The nervous system must also be aroused by the application of the above liniment to the spinal column. At another stage of the malady, diaphoresis (increased perspiration) must be excited by clothing the body with flannel, and drenching with a preparation known as solution of acetate of ammonia, to the amount of six or eight ounces per day, until the skin feels soft and warm. In case œdema (local dropsy) of the body or extremities sets in, the patient then requires diuretics. One or two ounces of sweet spirits of niter, or two ounces fluid extract of buchu, may be given in linseed tea, until the kidneys respond, and the secretion of urine becomes augmented. The bowels are sometimes inactive, but that inactivity may be corrected by adding to a bran-mash from four to six drachms of fluid extract of mandrake. Occasionally the patient becomes uneasy, paws with his fore-feet, and evinces signs of abdominal pain. In that event, he should have two or three drachms of fluid extract of Indian hemp. These latter symptoms indicate danger, showing that the bowels are congested, and the result may be mortification of the same.

The following case, occurring in the author's practice, goes to show that influenza may sometimes be followed by an acute disease of the brain, pleurisy, etc.: The patient was a bay gelding, aged about eight years. The groom informed me that the disease was ushered in by a shivering fit. Soon a discharge from both nostrils ensued. The throat became sore; cough and laborious respiration followed, and the animal became so weak that it was almost impossible to back him out of the stall. He ate but little, and in the act of drinking, the water returned through his nostrils. The treatment was commenced in the usual manner, by administering tonics and stimulants, which seemed to have a good effect on him. He also received enemas of soap-suds, yet, notwithstanding, symptoms of pleurisy supervened. He appeared to suffer much pain, and was very unwilling to have the region of the chest explored. The least pressure on the region of the intercostal spaces would cause him to grunt or groan with pain. The pulse was active and wiry, and the visible surfaces were much reddened. I administered a full dose of powdered niter, and twenty drops of aconite; and fearing effusion into the chest, I plastered the walls of the same with hot vinegar and mustard. The latter made the patient very uneasy, and he cut up all kinds of capers. In the course of a few hours I repeated the dose, and left the patient for the night. Next morning I again visited him, and was informed that he had made several attempts to get up in the manger. I noticed that his head was elevated, and the eyes fiery red. He acted wildly and showed decided symptoms of phrenzy. On attempting to administer a sedative, the animal went into convulsions, and I was compelled to make my escape from the stall as quick as possible. He soon got his fore-feet over the tops of an adjoining stall, five feet in height, remained there for a short time, when he scrambled over and fell like a log right into the next stall. I never expected to see him rise again; but after a few minutes, with some assistance, he got up. I gave him twenty drops of aconite. During this day the frantic animal went over a stall twice in succession, and fell very heavily each time on his back, yet did not appear to have suffered any injury

Most of the time he appeared frantic, yet unconscious; and just before going into a convulsive fit, he would stamp with his fore-feet, continuously toss up his head, his neck being the seat of spasmodic twistings. Next day the symptoms had not altered much, and, during my examination, he pulled back, broke a por-tion of his manger to which he was tied, and fell head over heels on the floor. During the forenoon he became more calm, and lay on the floor, yet would occasionally jerk his head backward, and have slight convulsions. I continued the prostrating plan of treatment, only exhibiting smaller doses, and soon the patient began to improve. The membranes of the eyes, mouth, and nostrils as-sumed a more natural appearance; then the plan of treatment was altered, and the patient received tonics and stimulants, which soon resulted in convalescence.

Among all the cases that have come under my care and observa-tion, I have neither found it advisable nor necessary to practice the antiphlogistic treatment, in so far as it relates to bleeding and purging; and I can not conceive a case, so depressing as influenza is known to be, that would require any such heroic treatment. The safety of our patient depends on the judicious application of remedies such as are here named, and, therefore, I would warn the farmer against the use of fleam and cathartic in the treatment of influenza. Yet, after all, the best treatment may fail in restoring a patient. We require aid in the form of pure air, suitable diet, and good nursing. Failing in the latter important adjuncts, our treatment avails but little.

Influenza is a disease accompanied by a discharge from the res-piratory, conjunctival, and schneiderian membranes; and the organs of respiration themselves are more or less involved, as shown by cough and soreness of the throat, lassitude, fever, thirst, and loss of appetite. Now, if the term influenza means any thing, it signifies epidemic catarrh, and the cases alluded to were not of this character, but quite diverse, indicating a low congestive state. Still, the majority of cases that we have seen do appear to be noth-ing more nor less than influenza; but if the above diseases appear simultaneous with the latter, there is danger of persons making great mistakes in the treatment, for it is very apt to run into typhoid fever.

TYPHUS OR TYPHOID AFFECTIONS.

The veterinary surgeons of England have hitherto maintained a remarkable silence on the subject of typhoid affections occurring among horses; and so late as the year 1850, Mr. PERCIVALL informs the world, through the pages of the "Veterinarian," that in London very little is known about such disease. In view, therefore, of lighting up the dark spots that exist in our department on this side of the water, we furnish a translation from the French, by Mr. PERCIVALL. The article is a selection from a prize memoir written by a distinguished surgeon. It is preceded, as the reader will perceive, by a review from the pen of the translator, who says:

"In the 'Collection of Memoirs and Observations on (French) Military Veterinarian Hygiene and Medicine,' which we have so lately been engaged in examining, are contained two prize memoirs—one on FARCY, the other on TYPHOID AFFECTIONS in horses.* On the latter of these we would make a few remarks, if it were only for the reason of showing what is meant to be understood by such imposing titles. Typhus and typhoid are words but rarely heard in our own country in connection with veterinary, or at least with hippiatric, medicine. Our old writers on farriery described fevers in horses as very destructive in their character requiring antiphlogistic treatment:

"'*Typhus Fever.*—A disease touching which we (the author) are in possession of but few observations, and one that has been, and still is, in our opinion, mistaken for and confounded with either enteritis or gastro-enteritis—in cases, for example, in which its consequences are of little importance—though, perhaps, with putrid fever, when, on the other hand, malignant and exhibiting extraordinary violence, its progress is rapid and its termination fatal. In its most benignant form, typhus fever, indeed, bears so great a resemblance to pure inflammation of the primary intestinal passages, that it is often difficult, very difficult, even to distinguish them. As for the cause which occasions it to be confounded with putrid fever, it is no matter of astonishment to us, since, in our opinion, one fever possesses, in many respects, so great an analogy

---

* The reader will find an article on Typhus Fever, but in an *enzoötic* form, in "The Veterinarian," vol. xxii, p. 462.

with the other; while typhus itself, through causes which remain latent up to the present hour, is, like the gastro-enteritis of 1825, dangerous, fatal, and epizoötic. In general, typhus fever begins without any warning, though there are times in which its approach is marked by dullness and weakness, and a manifestation of being easily fatigued, sweating through little exertion, etc., with loss of appetite.

Of the first stage, the most remarkable symptoms are either a yellow redness of the pituitary membrane, or a dryness or a discharge of viscous matter, which adheres in thick incrustations around the alæ of the nostrils, with, occasionally, an appearance of drops of blood. The nostrils are more or less dilated, the respiration more or less frequent, according to the sharpness of the abdominal pains, or the existence of a lamentable complication of thoracic with abdominal disease. The ear applied to the windpipe or to the walls of the thorax, almost always detects a well-marked mucous rale; also there is cough, with full, strong, and frequent pulse.

In the second stage, the mouth becomes dry, and sometimes displays aphthæ or cracks of some depth. The tongue is red about its point and along its sides, but its middle and base are of a deep yellow tint. The breath is tainted. When abdominal symptoms are present, which is not always the case, they are sufficiently well marked. The dung is either very hard, or it is, more commonly, liquid; sometimes it is bloody, and issues a fetid odor, the debility frequently being such at this period as to cause constant opening of the anus. The pituitary membrane is dry, and covered with bloody points. The conjunctival membrane is reddened and infiltrated, and, like the pituitary, presents petechial patches. The pulse has not its force and fullness, etc. Some veterinarians have remarked, at this period, upon the insides of the thighs those little vesicles to which the name of sudamina has been given. Urine scanty, possessing a remarkable fetor, and voided occasionally with great pain; great dejection, and occasional stupor; and, ordinarily, at this period it is that the ataxic or adynamic symptoms set in.

In the third stage, the parotids tumefy and ulcerate. Any setons or rowels that may have been introduced, or blisters, become so many causes of mortification of the parts in which they have been applied. And now the mouth becomes quite dried up; the tongue acquires a horny feel, and turns brown; the gums and teeth are covered with a dingy slime. Bleeding from the nose, which mostly appears during the inflammatory stage, now returns. The pulse becomes irregular and more feeble; œdema, sometimes considerable, of the extremities; great debility. The temperature of the body sinks. The neck, belly, flanks, axilla, and groins become covered with cold and clammy sweats; the tympanitis increases, the evacuations being bloody and fetid to a remarkable degree. The animal, continually in pain, exhibits symptoms of colic, often seeking to lie down; but if he does, he soon rises again. Respiration greatly accelerated; nostrils widely dilated; pulsations of the heart tumultuous. At length, beginning to stagger, he falls, either to die an easy death or to expire amid agonizing convulsions. When, however, recovery in place of death follows, such symptoms as we have last detailed do not occur, or with only modified force and character. When blood has been drawn for experiment, under such circumstances, at a time when adynamic and biliary symptoms prevailed, it has been found to contain but little hematosine, but, on the contrary, a large proportion of serum, of a light greenish hue. The disease may last from four, seven, or eight to thirty days. Relapse is extremely likely, even under every prospect of convalescence, unless great care be taken. The diagnosis at the commencement is difficult, even impossible, when the characteristic symptoms are but scantily and imperfectly developed.

The affections whose symptoms most resemble those of the disease before us are plegmasia of the intestinal tube, and particular of the brain. Should sudamina be detected, they are, as in human medicine, to be regarded as definitely characteristic of typhoid fever. Prognosis, for the most part, unfavorable; and the more so according as the nervous accompaniments are more numerous, more violent, and more early in showing themselves; the stupor great, the strength failing more, the pulse faltering, the diarrhea greater, more bloody, fetid, etc.

Our etiology is all at fault. In truth, we know not the determining cause of typhus. The post-mortem states are by no means agreed upon by those who have reported them. Most agree that the digestive canal exhibits the most constant and marked symptoms of disorder. The mouth presents marks of inflammation, and, on occasions, aphthae or ulcerations, though commonly too superficial to destroy the living membrane. The pharynx sometimes the same as the mouth; at others, nothing. The stomach is frequently the seat of inflammation of a marked character. The intestines also reddened, to more or less extent. The cavity of the peritoneum, containing a citron-colored fluid, reddened, perhaps, the omentum being at times almost entirely destroyed. A liquid matter—a sort of bloody corruption of more or less consistence—is often found within the cavities of the small guts; and in those of the large, the cæcum and colon in particular, are sometimes to be remarked red and black spots, or *ecchymoses*, buds of smaller or larger size (*glandes de Bruner*) indistinctly planted over the mucous surface.

Malignant or Putrid Typhus—(Typhus Charbonneaux.)—These malignant and putrid affections, either constitutional, or, in some instances, but local in their origin, ever of a highly acute nature and of unequaled rapidity of progress, since not unfrequently they cause death in a few hours, are of more consequence for us to become acquainted with, as they are not only susceptible of being communicated from one animal to another, but even from animal to man, in whom, losing none of their malignity, they are likewise most commonly mortal. Frequently sporadic, sometimes epizoötic, but most commonly enzoötic, these diseases at times assume so redoubtable a form that nothing can arrest their progress. Invading farms or other extensive localities, they produce therein terrific ravages; for, being susceptible under various forms, almost equally fatal one with another, of prostrating a vast number of animals of the same or different species, these affections determine, under varying circumstances, losses which, on occasions, desolate and completely ruin the country where they have broken out.

Such are the two especial forms of typhus fever which have been observed among horse-kind—one of which would seem to occur at times in practice without being sufficiently heeded or recognized by veterinary surgeons in this country, while the other may be said to comprehend those malignant epidemics by which our cattle, if not our horses, have, of late years, been too fearfully visited.'"

In the treatment of typhoid affections, we endeavor to give tone to the system by furnishing nutritious diet, tonic, and stimulating medicines. Ginger and golden seal, with a small portion of hyposulphite of soda, are the best remedies.

PLAN OF A VAPOR BATH USED IN THE TREATMENT OF LUNG AFFECTIONS.

EXPLANATION.—*a*, represents a boiler originally erected for the purpose of supplying the infirmary with hot water; *c*, is the main pipe issuing from the top of the boiler, receiving the steam, and conducting it, when not required for other purposes, into either a flue or the open air at *b*; *e* and *f* are branch pipes from the main one (*c*), *f* being that which conducts the steam into a worm *h*, winding through a condensing trough *g*; *e*, the branch pipe which conducts the steam, prevented by a stop-cock from going in the other direction into the bath, the place of admission *w*, being on one side, close to the floor, at a point intermediate between the horse's fore and hind feet while standing in the bath, with his head outside; *m* is the bath, being a horse-box, such as is used for embarking horses on board of ship, with the addition of a lining of flannel, a roofing of hoops and tilting, and curtains over the doors, front and back, to prevent the escape of steam. The box, being placed upon wheels, serves, besides being used as a bath, for the transport of sick or lame horses; and, having doors at both ends, and a moveable platform for the horse to walk in upon, is, in general, entered without any great deal of unwillingness.

PNEUMONIA (INFLAMMATION OF THE LUNGS).

Inflammation of the lungs, known to veterinarians as pneumonia, is not apt to be so prevalent among horned creatures as among horses, except, however milch cows, when they are located in filthy, unventilated milking establishments. In such places diseases of the lungs are often fearfully prevalent, raging as an epizoötic, as it did a few years ago in the swill-milk establishments of New York. It makes sad havoc when prevalent among a large herd that may be confined in a barn or stable of contracted proportions. All domestic animals require plenty of room, as well as light and pure air; for it is now pretty clearly demonstrated that the pleuro-pneumonia, which appeared in Massachusetts, in 1860, on the premises of Mr. Chenery, arose there spontaneously. His barn, or rather the basement, where the cattle lived, was a room fifty feet square and only eight feet in height. The walls on the north, and a portion of the east and west sides, were composed of brick, and the remainder of boards, with joints battened. On the south, east, and west sides were glass windows; on the north side a door opened into the vegetable cellar, and on the roof there was a sky-light. A stairway led to the hay-loft above, another to the manure cellar below, both closed by doors. The manure cellar was eight feet deep, and extended under the whole room. In the winter and spring of 1859 were confined forty head of cattle, arranged on three sides, with their heads toward the center; and within that center was another square, containing animals, so arranged that almost all were brought face to face. The manure cellar, about this time, contained from fifty to one hundred cords of manure, with from ten to twenty hogs in it. "This," says Mr. CHENERY, " was my condition in the beginning of the year 1860. I had buried nearly half of my herd. I had experimentally acquired a knowledge of the fact that, in order to keep animals alive and in health, it was absolutely necessary that they should be supplied with pure air as well as with good food and pure water."

So far as the direct or indirect causes of pneumonia are concerned, we may safely infer that crowding and a bad system of ventilation includes them all. Yet, among horses, this disease often originates as the consequence of laborious work and feats of speed, which produces rapid and sometimes distressing respirations; but among cattle, whose powers of speed and endurance are not often put to the test, and whose natural respirations are slower, we infer that impure air, and perhaps exposure, too, are more operative than action or ill-usage.

The stimulating and morbid action of an impure atmosphere may produce a disease of this character by its irritating effects on the highly vascular membrane which lines the bronchi and air-cells. But then we all know that impure air fails to decarbonize the blood; hence it is rendered unfit to enter the system. It has been noticed, however, by the drovers of the East, that when cattle have been driven a long distance without food, and in tempestuous weather, they are apt to become the subjects of diseased lungs. Fortunately for the poor brute, pneumonia is not so painful as bronchitis, pleurisy, or laryngitis; and having passed through the acute stage, which is usually brief, it assumes a mild or sub-acute form, and ends in altered structure of the lungs, known as induration (hardening) or hepatization (liver-like), or it may end in pleurisy, and ultimately dropsy of the chest. I should judge that cattle, when the subjects of pneumonia, suffer less than horses, from the fact that the former will lie down during the progress of the disease, and the horse is scarcely, if ever, known to do so until shortly before he dies. In the horse, this malady occasionally takes a metastatic turn; that is, the disease is translated to the feet or brain, producing an affection of a very formidable and acute character. Among cattle, however, such translation is very rare.

Forms of Pneumonia.—Pneumonia is divided into several stages and forms, but as such refer more to its degree or intensity, I shall only inform the reader that pneumonia may exist either as a condition of congestion or of inflammation. Congestion signifies a distended or plethoric state of the blood-vessels of the substance of the lungs, and slow motion of the blood; and it sometimes sets in as suddenly as that form which is termed inflammatory; among cattle the former form is more prevalent. In the congestive stage the symptoms are those of embarrassment—the blood courses through its vessels sluggishly, and there is not that activity of the heart and lungs which is noticed in inflammatory pneumonia.

Symptoms of Inflammation of the Lungs.—There are many symptoms attending lung disease which are common to various other forms or affections. These must be inferred; for, by detailing them, I might only confound the non-medical reader, and render the diagnosis difficult, and, perhaps, impossible. The symptoms, in the early stages, are such as are usually observed at the commencement of febrile diseases, viz.: coldness of the extremities, and slight shivering fits; loss of appetite; labored respiration: pulse variable, and the mouth hot and clammy; the animal heaves at the flanks, is rather unwilling to move, and the fore-legs are widely separated, while the head is held in a drooping position. As the disease progresses, these symptoms vary more or less, and the appearance of the membranes of the mouth, nose, and eyes may vary from the color of bright scarlet to that of a leaden hue. In congestive pneumonia, the pulse is more voluminous, yet less active, than in health, or in acute disease. A cough, slight or active, as the case may be, is usually noticed; it is a sort of deep-seated, half-suppressed one, and sometimes it is the first symptom which calls the owner's attention to the ailing animal.

Treatment.—It is important, at the commencement of the treatment, that the patient shall be placed in a comfortable shed or barn, where pure air abounds; for, under such circumstances, the coöperation of Nature in the cure of the malady is secured. Should the disease be of an acute character, it will be accompanied by quick or labored respiration, and a strong, wiry pulse. In that event, I should give two drachms of the tincture of gelseminum. The brisket and sides of the chest are to be rubbed occasionally with mustard and vinegar. This acts as a counterirritant, and diverts the blood from the lungs to the surface. It was customary, in former years, to bleed and purge almost all animals when attacked with acute pneumonia; but as the mortality was then very great, and less under a more rational system, we are led to believe that bad treatment was the cause of our want of success. Two or three doses of gelseminum, given at intervals of six or eight hours, will, together with the elapsed time, have a tendency to modify the affection. Then active medication is to be suspended, and we immediately give life-sustaining agents, which consist of pure air, water, proper food, tonics, and stimulants. The best tonics and stimulants that I know of are powdered golden seal and ginger, equal parts. Dose, half an ounce night and morning, as a drench; or two ounces, night and morning, of the fluid extract of resin weed may be substituted. The animal should have one ounce of powdered chlorate of potass every twelve hours, until it is evident that convalescence is approaching, or has actually set in, when it may be discontinued. The best way to administer the chlorate of potass, is to dissolve it in a few quarts of linseed tea, or an infusion of slippery elm, which should be kept in a bucket before the animal until he has partaken of the whole of it. The chlorate of potass is also indicated as a valuable agent in the treatment of all lung affections, including pleuro-pneumonia and pleurisy; its use is to be discontinued, however, when the urinary secretion becomes much augmented, or it may overwork the kidneys. Chlorate of potass acts as a sedative to the nervous system and to the circulation, and is a stimulant to the digestive organs and kidneys. For all cases of congestive pneumonia, or when the surface of the body is cold, as well as the limbs and ears, warm stimulants are indicated—ginger and golden seal; and, if possible, the body and limbs should be clothed, for the purpose of maintaining the natural heat of the skin, and of producing an equilibrium of the circulation; for when an equilibrium in the circulation of the blood is secured, then congestion ceases to exist.

I would caution my readers not to place too much reliance in art and medicine for the cure of disease. The physician does much good when he practices rationally—constitutes himself the servant of Nature, does her bidding, and does the system no harm in the use of poisons and meddlesome medication. In the treatment of many forms of curable disease it is often a question of time; they will "have their run." Hence, what we most need is patience, great faith in Nature, and a moderate degree of faith in drugs. The great trouble is, our instruction derived from text-books and veterinary institutions of learning tends to develop heroic notions of medicine and its powers, and to depress those more potent remedies, the vitalizing and reanimating forces of Nature. A case of heroism of this kind, as an example, I must now relate, in order to illustrate my position:

Mr. S——, of Chicago, bought of Mr. P—— a fine, promising young gelding, aged about eight years. The purchaser had used him during a period of five weeks with perfect satisfaction. Suddenly the newly-purchased animal became sick, and the services of a noted blacksmith were secured. The man of "tongs and hammer," realizing that the patient was the subject of the prevailing influenza, gave the patient "fits," in the form of the administration of a physic ball. He then abstracted two gallons of blood. The horse grew no better very fast. In consequence, he got more medicine. Purgation set in, which continued for twenty-four hours, when the horse died. The lives of many valuable animals are thus sacrificed. Finally, those men generally become the most distinguished who soonest emancipate themselves from the theory of the heroic schools of medicine.

The plan of treatment here recommended may appear to some of our readers as very simple. That is just what we claim for it. It is the simplicity of the practice which makes it really valuable, because such a system harmonizes with the law of Nature, and does not oppose her forces, which are always operating physiologically for the restoration of the sick and infirm; and it is our business, as rational physicians, to aid Nature in all possible ways. This is the kind of knowledge that we require—how to aid Nature, or give her the required assistance; and, in view of obtaining that information, we must read the great book of Nature, as its leaves are unfolded to us in the various departments of her great domain. We must altogether abandon a blind faith in doses of poisonous medicines and bleedings; otherwise, all our efforts are as but a withering simoom. For example, over seventy thousand head of cattle have lately perished of what has been called the " rinderpest," in and around the great city of London, which is claimed to be the model seat of veterinary learning for all British subjects It is probable that the heroic practice there taught has something to do with this tremendous mortality; for we learn that a more simple plan of treatment has been inaugurated by another class of physicians, whose medicines are almost inert, and they have been more successful than the advocates of setons, blisters, sedatives, and narcotics, not forgetting their " sheet-anchor," calomel—as Napoleon observed, in derision, when on the island of St. Helena, " Bleeding and calomel for ever!" Simple, sanative medicines act potently, yet silently, after the fashion of Nature's workings; so that we should never attempt to force her to adopt our system of cure for that of her own.

PLEURISY.

Pleurisy signifies inflammation of the pleura. This membrane lines the internal surface of the chest, and furnishes a covering to the organs within the chest. It also serves to divide the chest into two cavities, which are known as the right and left. This membrane is similar in its texture and function to that which lines the cavity of the abdomen. Both are serous membranes, secreting, on their inner surface, a lubricating fluid, which guards against friction. This membrane is the seat of pleurisy, and pleurisy consists of inflammation of it.

Causes.—It may be presumed that the ordinary causes which produce affections of the lungs are operative in the production of this malady, viz.: exposure, by which the surface of the body becomes chilled; injuries in the thoracic region; over-work; an impure atmosphere; a plethoric state of the system; in short, any cause which is likely to produce an inflammatory affection in other parts of the body. I have known horses become the subjects of this malady that have been confined to the stable for weeks. Such cases are clearly traceable to plethora, want of exercise, and impure air. It is possible that some animals inherit a predisposition to pleurisy. It has been noticed that animals of the sanguine temperament are more prone to the malady than others of different temperaments. Hence, the only way to prevent pleurisy is to try and keep up a good standard of health in horses, by affording them a chance to have regular exercise, pure air in the stable, and a fair allowance of good wholesome food. Pleurisy in the horse is very apt to end in hydro-thorax, or dropsy of the chest; and the case that I am now about to offer will illustrate this, as well as the symptoms and treatment.

Case of Pleurisy and Dropsy of the Chest.—A few months ago I was requested to visit a gray gelding, the property of a Mr. Waldo. The animal had been out of health for a week or more, suffering from a slight attack of influenza, but was improving—so thought the person who was in medical attendance; but the owner thought otherwise, and decided to secure my services.

Symptoms.—On examination, the following symptoms were found present: Pulse, wiry; respirations, quick, laborious, and accompanied with a slight grunt, indicating pain within the chest. When pressure, ever so slight, was made on the sides of the chest, it elicited very decided symptoms of pain. The animal would occasionally turn its head toward the sides, as if to inform us of the seat of his sufferings. Dropsical swellings were observed under the chest and belly, and the hind legs were also dropsical, and the vascular surfaces of the mouth and nose were tinged slightly yellow; the feces were scanty, and the urine was high-colored; the appetite was much impaired, but the patient was very thirsty. On the strength of the condition of the animal, and the presentation of the above symptoms, and percussion revealing water in the chest, I informed my employer that the case was one of pleurisy and dropsy of the chest. It is probable that water had been accumulating in the chest for several days, and that the pleura was involved at the commencement of the malady.

Treatment.—Dropsy of the chest is a very formidable disease to treat, and there are but few cases of cure on record; yet, I think if such cases came early into the hands of educated surgeons, and they use life-sustaining agents in view of husbanding the powers of the vital forces, more cures will hereafter be recorded. The treatment of this case was as follows: I rubbed the sides of the chest occasionally with a portion of camphor dissolved in olive oil, and gave, morning and evening, during a period of seventeen days, thirty grains of iodide of potass in two ounces of fluid extract of resin-weed root; an occasional dose of golden seal was given as a tonic. The iodide of potass acted on the system as a glandular stimulant, and thus increased the power of the absorbents and ex-crementitious vessels, to rid the chest of its fluid contents, while the resin-root acted as a diuretic, thereby carrying off some of the fluid by way of the kidneys. During the period of the sickness of this animal, the resonance of the chest gradually became clear, so that at the end of three weeks not more than a couple of pints of serum remained in the chest. The animal rapidly convalesced.

During the treatment of such a case as the above, I usually feed the animal liberally on oats and sweet hay; and, in order to obviate constipation, I order an occasional bran-mash. The animal, however, may not, at the commencement, have much relish for food, but after a few doses of the medicine here recommended have been given, the appetite will soon be restored. It will be noticed, during the progress of this malady, that the animal does not lie down, but stands with the fore-legs widely apart, although, at the very commencement of acute pleurisy, the subject will often get down, in view of mitigating the lancinating pain from which he suffers, in consequence of the distension of the vessels of the pleura. Pleurisy sometimes sets in as an accompaniment of influenza, which almost always ends in dropsy of the chest; and as a case of this kind is a good one to place on record, I here introduce one from my note-book:

Pleurisy ending in Dropsy and Death.—The horse, the subject of the above-named maladies, was the property of a milkman of Chicago. The animal had been previously treated for the prevailing influenza, accompanied by pleurisy, and had about recovered so as to take daily exercise; but finally the owner discovered that the horse was suddenly taken with a relapse. Feeling alarmed, he procured one pint of linseed oil, with which he drenched the animal. Still, the horse rapidly grew worse. At this period, my services were sought. On auscultating and percussing the chest, I found that the right cavity was occupied by a large quantity of water. Dropsical swellings were found in various parts of the body, viz.: the sheath, limbs, and region of the pectoral muscles. On applying the ear to the trachea, the respiration was of a mucous character, and tubular, showing that the lungs were much diseased. The examination revealed the fact that about one-half the left lung was involved in disease. The animal had no appetite, yet was thirsty; and when urged to move, would utter a grunt, indicative of pain. It was evident that the relapse was, or might be, considered a case of pleurisy, ending in effusion into the chest. The worst feature of the case was that the animal purged violently (liquid stools), having a very strong odor of linseed oil; it covered a large space on the floor, and ran down his hind extremities. This seemed to me to render the case hopeless; for super-purgation of itself will often cause the death of an otherwise well animal. But this poor creature was laboring under a malady from which few ever recover. However, I thought that while "there is life there is hope," and I commenced the treatment by performing the operation of tapping for dropsy of the chest. I drew off three and a half gallons of fluid, of a light straw color, from the right cavity of the chest. On auscultating the chest, both sides appeared resonant, showing that the waters must have occupied both cavities; hence, the mediastinum (the membrane which divides the thorax into two equal cavities) must have been ruptured, thus permitting an influx and reflux of the fluid. Next, counter-irritants were applied to both sides of the chest. The patient was properly clothed, tonics and astringents were given, and the case was left in the care of an attendant for the night. Next day I found the patient was respiring very comfortably. The tapping of the chest seemed to have done much good, yet the purging still continued; so I ordered scalded milk and charcoal, and administered tonics and stimulants, as before. But it was very evident that the animal must die of super-purgation, and next morning I received word that death had taken place. It was unfortunate, under the circumstances, that the owner should have given linseed oil; for it is a very powerful irritant and cathartic, and produces much irritation and inflammation on the surface of the interior of the stomach and intestines, as well as uncontrollable purgation. It is a remedy which is very quickly absorbed, and finds its way into the blood in a very few minutes, as the following case will show: I once gave a horse a dose of linseed oil for the purpose of experiment. (He happened to be in good health, or I should probably have lost him.) Ten minutes after administering the dose, I applied a linen handkerchief to the highly organized membrane lining the eyelid, and both the odor and color of the oil could easily be detected. On exposing the handkerchief to a moderate degree of heat, for the purpose of evaporation, a grease spot remained on the same; hence it is an agent that should never be used in pleurisy, nor in any disease occurring in the organs of respiration. Dropsical diseases, supervening on influenza, are never benefited by cathartics, for they always lead to debility and prostration. Influenza is a disease of very prostrating character; hence the proper plan of treatment is to sustain the vital power, by nourishing diet, tonics, and stimulants—ginger and golden seal. While commenting, however, on the linseed oil treatment, I do not wish to censure the owner of the animal; for I should be throwing stones into glass houses, having myself, in former years, used and recommended this kind of oil for various diseases; but, after many mishaps, I am satisfied that it is unsafe in the treatment of equine affections.

PERICARDITIS (INFLAMMATION OF PERICARDIUM),
(SOMETIMES CALLED "HEART DISEASE.")

The pericardium is a membranous sac, which surrounds and incloses the heart. It is composed of two layers, united by cellular tissue. The external layer is dense and fibrous, and is attached, by ligaments, to the sternum, or breast-bone and diaphragm, and above to the roots of the large blood-vessels. The internal layer exhibits internally a smooth surface, like the lining membrane of the cavity of the chest. From this surface is secreted a serous fluid, which lubricates the heart, and protects it from the friction which would otherwise occur. The pericardium is a wall of defense for the heart, and keeps it in its proper position.

Pericarditis signifies inflammation of the inner membrane of the pericardium. It is a disease which is almost unknown to the ordinary practitioner in this country, and entirely unknown to the husbandman; yet it is a disease that carries off a large number of horses and cattle, and often the parties concerned in the treatment and ownership of the animals do not even suspect the true nature of the malady. After the inflammatory action of the pericardium has reigned for several days, unsuspected and unchecked, an inordinate effusion from the serous lining takes place. We then have dropsy of the "heart-bag." Water accumulates around the heart, which interferes with its action so as to put a stop to circulation, and this results in the death of the animal. Pericarditis often accompanies, or is the sequel of, pleurisy and other diseases of the serous membranes; yet, fortunately for the patient as well as the practitioner, it does not require any special treatment other than that laid down for pleurisy and disease of other serous membranes.

Causes.—The causes of pericarditis are the same as those which are said to be operative in the production of pleurisy and disease in the serous membrane, viz.: sudden impressions of cold upon the external surface of the body; injuries; overtasking the powers of the animal by laborious and rapid work. Sometimes foreign bodies, such as pins and needles, have been found within the pericardium, which must have been taken into the stomach with the food, and have worked their way so as to penetrate the pericardium and heart. Such cases almost always prove fatal.

Symptoms.—The general symptoms of the disease are anxiety, shivering, coldness of the skin and mouth, loss of appetite; the jugular veins are congested, and they pulsate like an artery; there is a sort of tinkling or metallic sound, when the ear is applied to the chest and lower part of the neck. In the latter stages the respiratory murmur becomes very indistinct, and there is a sort of double expiration, and the ribs become contorted, the same as in dropsy of the chest.

Treatment.—The heat of the external surface of the body must be restored, and kept uniform by means of leggings and body clothing. Both sides of the chest are to be briskly rubbed, two or three times daily, with hot vinegar and strong mustard. One drachm of iodide of potass, dissolved in five ounces of warm water, should be given every four hours; but, should the case be seen, and the true nature of the malady discovered in its early stages, omit the iodide of potass, and substitute two fluid drachms of tincture of gelseminum. Two or three doses, at intervals of four hours, will suffice; then resort to the iodide of potass, or substitute for the same four drachms of chlorate of potass. So soon as the animal appears to improve, discontinue the above-named medicine, and give tonics. Four drachms of fluid extract of golden seal, twice daily, wi suffice.

A fatal case of pure pericarditis is related by Mr. WOODGER, V. S., in the London "Veterinarian." The autopsy revealed a healthy state of the lungs and pleura; yet the pericardium contained about three quarts of serum, highly colored. The walls of the pericardium were considerably thickened, and studded with merous bands of lymph, on the inner surface, of a bright yellow color. Several small patches of inflammation were likewise observed on the ventricles of the heart, near the valves of this organ. The heart itself was also larger than usual, and much softened.

HEAVES.

A disease prevails among horses, more particularly in the Eastern States, known as heaves. The affected animal heaves at the flanks, or, rather, performs what is known as abdominal respiration, by bringing into play, at the moment of expiration, the abdominal muscles, for the purpose of aiding the lungs and diaphragm in the function of expiration and respiration. The disease somewhat resembles asthma, it being of a spasmodic character, and, after continuing for some time, will suddenly disappear. The disease may be defined as being great difficulty in breathing; continuous; aggravated, by dust on the road, musty hay, improper provender, impure air, and sudden atmospheric changes. On applying the ear in the region of bronchi, or to the upper sides of the chest, a wheezing sound of respiration is heard. These symptoms, together with the heaving motion at the flanks, constitutes heaves. When the disease is merely functional—that is to say, brought on by feeding musty hay and musty clover—a cure is affected by feeding prairie hay, which contains the well-known silphyum, "resin weed." The husbandmen who reside in the vicinity of where the resin weed grows are well acquainted with the properties of this plant, and they declare that it is a specific for the treatment of asthma or heaves. I have used the article in the form of fluid extract, prepared from the root, and I find it to be a very valuable remedy. The dose of the fluid extract is two ounces, morning and evening. There are some cases, however, which are incurable, owing to organic disease of the bronchi, or rupture of air-cells in the lungs. Such cases are known by the unnatural respiration when the horse is drawing a heavy load or traveling up hill. His breathing is then very distressing, and he not only brings into play the abdominal muscles, but also the muscles of the chest (intercostals). Such cases, although considered incurable, may be palliated by the daily use of fluid extract of resin weed. Provided prairie hay can not be obtained as food, the ordinary hay used should be sprinkled

with salted water, in the proportion of a handful of salt to one
bucket of water. Sprinkling the hay with weak lime-water has
also a good effect. Sliced carrots, parsnips, or potatoes, fed occasionally, are also beneficial. But, of all our remedies, none equals
a run at grass on prairie lands.

MAIN ARTERIES ON THE INSIDE OF ONE OF THE FORE-LEGS, FROM
SHOULDER TO KNEE.

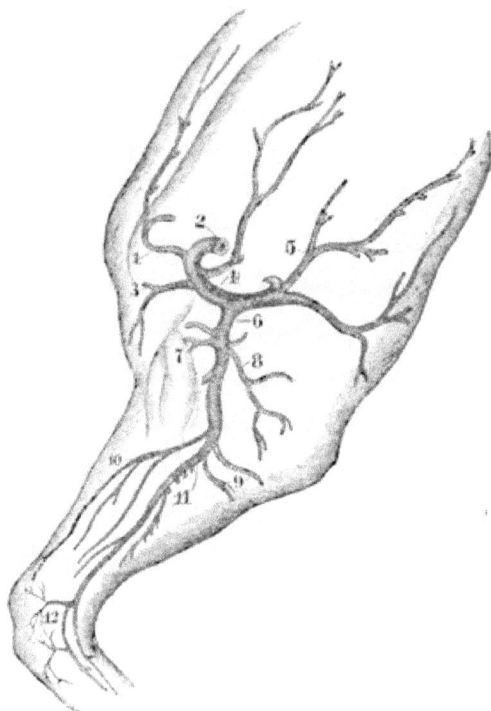

EXPLANATION OF PLATE.

1. Humeral thoracic.
2. Axillary.
3. Internal thoracic.
4. Dorsal scapular.
5. External thoracic.
6. Humeral.
7. Brachial.
8. Branches of the triceps.
9. Ulnar branches.
10. Spiral branches.
11. Radial branches.
12. Metacarpal arteries, large and small.

SECTION V.

DENTITION AND DIET.

THE HORSE'S AGE AS SHOWN BY HIS TEETH—ON THE DIET OF HORSES—THE EFFECTS
OF VARIOUS KINDS OF FOOD—AS REGARDS THE QUANTITY OF FOOD REQUIRED—
AS REGARDS CHANGES IN DIET—CONCLUSION.

THE HORSE'S AGE AS SHOWN BY HIS TEETH.

FIGURE 1 is a representation of a foal's " *mouth*," as it appears at the age of twelve months; the *temporary* teeth are
shown, and all, except the corner ones, are worn quite smooth.
(See remarks on *temporary* teeth.) Figure 2, a two-years old;
the two *anterior*, or front *temporary* teeth, are shed, and the permanent ones, with their deep *pit* in the center, and unworn edges,

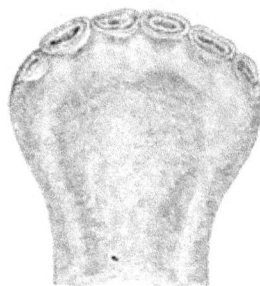

Fig. 1.

TEMPORARY TEETH OF THE COLT.

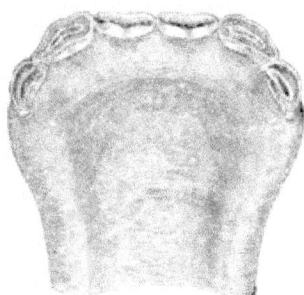

Fig. 2.

TWO YEARS OLD.

are approaching a level with the remaining *temporary*. Figure 3,
a three-years old; the two *permanent middle* teeth are seen, with
their deep *pits*, etc., and are nearly on a level with the two *front
permanent* teeth, which are smoothed down. Figure 4, a four-
years old; the permanent corner teeth are now " cut," and the
tushes have appeared, although the latter may not appear until

(127)

five. Figure 5, a five-years old; the *corner* teeth are well up; the faces of the *front* and *middle* are worn more or less, and the tushes are more developed. Figure 6, a six-years old; the *black* marks have disappeared from the two front teeth, and a

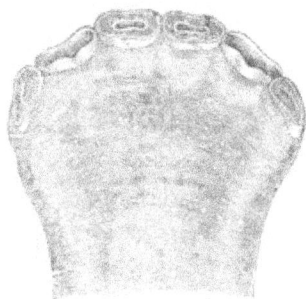

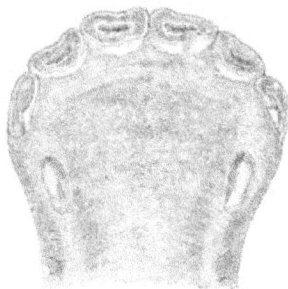

THREE YEARS OLD.

FOUR YEARS OLD.

brownish mark is perceivable. Figure 7, a seven-years old; the black marks have now disappeared from the two *middle* teeth, and a brown spot is seen in the center. Figure 8, an eight-years old; the *marks* and cavities of the *corner* teeth are now effaced. Figure 9 is a horizontal view of "Black Hawk's" incisors, and, con-

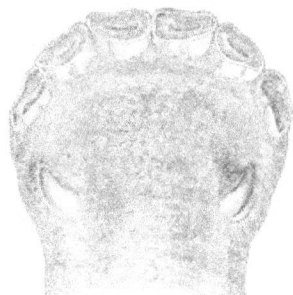

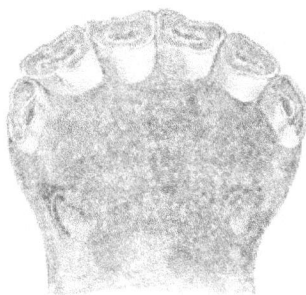

FIVE YEARS OLD.

SIX YEARS OLD.

trasted with figure 10, (a young mouth,) shows the degree of inclination existing between the incisors of the aged and young animals.

THE COLT'S MOUTH.—*Remarks on the Temporary Teeth.* (*Incisors.*) In the course of from ten to fifteen days after birth (varying occasionally), the *front* nippers in the upper and lower jaw make their appearance, and, during the period, which occurs between the third and fifth week, the *middle* incisors appear. Between the seventh and ninth month the lateral or *corner* teeth are cut. The

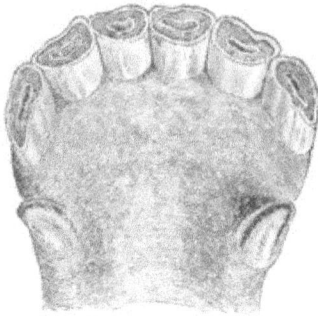

Fig. 7.

SIX YEARS OLD.

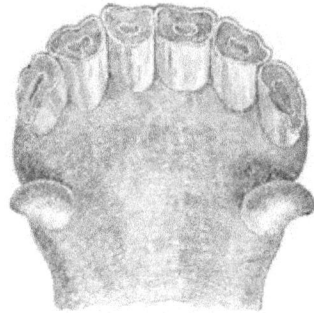

Fig. 8

EIGHT YEARS OLD.

colt is now in possession of a full set of *temporary* incisors. These *temporary*, or " milk " teeth, differ from the *permanent* ones by being much smaller and (previous to shedding) whiter, having on their anterior or *front* surfaces grooves or furrows; also a well-defined body, neck, and slender fang. They also present oval surfaces *anteriorly*, or in front; and within the mouth they are concave. Their surfaces are generally uniform and smooth, and resemble the eye of an English horse-bean.

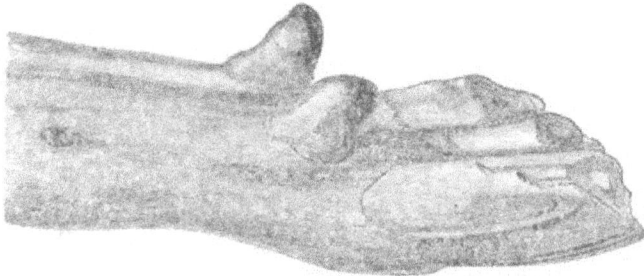

Fig. 9.

HORIZONTAL VIEW OF OLD BLACK HAWK'S INCISORS AND TUSHES.

THE PERMANENT TEETH.—Constitutional idiosyncrasies are often operative in hastening or retarding dentition, and the character of the food, whether it be *natural* or *artificial;* and the

9

made of obtaining the one or the other makes considerable difference in the wear and tear of the same, consequently we must
bear in mind that the popular theory of " age " is sometimes more
arbitrary than truthful, and allowances for the same must be
made. At the age of *two*, or a few months over, the *front permanent* teeth ought to appear, and, consequently, at the age of *three*,
varying a few months, the *middle* teeth are up. At *four*, varying
as in the preceding cases, the *corner* teeth begin to show themselves, and are not up *square* with the others until the fifth year.
The animal now emerges from colthood, and is known as a horse.
As regards the female, during her minority she is termed " *filly;* "
at *five* she enters the adult stage, and, consequently, is denomi-

Fig. 10.

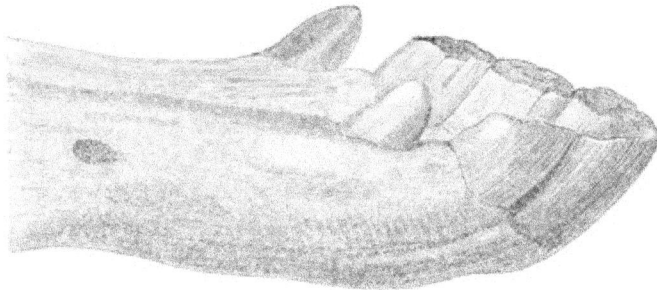

A YOUNG HORSE'S MOUTH.

nated a mare. Supposing the horse to have a full set of nippers,
or incisors, at the age of five, the *marks*, or black incrustation,
will have so worn away at six as to leave a faint brown mark
At *seven*, similar appearances have taken place in the two next,
or middle, teeth. At *eight* the marks, or rather the cavities, of
the two next, or corner teeth, are now about obliterated, and the
face of the teeth are nearly level, and the central enamel is nearer
the inward than the outward border. After the marks have all
disappeared in the lower jaw, the exact age can not clearly be
shown. At the age of *nine* the front teeth appear somewhat
rounded; the middle and corner ones contract their oval faces,
and the central enamel diminishes and approaches the inward
border. At *ten* the *middle* incisors resemble those at nine, and
the central enamel has approximated the inward border and is
rounded. At *eleven* the middle teeth resemble those at ten, and

the central enamel is almost worn off. At *twelve* the corner teeth take on the triangular form, and their central enamel is very contracted, and a yellow mark is left. At *thirteen* the sides of the front teeth thicken, and they begin to assume a triangular appearance, and the whole incisors of the lower jaw complete the triangular form at the age of *seventeen*. At *eighteen* the front teeth are somewhat flattened from side to side; the flattening goes on in the *middle* and *corner* teeth, and is completed at the age of twenty-one, giving them the appearance of a reversed oval. As regards the *marks* in the upper incisors, they disappear from the *front* teeth in the course of the *ninth* year, from the *middle* in the *tenth*, and from the *corner* or lateral ones in the *eleventh* year;

Fig. 11.

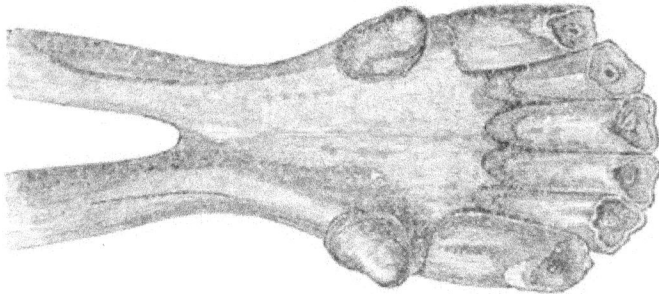

A PORTION OF THE LOWER JAW OF OLD BLACK HAWK,
(Aged twenty-three years and eight months.)

and as the animal increases in years, the upper corner teeth become indented or notched. The *inclination* of the teeth enables us to form some idea of the horse's age. In a young horse, the teeth are upright (see fig. 10); after the age of eight they gradually become horizontal, (see " Black Hawk's " mouth,) and the upper teeth overlap the under ones, and thus wear off their outer edge.

THE TUSHES, OR CANINE TEETH.—Between the ages of four and six the tushes, which, in the male, are four in number, make their appearance. In shape they are conical, with a sharp point, and curved. On the inside there are two furrows, which commence at the base and meet at the apex, leaving a triangular eminence between them. The sharpness of the apex, the degree of curvature, and the distinctness of the furrows are good criterions of youth; for, as the animal advances in years, the tushes become blunt, less angular, and more rounded, and the furrows have disappeared. Supposing that the tushes are not completely evolved until the age of *five;* in the sixth year the apex of the cone is worn some; in the *seventh* the furrows grow shallow; in the *eighth* year they are obliterated, after which period the apex gradually wears away, and the body of the same becomes rounded and pointed, or blunt, according to circumstances. Aside from the teeth, an aged horse may be known by the deep pits above the orbital processes; the sunken eye; by the prominence of the joints and loss of plumpness in the muscles; the lips are somewhat pendulous; the withers sharp; the back becomes arched; the teeth are lengthened, and become yellow.

Fig. 12.

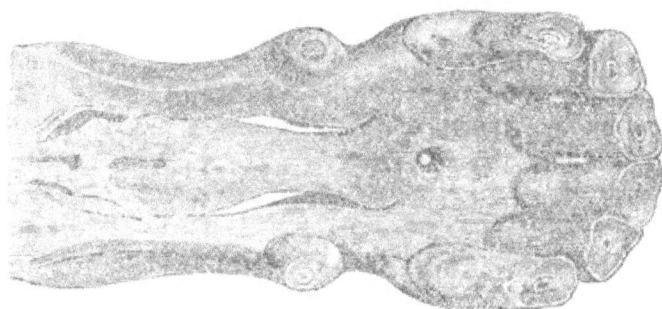

A PORTION OF THE UPPER JAW OF OLD BLACK HAWK,
(Aged twenty-three years and eight months.)

THE GRINDERS, or MOLARS, afford but very little information as regards the precise age of a horse. As he advances in years, however, the outer edges become sharpened, so that it often becomes necessary to rasp them. As regards their development, it is understood that the foal is born with two upper and lower grinders in each jaw. At the end of a month, sometimes more, a third appears. At the completion of the first year, or thereabouts, a fourth grinder in each jaw appears. Thus the yearling has sixteen grinders. At the age of *two,* a fifth grinder appears, and at the age of *three* the sixth and last appear. It must be remembered that cribbers and voracious feeders are apt to deface their teeth, and thus some persons are apt to be deceived as regards the exact age; but a good judge, who takes into consideration not only the appearances of the surfaces, but also the marks, points, and the inclination of the teeth, will not be apt to make any grave mistake.

On the Diet of Horses—(Men often Dig their Graves with their Teeth, and the same is true as regards Horses)—Variety of Food necessary—Effects of certain kinds of Food—As regards the requisite quantity, etc.

It is admitted by medical men, and the experience of the reader may possibly have confirmed the fact, that a great proportion of the diseases occurring among live stock are produced, either directly or indirectly, by errors in diet; consequently the means of prevention should be known, and used accordingly.

No special form of diet can, for any great length of time, preserve the integrity of the animal system. In order to substantiate this assertion, we shall show the effect of simple elementary principles on the body of man, inferring, at the same time, that man's servant, the horse, is included. The reader has probably heard of the experiment made by Napoleon Bonaparte, which was that of trying to supply the nutritive wants of his system by living on a concentrated preparation of beef, in the form of jelly. He gave it a fair trial, yet came very near starving on the same, and he was led to remark that the stomach was a sort of scavenger, which required a given amount of rubbish to sort from, and select the necessary elements for renovating the tissues. He probably selected jelly because it is highly nutritious, containing more of the flesh-making principle than most articles of food. It was rich, not only in nitrogen, but also in oxygen, hydrogen, and carbon, four of the principal elements which compose the animal fabric.

The fact is, no single article of diet, let it be ever so rich in the elements of organized tissues, can long support life. A nation of men subsisting long on a simple form of diet, without variety would soon become emaciated, and die of inanition. Let the Grahamites abstain from milk, and live on bread and water, and they would soon exterminate themselves. Milk is a highly nutritious compound, and furnishes their systems with the equivalents that we obtain from beef and mutton. The inhabitants of the "Celestial Empire" (Chinese) are great consumers of rice. It is the principal article of diet among the poorer classes, many of whom are lank and lean, and would make very respectable walking lanterns, (if lighted candles were placed within their abdomens), only they happen, occasionally, to catch a stray dog or pig, on which they make a savory meal, and thus furnish the material for the formation of muscle and fat.

MAGENDIE has proved that even the canine race can not live more than forty days on any single article of diet, let it be ever so nutritious, for it is either followed by starvation or disease; hence the necessity for variety in food. In allusion to disease being produced by the long-continued use of a single article of diet, I would mention that the Scotch peasants are great consumers of oat-meal. This article is little inferior to wheat in the flesh-making principle, and we might naturally infer that an article of diet so valuable and palatable, when properly cooked, should tend to promote health. This, however, is not the case. Those who eat the most oat-meal are, according to medical testimony, the notorious subjects of intestinal concretions, and in the Edinburg Anatomical Museum is to be seen a vast and valuable collection of intestinal calculi, most of which caused the deaths of confirmed oat-meal consumers.

Dr. CARPENTER, an eminent physiologist, says that "no fact in dietetics is better established than that concerning the impossibility of long sustaining health and life on a single alimentary principle. Neither pure albumen, fibrine, gelatine, gum, sugar, starch, fat, nor oil, taken alone, can serve for the due nutrition of the body. This is partly due to their failing in supplying the waste of the tissues, and partly to the fact that single alimentary substances, long continued, excite such a feeling of disgust that the animals experimented on seem to prefer the endurance of starvation to the ingestion of the same."

The reader is probably aware that when a person has long been confined to any particular article of diet, a craving for something else is experienced, which very few persons can resist. This teaches us that, in order to preserve the health of live stock, we must vary the diet, and are not to be over-particular in selecting the most nutritious articles. But we want, as Napoleon says, a little rubbish—coarse rubbish The internal surface of the stomach and bowels require to be irritated once in awhile, and this probably was the idea which Graham had when he first recommended coarse food. The stomach must be made to labor hard at times, or its function will deteriorate. Persons who complain of weak stomachs and dyspepsia are those who live on dainty viands, and seldom, if ever, distend that organ to a healthy capacity with coarse material. Consider, for a moment, the condition of Spanish and Italian peasants. They have not much of our national disease (dyspepsia) among them. The bread that they eat is made of coarse material; yet with that, and the addition of a little oil, wine, and a few vegetables, they can indure greater fatigue, and often carry a heavier burden than he who lives on more concentrated food. We may distend the horse's stomach with coarse food, and, perhaps, not impair its function so much as when overburdened with meal and concentrated food.

The stomach must be made to work for a living once in awhile. Hard work agrees with it, and coarse fodder stimulates and develops its latent powers, and augments the gastric secretion, which is the active solvent of the food. Labor operates on the stomach in the same manner and in the same ratio that it does on the brain or muscles—increases their capacity. Compare, for example, the brawny arm of the mechanic with that of the counting-house clerk, or the powerful muscles of the truck and farm-horse with those of the pet saddle-horse. This comes of work, hard work. See the king of birds, the eagle, towering above and beyond the sight and ken of man, darting, with almost lightning speed, from mountain to valley, buffeting the rude shocks of heaven's artillery. He acquires strength of muscle and wing by extraordinary exertion and desperate feats of flight. After the same fashion we develop the mental faculties, augment protracted mental labor, close thought, and study, light up the intellectual nature of man, and develop the latent powers of his brain; and the more his mind acquires, the greater and more varied are its powers.

It must be borne in mind, however, that the various functions of the body require periods of rest; for, should a horse be permitted to stand up to a full crib, and spend the greater part of the day and night in cramming his stomach, disease, sooner or later, must surely occur. The same is true as regards man. Let an individual gormandize through the day, and then indulge in a late supper, and continue the practice, he soon acquires a sympathetic headache, or the stomach grows refractory, and casts up the burden, for the simple reason that its function is overtaxed. It must have rest. The same rule applies to the muscular and mental natures. If, therefore, hard labor develops the function of either, rest, at suitable intervals, preserves their integrity, and, therefore, must not be disregarded.

In selecting food for working animals, we must remember that they require certain inorganic equivalents, which seem to be as necessary for the support of the system as nutriment. Salt, for example, is not nutritious; yet neither man nor brute can long exist without it. Common salt, chloride of sodium, is decomposed in the stomach, and is there found in the form of muriatic acid and soda. The former is supposed to aid digestion, and the latter eliminates bile. Neither is phosphorus (found in straw) nutritious, yet that article is absolutely necessary for the support of animal life. It is an element of both vegetable and animal organization. The former absorbs it from the soil, and, in turn, yield it to animals, by the process of digestion. Oats and beans are nitrogenous compounds, flesh-making equivalents, yet they furnish only one part in a thousand of the article we need—phosphorus; while cut straw, potatoes, and several other "inferior" vegetables, contain more than double the quantity of the same; so that a horse must eat such rubbish as straw, potatoes, carrots, beets, and "stubble," in order to supply the necessary material. Then consider that sulphur, iron, chlorine, lime, potassium, magnesium, and several other mineral substances, not in the least nutritious, are alike necessary for the support and integrity of the living organism, and, therefore, should be the elements of food. Some articles furnish the needful in abundance; in others there is a deficiency. This supplies another argument in favor of variations in diet.

A lecturer on physiology has remarked that "there exists a peculiar analogy between vegetable productions and living animals. Animal and vegetable fibrine—albumen of eggs and the gluten of wheat—contain about 15 per cent. of nitrogen, so that they are somewhat identical. If you take 100 lbs. of flour and wash it in water, frequently changing the same, you get 15 lbs. of gluten. This is the flesh-making principle, and represents 15 lbs. of the albumen of flesh. The gluten of flour, caseine of cheese and peas, albumen of eggs, and the flesh of an animal contain also a relative amount of carbon, hydrogen, and oxygen; so that the flesh of animals is already prepared for them in the vegetable world. The digestive organs of animals merely change the mechanical form and condition of the former; their chemical composition remain about the same."

THE EFFECTS OF VARIOUS KINDS OF FOOD.

It is customary, in some stables, to feed horses, nearly all the year round, with what is known as "cut feed," which is composed of cut hay, meal, shorts, salt, and considerable water; the whole is mixed together, and served out, sometimes, without regard to quantity. This kind of food might, and sometimes does, agree with horses, but it is not right to feed them, year after year, on the same, for the reason just set forth. Another reason for objecting to this food is, that, in the stables alluded to, we hear of a great many cases of tympanitis and flatulent colic (diseases somewhat identical), arising, no doubt, from the presence of so large a quantity of water as some persons are in the habit of using. It saturates the food, and retards digestion. Not only this, but when dry food, highly charged with water, enters the stomach, the temperature of the latter causes the food to swell—increase in bulk—and distends that organ, and also favors fermentation instead of digestion; hence arises flatulency. We do not, however, mean to contend that such food is at all times the direct cause of indigestion, colic, etc., because many stablers are ready to testify that they have fed the same for many years without any apparent inconvenience to their horses; but we contend that it acts indirectly in the manner alluded to; and, although some horses may "get used to it," and others, having wonderful digestive organs, assimilate it, yet the day of reckoning may not be far off. We contend that water taken with food always retards digestion. The proper solvents of the food are the gastric fluids, and the horse has abundant facilities for supplying the requisite quantity. An ordinary horse is said to secrete, while feeding, fluid, of salivial and gastric characters, at the rate of one gallon per hour—enough, we should judge, to saturate a common meal; therefore the water is not needed. We urge no objection against the more rational custom of merely sprinkling the food with salted water, in view of absorbing dust, which often abounds in inferior hay, but do seriously object to the practice of using a large quantity of cold water in the preparation of food for horses.

From experiments made by scientific men, it has been ascer-

tained that parsnips, carrots, turnips, and cabbage, which contain
from 80 to 90 per cent. of water, require over twice as much time to
digest as when the food is free from water. Cabbage, for example,
requires twenty hours, and broiled beef-steak only eight, to digest.

Turn a cow into a luxuriant pasture of grass or clover, and, after
partaking of one or the other, she is liable to become "blown" or
"hoven"—tympanitic; the abdomen becomes enormously distended with gas, (either carbonic acid gas, or sulphureted hydrogen,) and, unless the same be condensed or evacuated, rupture and
death are sure to follow. This imperfect digestion and consequent
generation of gas is due to the presence of vegetable fluids found in
green fodder. Therefore, animals having weak digestive organs,
predisposed to flatulency, should have the privilege of watering
their own food with salivial fluid. The best diet for such an animal would be "dry feed," composed of ground oats, cracked corn,
"fine feed," and a small quantity of sweet hay. On the other
hand, a constipated state of the bowels always indicates coarse
food; and in this view the English use chopped straw and coarse
bran, with decided advantage. Animals should never be watered
immediately before nor after meals; after the lapse of an hour
from feeding time is the best.

As regards the Quantity of Food required.

The adult horse does not require so much of the flesh-making
principle as the young and growing animal, but he seems to require
a greater variety. The adult merely requires enough to replace
the waste—the wear and tear of his system. If he obtains more
than this, the surplus is either excreted from the body, or else
stored up within the same in the form of fat; and every body
knows that a fat horse or fat man are not best adapted for a race
nor hard labor, but of all others, (except those in a state of debility,) they are most subject to acute disease. With the young
and growing animal the case is different. Here we require bone,
muscle, and nerve. Oats, corn, and pollard furnish the same.
The colt obtains from its mother's milk all the elements of its own
organization in a concentrated form—all that seems necessary for
developing bodily proportions and hereditary traits; therefore,
when weaned, the colt must be furnished with the same equivalents
in the form of fodder: ground oats, wheat bran, and meal furnish

the same. It is the young and growing animal that requires our greatest attention. If our readers desire to raise colts that shall remunerate them for the trouble and expense incurred, they must feed the same, during their minority, with a liberal hand. Any neglect at this period can never be made up in after life. The subject will always remain lank and lean—living monuments of their master's folly or ignorance, as the case may be. In addition to the food required for the colt's growth, we must also furnish enough to supply the waste incurred by expenditure of muscular power. We all know that the young are very active and playful. Every muscular movement involves an expenditure of vital force, and thus exhausts the system; therefore, in view of developing their full proportions, and promoting the integrity of the living mechanism, they must have nutritious food, and plenty of it. They are not, however, to have a large quantity at a time, but little and often. Their stomach is small, not larger than that of a man's. Should it be overdistended with innutritious food, the organs of respiration and circulation become embarrassed, and the blood loaded with carbon. They require food often, because the digestive organs are very active, and soon dispose of an ordinary meal. Then comes the sensation of hunger, which every one knows is hard to bear.

The climate or temperature of the surrounding atmosphere has a wonderful effect on the animal machine. Let two horses be located in different stables, one of which shall be, like "Jack Straw's' house, neither wind-tight nor water-proof; the other built on the air-tight principle. The occupant of the former will require more food than he of the latter, because cold air has a depressing influence on the body, exhausting superficial heat. Our readers are probably aware that if a hot brick be placed in contact with a cold one, the caloric radiated from the first is absorbed by the latter, until a sort of equilibrium be established. The same is true as regards the body of a horse. He being in an atmosphere many degrees less than that of his own body, gives off the heat of the same to the surrounding medium. The heat thus given off has to be replaced by food, which is the combustible material. But it often happens that the digestive organs are deranged, incapable of assimilating a sufficiency of carbonaceous material to maintain even the normal temperature of the body. The consequences are, loss of flesh and health. On the other hand, a fat horse can endure the depressing influences of a cold atmosphere, because he has within the body a vast generator and non-conductor of heat in the form of adipose matter. It has been proved that the immediate cause of death in warm-blooded animals, when food has been withheld, was their inability to keep up that temperature necessary for the integrity of vital operations. The animal located in an air-tight stable is, probably, surrounded by an atmosphere almost equal in temperature to that of his own body, and, consequently, he does not require so much food as the former. Hence, the amount of food necessary in the one case might be too much for another, and, consequently, operate injuriously. A cold, bracing wind is said to "sharpen the appetite." The inhabitants of northern regions require more food than those of the southern. A horse, therefore, of weak muscular organization, the subject of debility, requires a comfortable stable, and food that will develop muscles. Now, it has been ascertained that horse-beans contain more of the nitro-albuminous principles than any other article of diet. English horses are very partial to the same, and in such the development of muscle is very remarkable; consequently, a fat horse requires less of the nitro-albuminous principle, and more of the aqueous, in the form of carrots, potatoes, and beets.

It must never be forgotten that some horses will keep in fair working order on a moderate supply of aliment. In such cases, we infer that the digestive organs are in a state of activity capable of extracting all the nutrient properties from the same. On the other hand, a horse performs less work than the former, consumes twice as much food, yet actually loses flesh. The digestive organs are then at fault. A horse, however, may lose flesh, become weak in the legs, etc., in consequence of the laborious nature of his employment. Therefore, any derangement occasioned in this way prevents the food from being converted into healthy chyme, chyle, or blood, and thus the necessary deposit of new matter is prevented.

Most voracious feeders are dyspeptic, and such are almost never free from intestinal worms. Their breath becomes feted; saliva, thick and tenacious; excrement, slimy; and should the subject be fed on oats, the same would be found, after traversing the alimentary canal, unmasticated. The principal remedy for indigestion is change in diet. Sometimes it may be proper to allow scalded oats, although cooking does not add to their nutritive quality, yet, more

or less, completes the destruction of organization, and better prepares them for the action of weak solvents.

Changes in diet will sometimes work wonders in the restoration
of a dyspeptic; yet he will require, also, suitable medicinal agents,
in view of augmenting the digestive function, for which purpose
the following is recommended:

> No. 17. Fluid extract of chamomile.............. **4 oz.**
> Fluid extract of ginger................. **3 oz.**
> Powdered hyposulphite of soda.......... **2 oz.**
> Water............................... **5 oz.**
> Mix.

Dose, a wine-glassful night and morning.

CHANGES IN DIET.

Sudden changes in diet are not recommended. For example,
should an animal have been previously fed on corn, meal, or oats,
he should not be turned out to shift for himself, as the saying is,
and depend entirely on grass for a living. Grass acts on horses
unaccustomed to it as an aperient—scours them—which is a debilitating process. Grass may improve the health of a fat, lazy, or
humory horse—reduce flesh and purify his blood; but the emaciated horse requires, in addition, a daily feed of oats or cracked
corn, to make up for the deficiency of carbon in the former. For
these, and other reasons that we might urge, the reader will perceive that changes in diet can only be made valuable in proportion
to our knowledge of the wants of the animal economy.

A great proportion of our horses are too well fed, obtaining more
food than they require. In this land of plenty, most of our valuable horses are overfed; and more especially does this happen
among animals owned by wealthy and liberal individuals. The
impression we wish to convey to the mind of the reader is, that
the food of such is not proportioned to labor; in other words,
there exists a disproportion between the amount of carbon taken
in the form of food, and the oxygen received, in the process of
respiration. Now, to illustrate this, we will suppose that a man
engaged in mercantile pursuits owns one or more horses. He has
not the time nor inclination to give the one or the other the necessary amount of exercise. They stand up to a full crib, from day
to day, enjoying or, rather, gorging themselves with, a certain

amount of fodder over and above what they actually require, and much more than they really need. The surplus is often stored up in the form of fat, and this induces acute diseases, and they die of too much food and care. It is very rare that we have occasion to recommend a man to feed his horse more liberally, but almost always the reverse. Starvation is said to be the cause for many equine diseases; but, so far as our experience goes, such cases are, in this country, very rare. We conceive the term starvation to be a libel upon civilization; and so unfrequent is its application among a nation of husbandmen, that it is omitted in our dictionaries. A man on a barren rock, or a horse in the deserts of Arabia, might probably starve; but the idea of the latter starving in the vicinity of a well-stocked barn or stable, within striking distance of a land of plenty, seems to us a very absurd conclusion. There are enough horses to be found dressed up in the garb of starvation, having tight skins, prominent ribs, and a cadaverous countenance, living, yet half dead. But they know nothing of the "famine in Egypt;" they get not only enough, but too much of the same kind. They probably require a change in diet, or else they are confirmed dyspeptics, laboring under a chronic form of indigestion; and if such should be the case, quantity is objectionable, and good quality more desirable. The stomach, however, is not at fault, its function being deranged. Or they probably get enough, and perhaps too much, for a weak stomach: hence loss of flesh, etc.

Some men are in the habit of bleeding their horses every spring. This is done in view of reducing fat and flesh, the subjects being humory (plethoric), their systems abounding in highly carbonized blood, which is proof positive that the same have been overfed. The racer, before he can perform a feat of speed, must be prepared, as the saying is. This implies bleeding and physicking. Some works on farriery lay down regular rules for putting a horse in racing condition, and the remedies are fleam, physic, and bran, proof positive that such animals have had too bounteous a supply of nutriment.

Some horses—and the same is true of man—grow poor in consequence of having to carry about a juvenile restaurant within their digestive organs. They probably become exhausted, or plethoric, as the case may be, in consequence of an overburdened stomach. A stomach overburdened reacts on the nervous system, deranges the physiological condition of the subject, and lays the foundation for hypertrophy (which is an abnormal increase of fat or tissue), or the opposite, which condition is known as atrophy—a wasting of the same. But most frequently an overburdened stomach induces diseases known as staggers, cerebral congestion, softening of brain, etc.

Many horse owners are continually devising means to excite the appetite of their horses, in order to get as much food as possible into their stomachs. Some men seem to think that an error in this direction can never occur; and should the animal refuse to consume the abundance thus placed before him, the liberal owner is apt to consider his pet sick, or think that he must have a poor appetite. Now, it were far better, for both man and horse if the former would only experiment in the opposite direction, and ascertain how small a quantity a horse may subsist on. A small quantity of good food, well digested, answers the purpose of nutrition much better than a large quantity, imperfectly so. It is very interesting to contemplate how efficient a little food proves for the promotion of health and longevity. For example, a little barley and coarse fodder will suffice for the "courser of the desert." Shetland and Welsh ponies will live and grow fat on the mere vestiges of vegetation. The best cow in the world—the property of J. H. Kelly, of Cleveland, Ohio, weighing 1,350 pounds—is fed exclusively on hay and straw. Very poor fodder some may exclaim; yet in the course of ten months the same creature yielded 4,921 quarts of milk, and during a single month, after calving, she gave 620 quarts, and the least she gave during the winter months was 562 quarts. A pig will grow fat on small quantities of the right kind of food, and yet lose flesh when suffered to gorge himself with the same. Many swine, in the vicinity of coal mines, consume both coal and charcoal, and little else, yet they thrive well.

Evidences can be furnished going to show that both the superior and inferior orders of creation might subsist on much less food than they are in the habit of using, and without danger to themselves. Shipwrecked mariners have been known to exist several weeks without food; and there is a man now residing in Illinois, who, during a period of thirty-two days, never tasted food. The sleeping man of Rochester is another example; and a case is recorded in one of our medical journals showing that an individual once lived for several months on nothing but pure water. As regards horses, they form no exception to this peculiarity. We might introduce evidence, convincing and positive, of their ability to endure the privations of hunger, and, at the same time, show that they suffer but little from its effects. A single case will serve to illustrate this. We once treated a case of tetanus (lock-jaw). The subject never tasted food during a period of sixteen days; on the seventeenth the masseters relaxed, and the faculty of swallowing returned. At this period we might suppose him to be "hungry as a bear," yet, on offering him a few oats, he did not appear to be very ravenous, and partook of food subsequently offered him as if nothing had happened. These are extreme cases, yet they go to show that there is no cause for alarm because a horse happens to be "off his feed" once in awhile. Such condition may ultimately prove salutary, affording the stomach and its associates time to rest from their herculean labors.

The fact that most of our adult horses get more food than they need has been demonstrated by analysis of their excrement, which has been found to contain a large amount of nutritious material over and above what the animals actually need. We can develop the gormandizing powers of very many horses, by placing before them, from day to day, more than they require. Their appetites, like some of ours, are not proof against temptation; and the digestive organs may be trained to dispose of twice the quantity of food actually needed, and the habit, at first acquired, becomes permanent, and the creature is known as a voracious feeder—a glutton. Hence, through indiscretions of this character, we can augment both function and capacity of stomach. We remember examining the stomach of a horse, the property of a baker, who was in the habit of feeding the former on brown bread. The animal died of chronic indigestion, and his stomach exceeded in capacity that of two ordinary horses. Great care, therefore, is requisite in regard to the proper feeding of horses; for, in their domesticated state, they have lost those natural instincts which serve to inform the untamed animal of the necessary amount of food which his system needs, and they are in the condition of a thoughtless child that will eat all day, and, on retiring to bed, will crave and cry for more.

Our readers have probably heard of the gormandizing propensities of natives in the arctic regions. Some of them think nothing of bolting down twenty pounds of meat and oil per day, and making a good supper on tallow candles. A case is related, by Captain COCHRANE, of a Russian who ate, in the course of twenty-four hours, the hind-quarter of an ox, twenty pounds of fat, and drank a quantity of melted butter. He also states that he has seen three gluttons consume a deer at one meal. But we need not go beyond our own immediate vicinity to prove that the gormandizing powers of both men and horses are equally extraordinary. The corn-dealer's bill furnishes one illustration; and the lengthened meal which some of our young men indulge in, commencing in the morning and only ending at night, completes the evidence. Hence, with these facts before us, we may safely conclude that errors in diet are constantly occurring, and, consequently, a great many unnecessary diseases arise in consequence; therefore, we recommend our readers to make an experiment in the opposite direction, and ascertain how small a quantity of good food will answer the purpose of nutrition. Should the quantity be insufficient for the animal's wants, we shall soon be made aware of the fact by loss of flesh and other unmistakable signs. On the other hand, the error alluded to is not so easily corrected; for the animal may die, overburdened with fat, of an acute disease, before we can reduce his system.

CONCLUSION.

The reader will perceive that in the management and feeding of horses there opens a fine field of observation and improvement; yet, in order to apply that unlimited power which man seems to possess over his own organization and that of the inferior orders of creation, he must be conversant with animal physiology; for on this science alone do we base the problem of life.

Now, reader, after having presented this essay for your consideration, pray do not find fault with the stable-keeper because your horse does not look fat and sleek. You had better trust to the discretion of the man who, having been long in the stable business, is perhaps better qualified than yourself to judge of the effects of food under the states of rest and exercise, and knows how to graduate the same accordingly. We frequently have occasion to notice that horses owned by stable-keepers are never so fat as the boarders—a very good proof that the latter get more than they require. Some men are in the habit of ordering a given quantity, say six or twelve quarts, of oats at a feed, whenever they put up, and the feeder gives it. At the same time he is well aware that the animal does not need it, yet he must obey orders; for should he suggest that the quantity be too great, his motive may be questioned, and wrong inferences drawn. Depend upon it, therefore, that when the stabler, by a judicious system of feeding, prevents fat from accumulating on horses, he is conferring a favor on his patrons and benefiting the animal.

As regards the number of meals per day, our own observations satisfy us that working horses require three meals per day—a bountiful, yet very early breakfast; at noon, a light meal, composed of cut hay and oats; and at night the quantity may be increased equal to the morning meal. It is very poor policy to feed or water on the road when performing a short journey; yet, should a horse seem to stand in need of something, we should not object to a handful of oats and a few swallows of water occasionally.

SECTION VI.

GLANDERS AND FARCY.

GLANDERS—GLANDERS CAN BE COMMUNICATED TO MAN—CONTAGIOUSNESS OF GLANDERS—SUSPECTED GLANDERS—HOW CAME THE DISEASE TO BE CALLED GLANDERS?—DIAGNOSTIC SYMPTOM OF GLANDERS—TREATMENT OF GLANDERS—FARCY (DISEASE OF THE ABSORBENTS).

GLANDERS.

ONE of the greatest evils which farmers and horse-dealers have been subjected to, during the prevalence of our late national troubles, was the sale of condemned army horses, many of them being afflicted, either insidiously or actually, with that form of equine affection known as glanders; for it is a well-known fact that this terrible malady is, under certain circumstances or conditions of the animal economy, both infectious and contagious; hence the great evil.

For example, should a glandered horse be placed in a stable where the principles of ventilation and of the admission of light are entirely disregarded, the other inmates of the den, by continually breathing, over and over again, the contaminated atmosphere which must necessarily occupy the same, will (provided their systems are susceptible) take the disease by infection; that is to say, the miasmatic virus finds an introduction into the blood through the lungs and pulmonary circulation. On the other hand, glanders, under almost every circumstance, can be communicated by contagion or touch; that is, bringing the glandered nasal discharge in contact with an abraded surface, including cuts and scratches.

GLANDERS CAN BE COMMUNICATED TO MAN.

Within the last quarter of this century, two veterinary surgeons—one residing in Walworth, and the other in Wolverhampton—are reported as having died from inoculation of glanders. This terrible disease is not often seen in Scotland, but very frequently in England, and still more so in Ireland. From the latter circumstance, the malady is often found to be imported about the west coast of Scotland. London has always been rather renowned for the prevalence of glanders among omnibus, cab, and other

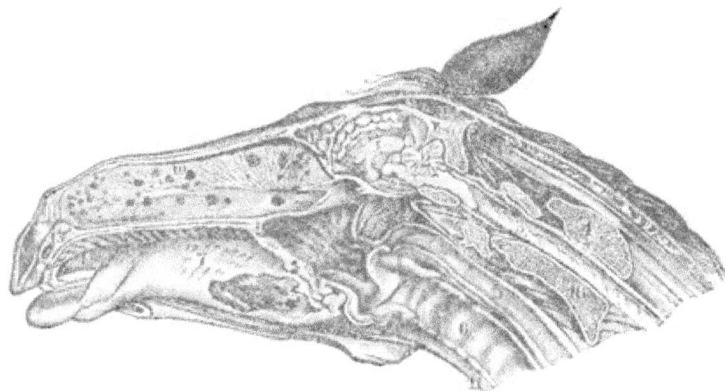

SECTION OF A HORSE'S HEAD SHOWING ULCERATION OF THE SCHNEIDERIAN MEMBRANE, WHICH CONSTITUTES THE DIAGNOSTIC SYMPTOM OF GLANDERS.

EXPLANATION.—On contrasting the above picture with the one on the opposite page, the reader will perceive, in the above, a number of dark spots on the lining membrane of the nose (schneiderian), showing the chancreous ulcerations which constitute the diagnostic symptoms of glanders.

1. The cerebrum, or anterior portion of the brain.
2. The cerebellum, or posterior portion of the brain.
3. The spinal cord.
4. Shows a section of the cervical vertebræ, or bones of the neck.
5. The œsophagus.
6. The windpipe.
7. The tongue.
9. Section of the occipital bone.
10. The schneiderian membrane.
11. A muscle known as the sterno maxillarius. It is located beneath the neck, and is inserted into the angle of the lower jaw and the front part of the breast bone. Its action is to aid in closing the mouth and drawing the head downward toward the breast; one muscle acting draws the head to one side.
12. The frontal sinuses.
13. Section of the anterior portion of the upper jaw.
14. The hard palate.
15. The strong ligament of the spine known as the ligamentum suffavium.
16. Section of the inferior portion of the cervical vertebræ

horses. A very strict supervision is maintained, and all glandered horses are destroyed when discovered; but nevertheless we can state, on good authority, that the omnibus horses of London have suffered very severely from this disease, and do so still. The partial measures adopted by companies are not sufficient to eradicate it, and the "glandered night train" is not altogether a thing of the past. The danger to human life is so great that we feel happy to seize any opportunity to urge the adoption of the most effectual measures for the suppression of any practice which tends to prolong the life of the glandered horse.

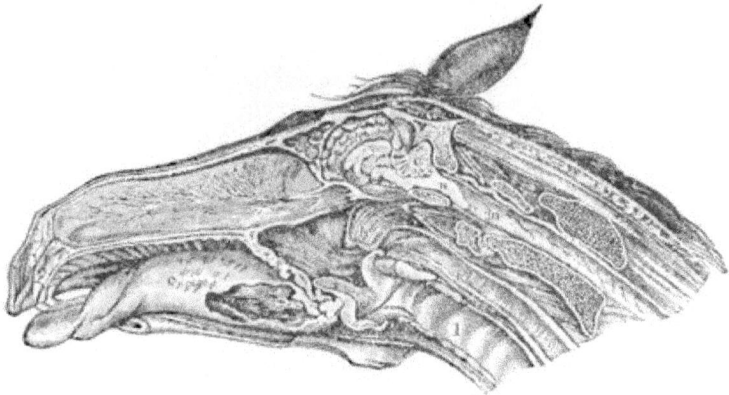

SECTION OF A HORSE'S HEAD IN A HEALTHY CONDITION—ITS ANATOMY.

DESCRIPTION.—a The lining membrane of the nose, known as the schneiderian membrane. On its surface are seen the vessels which furnish it with arterial blood.
b The anterior portion of the brain, known as the cerebrum.
c The posterior portion of the brain, known as the cerebellum.
d The frontal sinuses.
e The tongue.
f Showing the velum palate, or soft palate.
g The back part of the mouth.
h The pharynx, or top of the œsophagus.
i The epiglottis, or cartilage at the root of the tongue.
j The glottis, or opening into the windpipe.
k The œsophagus, or gullet.
l The trachea, or windpipe.
m The spinal cord.
n Junction of the cerebellum with the spinal cord.
o p Showing a section of the cervical vertebræ, or bones of the neck.
The dotted lines between a and g, show the direction leading to the trachea and œsophagus.

A fatal case of glander, reported in the Baltimore "Herald of Reform," is as follows: Mr. I. P. Burus, a grocer, died a horrible death in that city in consequence of poison communicated to his system from a horse afflicted with glanders. During the administration of medicine, Mr. Burus thrust into the animal's mouth his hand, a finger of which had been previously cut, and the flesh laid open. Through this wound the virus was absorbed, and mortification supervened. A surgeon was called upon to amputate the diseased member. Perceiving, however, that the poison had penetrated to every portion of the unfortunate man's system, he declined performing the operation, and stated that no earthly skill could save him. After lingering in great agony, death closed the scene.

Death of a Russian Lady from Glanders.—The awful death of Madame Palesikoff, one of the most charming among all that bevy of entertaining Russian ladies who sometimes gladden the winters of Paris, has created a terrible shock among the circles she so lately embellished by her presence. The unhappy lady left Paris but a short time ago on a summer tour to Germany. While stepping from the door of the opera-house in Berlin, to gain her carriage, she let fall one of her bracelets close to the pavement. Stooping to pick it up, she noticed, at the time, laughingly, that "one of the horses belonging to a carriage standing at hand, dropped his head so close to her face that he touched her, and left a moist kiss upon her cheek." In a few days the unfortunate lady was taken ill with that most horrible disease, glanders, and in a few days more, breathed her last, in spite of the attendance of the first physicians of Berlin, and every resource to be obtained by wealth, or by the ceaseless vigilance of friends.*

Still another case.—Sidney W. M., aged 23, a horse-slaughterer, residing at Plumstead, England, was admitted into Guy's Hospital, on March 13, 1861, under the care of Mr. Birkett. He had always enjoyed good health, but he lived freely. Six days previous to his admission he cut his right hand deeply over the dorsal aspect of the thumb. The wound bled freely, but he felt no inconvenience from it, of any consequence, until the 10th. He then had pain in the part, extending upward to the axilla, and also a numbing pain in the calf of the right leg. His appetite was good, and he had so little constitutional disturbance that he went as usual to superintend the work which the accident had prevented his carrying out. On the 11th the pain in the arm increased, and that in the leg, on the 12th, extended upward to the thigh. This had become considerably aggravated, and he then became generally indisposed, and was unable to stand. The following day he applied for admission. He stated that he was extremely ill, and that his limbs felt almost paralyzed. He was quite unable to stand. He had constant rigors, a burning skin, white tongue, and his pulse was 120. The wound in the hand was inflamed, and the calf of the right leg and the thigh were swollen, and evidently the seats of abscesses. These were freely opened, with relief. He was ordered effervescing medicine, and two pints of porter. The case was then regarded as one of pyæmia. On the 14th he was much relieved. He had slept tolerably. The abscesses discharged freely, and he was able to take his food. On the 15th he was seized with vomiting, bringing up a dark bilious fluid. His powers were less. Wine was given freely. On the 16th the vomiting continued, and he complained of a pain in the left thigh. An abscess was discovered, on examination, which was freely opened. The man, at this time, was physically in a very low condition. He took brandy, eggs, and other liquid nourishment with avidity. His manner was peculiar, and somewhat excited. On the 17th another abscess was opened in the left arm. On the 18th the left foot became the seat of an abscess, and on the 19th, for the first time, a pustular eruption was observed on the face and body, which suggested to Mr. Birkett the probability that some animal poison was the cause of all this mischief.

The eruption was peculiar, having no definite shape. It appeared to be more like small irregular blebs, containing pus, varying from the size of a pea to a sixpence. Some were round, others oval, and others of an irregular form. Toward evening, also, his breathing became much impaired, a bloody mucus obstructing the nostrils. Toward night he became delirious; his powers became less, the nasal discharge more profuse, and at 1.30, on the 20th, he died. The eruption, some hours before his death, in parts, had disappeared; but in others a fresh crop sprang up, particularly over the region of the sternum. These were, however, of the same character, but smaller. No post-mortem examination was made.*

CONTAGIOUSNESS OF GLANDERS.

Mr. PERCIVALL submits the following deductions, regarding the contagiousness of glanders, as the result of facts gleaned from his own experience:

" 1. That farcy and glanders, which constitute the same disease,

---

* Medical Times and Gazette

are propagated through the medium of stabling, and this we believe to be the more usual way in which the disease is communicated from horse to horse.

2. That infected stabling may harbor and retain the infection for months, or even years; and although, by thoroughly cleansing and making use of disinfecting means, the contagion might be destroyed, yet it would not be wise to occupy such stables immediately after such supposed or alleged disinfection.

3. That the virus, or poison of glanders, may lie for months, in a state of incubation, in the horse's constitution before the disease breaks out. Of this we have had the most positive evidence.

4. That when a stable of horses becomes contaminated, the disease often makes fearful ravages among them before it quits; and it is only after a period of several months exemption from all disease of the kind that a clean bill of health can be rendered."

From the preceding evidence, it is probable that the reader will entertain but little doubt of the contagious character of the malady; yet it is very important that every one, either directly or indirectly interested in horses, or having any regard for the welfare of mankind, should be familiar with all that is important and useful as regards the cause and nature of the awful malady now under consideration. It is often mistaken for other diseases that (in so far as contagion is concerned) are perfectly harmless; yet many valuable human lives have paid the forfeit, and many priceless animals have been sacrificed on an altar of ignorance which the light of science has but recently illuminated. Hence, correct information is what the people require, in consequence of the emergency of the peril; and this is what the author aims at in offering this article for the consideration of the reader.

Many hundreds of times, during the professional career of the author, have horses been brought to him for treatment, declared by their owners to be the subjects of glanders, simply because they had enlarged glands under the jaw, and a nasal discharge; and by the successful treatment of the same, he has got the credit of curing glanders, a feat which he never pretends to have accomplished. Mr. GAMGEE has very lucidly illustrated this part of the argument, as regards the mistakes made in diagnosing glanders He states that glanders may be suspected, instead of being a reality; hence, many supposed cures are on record. The following are his remarks, under the caption of "suspected glanders":

" Glanders is, fortunately, a rare disease in this country (Scotland), thanks to the pole-ax. Englishmen have long since advocated and practiced the shooting of suspected animals, rather than trusting the lives of men and horses to the chances of escape, wherever cases of chronic nasal discharge are met with. The command officer and veterinarian of a British cavalry regiment would consider it a great disgrace if such a disease acquired any firm hold in their stables ; and in spite of occasional introductions of the disease when a number of remounts may be purchased, the unrelenting order to kill rather than attempt to cure, saves the public purse and the reputation of those responsible for the health and condition of our troop horses. I am as great an advocate for the slaughter of glandered horses as I am for the slaughter of cattle affected with rinderpest. Glanders is more incurable than the cattle plague, as not even ten per cent. recover, but its communication is less certain and swift. It never could and never did destroy its tens of thousands over a country in the short space of time in which the steppe murrain spreads over the land, but it is, nevertheless, wise and proper to stamp it out. We have not indulged, as do our neighbors the French—who manage this matter, at all events, worse than we—in wild theories as to the transmissibility of acute and not of chronic glanders. We admit it to be always contagious and always deadly, and prevent it killing by shooting its victims.

NASAL GLEET.

We must not, however, forget that there are hundreds—nay, thousands—of cases of chronic nasal discharge which admit of some diagnosis on the part of skilled veterinarians, and which are erroneously set down as cases of glanders. Many of these cases are condemned because they baffle the attempts to restore them for a great length of time; and, unfortunately, in this country many forms of nasal disease have been rarely cured, simply because their nature has not been understood, and bold surgical operations have been dreaded. I could relate the histories of many cases which have yielded to radical measures after several veterinarians had pronounced the animals incurably glandered—more to get rid of them, perhaps, than from a conviction that they were suffering from the disease. I have seen as many as half a dozen animals, in a stable containing a score of horses, given up by practitioners because they had chronic discharge and swollen glands; and in ten days or a fortnight all the animals have recovered under the influence of nasal injections, and the internal administration of tonics.

It is impossible to enter into details as to the diagnosis of a great variety of cures. I may mention, however, that the curable discharges set out as incurable forms of glanders may be grouped under six heads.

1st. Fetid discharges from the nostril, owing to a carious tooth and caries of the upper jaw. The fetor is characteristic.

2d. Intermittent discharges from an abscess in one of the turbinated bones within the nasal chamber. The swelling of the nasal bones and flow of pus when the head is jerked upward are diagnostic.

3d. Continuous or intermittent discharge from one or both nostrils, from accumulations of pus in the frontal and superior maxillary sinuses, indicated by the shape of the forehead and absence of resonance on percussion.

4th. Irregular discharge from the guttural pouches, brought on by exercise, or seen when an animal is made to eat hay, oats, a turnip, or carrots off the ground. As the head is depressed and jerked, a somewhat fetid and often abundant purulent matter flows freely.

5th. Discharges kept up by foreign objects in the posterior nares, which are continuous and fetid.

6th. Chronic regular nasal discharge—pure ozena—dependent on constitutional causes and defective management of acute catarrh.

All the foregoing varieties include the innumerable cases of suspected glanders for which animals are wrongfully destroyed. It is not every practitioner who can or would venture to open the guttural pouches, or practice dissection on the bones of the face to such an extent as we find requisite in overcoming deformities and curing long-standing diseases. The longer these cases are treated by any but the right plan, the greater the difficulties encountered when surgical operations are determined upon; and for this reason even those who would not dare to cut should always strive correctly to diagnose the form of disease presenting itself. It is not necessary to refer at greater length to the method of dealing with the very various forms of disease above indicated."

How came the disease to be called Glanders.

PERCIVALL is our authority for the following explanation "The derivation of our word glanders is traceable through the French language, from which we appear to have borrowed it, to the Latin roots *glandula* and *glans*, the latter signifying any fruit kernel, such as a chestnut or acorn; the former, its diminutive, any small fruit kernel; and both afterward used in medicine to denote the glands of the body, many of which—such as were then so called—are small and comparable, both in shape and size, to acorns or other kernels. Celsus applies the term *glandula* to a swelling in the neck, supposed to be glandular; and Vegetius uses the same to denote swollen glands 'between the cheek-bones and lower jaws:' from his saying, however, that the *glandules* are 'especially troublesome to *foales*,' it would appear the disease he meant to describe was not glanders, but strangles. The French veterinarians, following the ancient phraseology, called a horse exhibiting any submaxillary tumor or enlargement, *glande;* not with any special reference to glanders, but simply because his glands or 'kernels,' as our farriers denominate them, had become enlarged; hence, with the French, a horse was said to be *glande de gourme*, as well as *glande de morve* and *glande de farcin*. It seems to have been our English writers on farriery who have restricted the application of the term to the foul and malignant disease now known under that appellation. Before then, glanders appears to have had no other meaning save that the horse had tumefied glands, or that, in the farrier's phrase, 'his kernels had come down.' The French call the disease *la morve*. A horse, however, in the estimation of Lafosse, is not to be regarded as having *la morve proprement dite*, unless he be *glande*, or have tumefaction of his glands."

Diagnostic Symptoms of Glanders.—Glanders consists in a discharge, from one or both nostrils, of matter which, by transfer or inoculation, will produce the same disease in another animal (of the equine or human species), and which discharge is, sooner or later, accompanied by vascular injection and chancrous ulceration of the schneiderian membrane of the nostrils, and tumefaction of the submaxillary lymphatic glands, and by farcy; so that a horse can not be considered as the subject of glanders until these symptoms are made manifest.

On consulting "Hippopathology," I find a paragraph, credited to a French surgeon, who very accurately describes the symptoms of glanders. It reads as follows: "The signs by which the disease may be known are, when a horse, already too old to be troubled with strangles, without a cough, voids matter by the nose, and has a kernel sticking to the bone; and, besides, in glanders the matter usually flows from one nostril, whereas, in a cold, it runs always out of both. Some cast the matter that is voided by the nostrils into water, and, if it swim on the top, they conclude the horse to be free of this distemper; but if it sink to the bottom, it is a sign of glanders, the principal use of this experiment being to distinguish the pus. But you must not depend on the certainty of this sign; for if the matter stick to the nostrils, like glue, it is a bad sign, and you may conclude the disease to be the glanders, though the matter do swim on the top. When either the breath or matter that comes out of the nostrils stinks, the disease is almost always incurable. I have seen horses troubled with this distemper without kernels, or, if there were any, they were small and move-able; and the only sign by which we could discover it to be glanders, was the glueyness of the matter discharged from the nasal outlet."

Treatment.—The author knows of no remedy for the cure of glanders. He considers it an incurable disease. In fact, most of our educated veterinarians contend that the disease, like pulmo-nary consumption, is incurable.

MAYHEW, one of the most intelligent veterinary teachers of the present period, informs us that "no medicine can restore the parts which disease has disorganized. There is no cure for glanders, which is essentially an ulcerative disorder." And this opinion is indorsed by others of equal eminence in the profession, who were employed lately, by the members of an agricultural society in England, to ascertain if there was any specific for the disease known as glanders, and the verdict was that no specific could be found. So soon as glanders is discovered in the horse, he should, by all means, be destroyed, and buried deep in the earth.

FARCY (DISEASE OF THE ABSORBENTS).

This disease is usually met with among horses of the scrofulous diathesis, which diathesis is known by a proneness to diseases of the skin, and does not readily yield to medical treatment. Such horses are noticed as belonging to the nervous temperament, having thin, spare muscles, wiry neck and limbs, very quick and active in their movements, and having a very narrow chest. Yet it should be remembered that the scrofulous diathesis may exist in any horse, without regard to his conformation or temperament, provided he inherit, from either sire or dam, the requisite predisposition.

The disease sometimes steals on in a slow and insidious manner, so that it is not known to exist until, all at once, the animal becomes lame, and, in the course of a few hours, his legs swell; large inflammatory farcy-buds appear just as suddenly; a stinking discharge takes place from the nostrils, and, unless the animal be destroyed, he soon dies. This form of farcy may, very properly, be denominated malignant farcy. When the disease manifests itself after this fashion, it is dangerous for a person to handle the subject. He should be destroyed and buried.

A case of this kind occurred very recently in a large stable, the facts of which are as follows: The author was requested to visit a bay gelding, of the nervous temperament, supposed to be the subject of a catarrhal affection. He did not show any very marked symptoms of ill-health, except a slight discharge from both nostrils. He had performed labor up to within a few hours of the time of my visit, and had partaken of his usual amount of food. This was toward evening. I directed the foreman to rub the throat with stimulating liniment. Two ounces of fluid extract of resin weed were given him; his legs being cold, they were bandaged, and a warm bran-mash was placed before him. Next day he was a pitiable sight to behold. His limbs and other parts of the body had become dropsical; his face and eyelids had become so tumefied that the eyelids were almost totally closed, and scalding tears ran profusely down his cheeks. In various parts of the body the lymphatics were tumefied, presenting unmistakable farcy-buds. A stinking discharge ran from his nostrils, and he was so lame that it was almost impossible to get him out of the stable. The case being diagnosed as farcy, the animal was shot.

Mr. PERCIVALL alludes to the lameness and rapid tumefaction in the following language: "I have known horses so lame from farcy before the disease had, in any local or characteristic form, declared itself, that shoes have been removed and feet searched, etc., to discover the seat and cause of lameness, no suspicion having existed at the time that farcy was present in the animal's system. It may so happen, however, that some of the preliminary symptoms are observed or observable; that, on the contrary, farcy at once develops itself in an attack on some locality—most probably one hind limb. Indeed, so sudden, sharp, and severe are attacks of farcy, in some instances, that, in the course of one night, the horse's limb will be swollen to a frightful size, so as to incapacitate him almost from turning in his stall and walking out of the stable. Ordinarily, the development of farcy plainly accounts for the halting or lameness, yet the lameness may appear without any ostensible cause."

Symptoms.—The ordinary symptoms of farcy are, slight impairment of the general health: feverish symptoms; a small, quick pulse; swelling or tumefaction of one of the hind legs, generally the left, with much lameness; tumefaction of the lymphatics, on various parts of the body, and in the groin. On the inside of the thighs, along the course of the femoral vein, corded, nodulous swellings will be found. Sometimes one or both hind limbs will swell below the hock, and an abscess may form in the vicinity of the fetlock, and isolated blotches will break out, suppurate, and terminate in farcy ulcer. No swelling of a hind limb, or any other part, constitutes a case of farcy apart from the unequivocal signs of lymphatic disease. There must be present corded, nodulated swellings, buds in some form or other, together with actual or approaching tumefaction of the lymphatic glands, or the case is not farcy. The disease evidently affects the absorbents. When it commences in those which are deep-seated, it usually ends in glanders. Glanders and farcy are considered the same disease; yet, unlike glanders, many cases of cure of mild farcy are on record, but the author knows of none. A disease known as lymphatitis is often mistaken for farcy, and, as the former is curable it is often recorded that the latter affection has been successfully treated.

Treatment.—I do not know that it is good policy for me to recommend any form of treatment for the disease known as true farcy. It is against the laws of this country, and the interests of society for any one to keep or treat a case of declared farcy; therefore, the advice offered in regard to the treatment of glanders applies also to farcy. To show the reader that the author is not regular in his opinion as regards the treatment of this disease, a quotation from MAYHEW is here introduced:

"Farcy is, by the generality of practitioners, regarded as a more tractable disease than glanders. Certainly the course of the disorder is arrested much easier; but, to cure the malady, there is a constitution to renovate and a virus to destroy. Is it in the power of medicine to restore the health and strength, which have been underfed, sapped by a foul atmosphere, and exhausted by overwork? Tonics may prop up or stimulate for a time; but the drunkard and the opium-eater, among human beings, can inform us that the potency of the best selected and the choicest drugs, most judiciously prescribed, and carefully prepared, is very limited. Sulphate of copper, iron, oak bark, cayenne pepper, and cantharides, probably, are the chief medicines the practitioner will give. With such the horse may be patched up; he may even return to work. But at what a risk! He carries about the seeds of a disorder contagious to the human species, and in man even more terrible than the quadruped. Is it lawful, is it right, to try to save an avaricious master the chance of a few shillings, and incur the risk of poisoning an innocent person? The author thinks not. Therefore he will give no directions how to arrest the progress of farcy. The horse once contaminated is, indeed, very rarely or never cured. The animal, after the veterinary surgeon has shaken hands with the proprietor and departed, too often bears about an enlarged limb, which impedes his utility, and, at any period, may break forth again with more than the virulence of the original affection."

SECTION VII.

DISEASES OF THE DIGESTIVE ORGANS.

Lampas—Spontaneous Salivation—Bots—Inflammation of Stomach—Rupture of Stomach—Gorofd, or over-distended Stomach—Meteorization—Flatulent Colic—Spasmodic Colic—Inflammation of the Peritonæum—Ascites—Inflammation of the Intestines—Strangulation of Intestines—Wounds penetrating the Abdominal Cavity—Splenic Apoplexy—Functional Disease of the Liver—Pancreas; its Function—Parasites which affect the Intestines.

LAMPAS.

LAMPAS is the name given to a slight tumefaction of the soft palate of the horse's mouth. It usually occurs during the period of dentition, at which time the mouth is hot and tender. At the time when the tushes are about making their appearance, the membrane over the crown of the tooth becomes tense, so as to cause the root of the tooth to press on the tental nerve. This causes much pain. In such cases we make a slight incision over the region of the tooth, which immediately relieves the animal. But most men overlook this difficulty, and, because the palate happens to be tumified, it is at once attacked with the barbarous firing-iron. This treatment was condemned by PERCIVALL, who, in writing of it, says:

"Are lampas disease? The complaints, frequent and grievous, which are daily reaching our ears, are enough to persuade us that they are disease. Every groom that has an unthriving horse, or one that does not feed, is sure to search for lampas; and, should he find any, the cause of loss of appetite, etc., is detected, and the remedy obvious—*burning them out*. Many a poor wight of a horse, even when suffering from a constitutional disease, has been subjected to this torturing operation, with a view of demonstrating the sagacity of the groom, and thereby has got added to his other

ailments a foul, sloughy, carious sore upon the roof of his mouth
This may be said to be the fruits of the removal of lampas.

Supposing that the existence of lampas is owing to the teeth,
surely the teeth should be removed, and not the bars of the mouth.
In cutting or burning away the lampas, we are mistaking the effect
for the cause. If it be contended that lampas do not owe their
production to the irritation of teething, then I should like to be
informed what does give rise to them; and, let what will give rise
to them, I do not imagine there is any veterinarian hardy enough

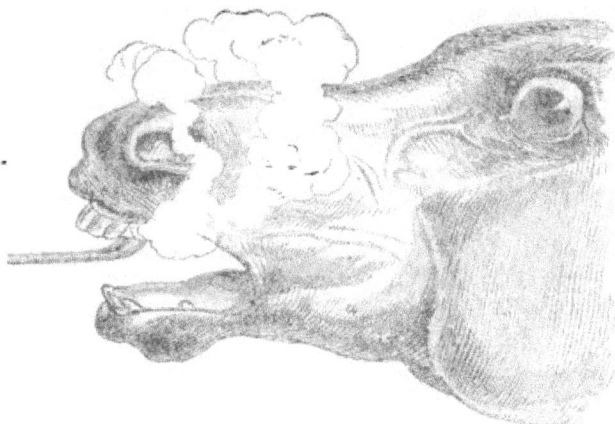

BURNING A HORSE'S MOUTH FOR LAMPAS.
(A cruel and unnecessary operation.)

to contend that the cause resides in the palate, and becomes removed by the hot iron. Those who are entering private practice,
and find themselves compelled, at all times, to belie their consciences by the performance of unnecessary operations to please
their employers, may be told that burning out lampas is, after all,
preferable to lancing or cutting the bars; for, unless the palative
artery is wounded, very little blood is obtained by stabbing the
mouth; and the wounding of this vessel, which will certainly take
place, should the punctures be made along the sides of the palate,
or extend forward beyond the fourth bar from the front teeth, is
not always a very safe proceeding."

It is an unfortunate occurrence that men can be found who will
perform barbarous and unnecessary operations just to please their
employers. If men were true to themselves and their profession,
neither threats nor promises of reward would tempt them to prac-

11

tice cruelty on animals. It is a fact that some men are often apparently compelled to perform unnecessary operations even when the inward monitor admonishes them that they are about to outrage the feelings of a poor dumb brute. Yet, for the honor of our profession, let it be recorded that there are but few, if any, of our educated surgeons that can be induced to perpetrate this outrage. If medical men will only take the pains to explain this matter, and prove to the people that the enormity here complained of is both absurd and barbarous, then the evil will soon cease; but so long as horse-owners believe lampas to be a disease, and men can be found ready to " burn them out," just so long will the evil exist. The more intelligent portion of husbandmen are now beginning to lend their influence to the work of rebuking ignorance, and of reforming the barbarous customs of bygone days, and, erelong, it will be a difficult job to convince them that their horses have lampas, and that the firing-iron is the remedy indicated. Surely, if the distinguished PERCIVALL calls lampas a " supposed malady," and sets his face against the means used for its removal, the wise men of the fleam should indorse his sentiments.

Treatment.—Should the colt's mouth, while teething, be hot and tender, it should be sponged often with a portion of equal parts of vinegar and water; and should any of the teeth appear imprisoned by the membrane of the mouth, a slight incision over the point of the tooth will give immediate relief.

SPONTANEOUS SALIVATION—" SLAVERING."

Persons who own and handle horses must have noticed, occasionally, a foaming at the mouth, and sometimes a constant and profuse discharge of saliva. This, when occurring without any assignable cause, is termed spontaneous salivation. The disease is usually occasioned by some irritation in the immediate vicinity of the salivary glands, or from perverted function in the glands themselves. The preparations of antimony and calomel, when given to horses for any great length of time, in the form of condition powders, are very apt to produce salivation. Musty clover, and hay that is intermixed with lobelia, will also produce it; so, also, will sharp, projecting teeth : and diseased teeth have the same effect. In such cases, the cause being discovered, the removal of the same is the first step toward the cure. Then drench the animal, twice daily, with the following:

N ,. 18. Powdered chlorate of potass........... 1 oz.
       Water......... .... ½ pint.

Should the animal show signs of debility or weakness, give one ounce of fluid extract of golden seal every morning, before feeding. Should the mouth appear to be sore, from the constant flow of saliva over its surfaces, or from any other cause, then the parts should be sponged two or three times daily with a portion of equal parts of powdered borax and honey, or with an infusion of bayberry bark.

When spontaneous salivation occurs in an aged horse, it denotes debility, and thus points out the propriety of sustaining the patient's strength by administering tonics and diffusible stimulants. The following is the best remedy for this state of the system:

No. 19. Fluid extract of chamomile........... 1 oz.
      Fluid extract of ginger............... ½ oz.

Give as a drench, night and morning. The patient should have a fair allowance of good nutritious diet.

Secretion of Saliva in Horses.—From experiments lately made, it appears that a healthy horse secretes from the paroted glands, which are situated beneath the ears at the posterior angles of the lower jaw, during active mastication, about one gallon of saliva per hour, and other glands in the vicinity of the throat and mouth secrete, in the same amount of time, three gallons, making in all four gallons per hour, which is mixed with the food during mastication. The following experiment was lately made on an aged horse, 15 hands 2 inches in height. The tubes leading from the paroted glands were divided on either side, and so fixed as to throw their contents into vessels held for that purpose; the animal was then fed on oats, which he masticated during a period of thirty minutes, at the end of which time half a gallon of saliva was caught. This calculation, however, will not hold good in all cases; for ravenous feeders, who do not thoroughly masticate their food, fail to secrete the necessary amount of saliva; hence the food is not properly masticated nor insalivated, and can not, therefore, undergo proper digestion; and when food is not thoroughly digested, it imparts but little of its nutriment to the body. Slow feeders, therefore, are more likely to accumulate flesh than the voracious ones.

Bots (Stomachic and Hemorrhoidal).

Some persons contend that bots are always injurious. The author dissents from this dictum. It is possible that, as in the case of intestinal worms, which are now recognized as the scavengers of Nature, that the bots are Nature's hirelings, created and commissioned to do her bidding, to maintain the integrity of her physiological laws. The parent of the bot, as Bracy Clark remarks, " selects her subjects," or, in other words, pounces on those who are not in rapport with Nature, and hence have no business to enjoy good health, nor even to live.

This was the case when the people of the great city of London were afflicted with the terrible plague, which ran riot and carried off about one-fourth of the inhabitants. The sanitary emissary of Nature, whose shield was emblazoned with the motto, " Thus far shalt thou go and no further," pounced upon selected subjects, the intemperate, licentious, and the gluttonous, and those who had violated Nature's laws by wallowing in filth in down-cellar locations, where the breath of life—*pure air*—scarcely ever entered. These were the selected sufferers. The same is true as regards the mortality attending the yellow fever, which made such sad havoc in the city of New Orleans some years ago. The medical authorities contend that the disease carried off over one thousand of the inhabitants, without touching a single sober or temperate person ; hence the plague, the yellow fever, and the cholera may be said to be the forces which Nature employs to maintain the integrity of her laws. Intestinal worms, found in the intestinal tube of the emaciated and the glutton, are said to be Nature's scavengers, and the same perhaps is true of bots. They may be the agents of Nature, employed to keep the balance of power within her own hand, for the purpose, sometimes, of preventing a too rapid multiplication of the species ; at others, to avenge her for crimes committed against the laws of physiology.

Let us, for example, inquire into the history and habits of some of the inferior orders of creation, and we may be led to infer that the presence of bots in the stomach of a horse is no deviation from the general rule which seems to pervade all creation. Our tenure of life depends on the use which we make of it, and the same is true as regards the horse.

In the study of physiology, we discover that animals and insects require the operation of certain forces, in order that their peculiar vital properties shall be manifested. They all require food, water, and oxygen—food for the development of organized tissues, water to maintain an equilibrium between the solids and fluids, and oxygen for promoting various changes, uniting some particles for the benefit of the whole fabric, and disengaging others destined for excretion. These have to be obtained under various circum- stances. The number of the different species of reptiles known to naturalists is about 1,300, and there are at least 160,000 species of insects. Among this vast assemblage of animate forms, a great proportion obtain food, water, and oxygen in a situation and at a temperature which is most congenial to each species, each one of which, as species, exhibit great varieties in physical organization and habits ; and hence the necessity for that diversity in geograph- ical distribution which seems to surprise some men. Each species of animal and insect carry about with them, in their own organi- zation, the fertile embryonic habitation for successive parasitic development, and all are, to a certain extent, dependent on each other for both food and life. It has been truly said that there " is life within life." Begin with the body of man, for example, and we shall find that it is infested with thirty-nine distinct spe- cies of entozoa. These are not confined to a single location, as the bots to the digestive cavity of the horse; but some are to be seen in the eye, bronchial glands, kidneys, liver, gall, bladder, in- testines, muscles, and even in the blood. There are several other species of entophyta, to the number of ten, inhabitants of the skin and mucous surfaces. So that the master can boast of a larger number of living parasites within and about his body than we have yet been able to find in his servant, the horse. And if the former can carry about in the living citadel such a myriad of liv- ing, active creatures, without inconvenience, and he being the weaker party, why should not the horse, who is stronger, be able to furnish nutriment for some half dozen or more bots that are occasionally found in his stomach, and to perform his 2.40 gait without inconvenience? Some of the inferior orders of creation are the receptacle of a still greater amount of parasites. The grasshopper, for example, is infested with a species of gordius, a sort of hair-worm, which some persons have erroneously supposed to be a transformed horse-hair. Several of these coil themselves into the digestive cavity of the former, at times penetrating the

abdomen, thorax, and cranium.   Their bulk and weight often ex-
ceeds that of an ordinary grasshopper ; still you see and hear them,
skipping, jumping, and chirruping, notwithstanding this immense
parasitic mass, just as freely as those not infested.

Then consider the condition of the hog.   We frequently find,
in his liver, a vesicle filled with fluid, apparently possessing no
further organization.   But examine it carefully, and we shall find
within its enveloping tunic others, the rudiments of successive
cells, in various stages of growth.   These are all young hydatids,
which contain still others, which increase in size until the parent
sac is so distended that it finally bursts, and thus liberates a mul-
titude of parasites, which, in their turn, undergo the same evolution, becoming each a parent hydatid, producing a subsequent
generation, which diffuse themselves over the whole body of the
pig; and hence arises that peculiar feature in pork known as
measles.

Examine the same animal after he is slaughtered, and you may
possibly find in the intestines a large number of the *ascaris lum-
bricoides* (the common worm of the human intestines); and they
are so prolific that naturalists have calculated sixty-four millions
of ova within the body of a single female, which are, at the proper
season, deposited within the intestinal tube of the pig, who, notwithstanding, grows fat, and furnishes a savory meal for the lovers
of pork.   Sheep, also, are infested with a species of entozoa termed
the "fluke," besides other different species; and it is only when
the latter become very numerous that they can be considered injurious.

We might go on to show that every living being is more or
less infested with parasites, and that parasites are, in their turn,
the local habitation for other parasites.   The very atmosphere we
breathe, and which is the purificator of the vital current (the
blood), teems with an innumerable host of living organized creatures.   The water which serves to quench the thirst, that plays
so important a part in our economy, and in the arts and sciences,
whether it be the ocean, lake, river, pond, or gully, all contains
crowds of parasites, or animalculæ, at times, so numerous that
several hundred thousand have, by means of a magnifying lens,
been discovered in a single drop of this fluid.   Yet such is good
and pleasant to the taste ; and the water is not injured thereby ;
neither is it, in turn, injurious to man.   Dr. LEIDY states that he

has, at various times, purposely swallowed large draughts of water containing myriads of animalculæ, without ever perceiving any effect; and he combats the notion that diseases are produced or propagated by these parasites taken into the body. The most curious feature in the history of parasites is their extraordinary powers of multiplication, which is doubted by some; but it is well known to others that some species are capable of producing a hundred repetitions of themselves, and the process can be repeated ten times in a season. The common white ant is capable of depositing eggs at the rate of 80,000 per day, for several successive weeks, and the common flesh-maggot can be generated by the million in the course of a few hours; and as regards growth and development, the common flesh-fly and the caterpillar increase in weight 200 times in the course of twenty-four hours.

But the bot is a creature that does not multiply nor increase in bulk at this rapid rate. He may be said to be a "slow coach," and when once located in the stomach of a horse, he generally makes it his abode for a season, at least. He is a sort of aristocratic entozoa. He lives in the upper region, the stomach. He seldom condescends to mix with the lower orders that infest the alimentary regions. The little creature seems to exercise considerable tact in selecting his location. Although he has but a squatter's title to it, yet it is the best and safest in the whole diggings. He is in the upper part of the stomach, where the fluids (poisons or medicines) with which you are about to coax or drive him off, are inoperative, (for they merely give him a sort of shower-bath,) and pass immediately through the stomach into the intestines, where all the fluid a horse drinks is generally found; therefore they can not act on the bot. Then, again, he is located on the cuticular coat of the stomach, a membrane as insensible as the horse's hoof, and, therefore, not liable to become diseased, nor to be acted on by either medicine or bot nostrums. You may kill the horse by the same, but the bot, being within his own castle, can refuse whatever you offer him.

We can not make medicine act on the external surface of the bot, for it does not absorb fluid; it is impervious. These creatures have been put into muriatic acid, and kept there for a time, without being injured. You may put them into new rum, and keep them for weeks, and, on taking them out and exposing them to the sun's rays, they will manifest vitality.

We often hear wonderful stories related of bots burrowing through the coats of the stomach. This, we think, rarely takes place while the horse is alive. That cavity is the home of the bot, its natural habitation; for we know of no other. Here it generally remains until it is capable of exercising an independent existence. In this situation the little creature is too comfortably located to burrow through the stomach into a cavity where it might perish for want of food. If the time has arrived for it to vacate its stronghold, instinct teaches it the most safe and expeditious route, which is through the alimentary canal. Turn a horse out to grass in the spring, or give him some green fodder in the stable, and the bots will soon leave him, if they are matured; otherwise they must remain until that period arrives, unless Nature has some work for them to perform. We shall not contend that bots are never found in the abdominal cavity, for some persons have testified to the fact; but, during a practice of several years, and having opportunities of making many post-mortem examinations, we have not yet been able to observe the phenomena, except in cases of ruptured stomach. Still, a few solitary cases are on record, and hence it remains for us to explain how they got there.

We all know that the moment a horse dies his whole body is subject to the common law of decomposition; but the central organs, where the greatest activity prevailed during life, are generally the first to succumb. Our business is with the stomach, the great chemical laboratory, the center of sympathies—an organ that is very seldom permitted to rest, consequently an active one. Its powerful solvents, during life, were busy in transforming hay and grain into chyme, chyle, and blood; but now that death has the victory, the gastric fluid acts on the coats of the stomach, and thus its decomposition is effected; so that what was previously good food for bots is now their bane, and, unless they escape, their death is sure and certain.

The peristaltic motion of the intestines, which favored the exit of the bots through that channel, has ceased; they are too well acquainted with its intricate labyrinthian outlet (their usual route) to even attempt its passage. No! the same energies of one Eternal Mind, "pervading and instructing all that live," suggests the only means of escape from threatening dangers. The stomach being partly decomposed, offers but little opposition to their encroachments. They burst their prison-house, and hence are found in the abdominal cavity; and when there, they may be said to have jumped from the *"frying-pan* into the fire !" Open a horse immediately after death, and, provided his stomach be in a healthy condition, we shall find that the bots have not penetrated beyond the cuticular coat of it; but if he shall not be examined until some hours have elapsed, the bots may be found to have passed through the walls of the decomposed stomach and its peritoneal tunic.

We can imagine, also, that a large number of bots might congregate at a given point in the stomach of a horse, and, aided by disease, occasion a loss of continuity in the fibers of that organ; then, on the slightest distension by wind, its walls might be ruptured and its function paralyzed, and thus the bot be involuntarily driven from its home, to seek shelter and food in another location.

We contend that the stomach of a horse is the natural habitation of the bot during its minority, and, at the proper season, the digestive canal is the usual channel for its introduction into the external world; and if these parasites are ever found in any other situation within a horse's body, they are there by the force of circumstances, owing to disease or rupture of the stomach, or from some morbid condition in the gastric fluids, which arouses a set of involuntary actions in response to a stimulus; because, during the whole period of their minority, that is, the larveal state, (a term which, in the language of entomology, applies to the bot from the time of emerging from the egg, or nit, up to that period when it vacates the horse and assumes the form of a gad-fly), they are in the same condition as a new-born babe or an idiot—the one imbibing its mother's milk, and the other performing unnatural antics, both appearing to lack that train of mental operations which implies knowledge, motive, or the consequences resulting from such actions. We very much doubt if the bot can, at any time, by voluntary act, vacate the body of the horse. Reasoning from analogy, we are led to the conclusion that the result is accomplished through their instinctive properties, which are common to many insects and parasites—a perfect adaptation of means to an end—by which they perform a certain set of operations without choice, purpose, or intention of their own, yet, in many cases, producing results which man, through the aid of his superior intellect, has not been able to surpass.

Persons who have paid any attention to the study of physiology, are aware that these instinctive or involuntary movements, performed without consciousness, are the birthright of a vast number of the inferior orders of creation; therefore, it is not likely that the bot would, even if he possessed the power, voluntarily vacate a location favorable to its growth and development.

Veterinary surgeons have long since discarded the absurd notion that bots are the cause of any pain or suffering to horses. In fact, some of the most distinguished of them assert that these little creatures, with their rough exterior, are rather beneficial than otherwise, and that, by friction and irritation, they arouse the sluggishness of the stomach, and thus promote digestion. Persons unacquainted with these facts are, therefore, apt to attribute effects, during life, to causes which happen after death, and, consequently, the poor horse has to be dosed with all sorts of nostrums.

So popular has been the belief that bots are injurious to horses, and, therefore, must be expelled at all hazards, that almost all the old works on farriery contained some favorite recipe for their expulsion. Popular opinion, too, has been so much in favor of the theory, that Mr. PERCIVALL thought it his duty, as a public teacher, to make use of the following language:

"You may boldly assert that bots are in nowise injurious. Still, you can not persuade the world so, and, therefore, you must be prepared to meet the complaints of those unbelievers, who will, now and then, declare that their horses have bots, which must be got rid of. But I know of no medicine that has the power of destroying; and even if we possessed such, I am not sure that we could, even when dead, detach them from the cuticular coat of the stomach, to which they are attached by small horns."

In allusion to the parasites which infest the human body, WATSON remarks:

"It is a curious fact that numerous parasites do crawl over the surface of our bodies, burrow beneath our skin, nestle in our entrails and riot and propagate their kind in every corner of our frame, producing ofttimes such molestation and disturbance as to require the interference of medicine. Nearly a score of animals that have their dwelling-place in the interior of the human body have been already discovered and described, and scarcely a tissue or an organ but is occasionally profaned by their inroads. Each, also, has its special or its favorite domicile. One species chooses the heart for its place of abode; another inhabits the arteries; a third, the kidney. Myriads of minute worms lie coiled up in the voluntary muscles or in the areolar tissue that connects the flesh fibers. The guinea-worm and chigoe bore through the skin, and reside in the subajacent reticular membrane. Hydatids infest various parts of the body, but especially the liver and brain. A little fluke, in general appearance much like a minature flounder, lives, steeped in gall, in the biliary vessels. If you squeeze from the skin of your nose, what is vulgarly called a maggot (the contents of one of the hair-pellicles), it is ten to one that you find in that small sebaceous cylinder several animalcules, extremely minute, yet exhibiting, under the microscope, a curious and complicated structure. Even the eye has its living inmates; but it is in the alimentary tube that we are most infested with these vermin."

It is evident, from competent testimony, that these, as well as other kinds of parasites, are always more or less injurious; hence the same may be true as regards the bot in a horse's stomach. The best authority we have for the origin and history of the bot is BRACY CLARK, V. S., a selection from whose works is here introduced:

"*The Œstrus Equi, or the Stomach Bot.*—When the female has been impregnated, and the eggs sufficiently matured, she seeks among the horses a subject for her purpose; and approaching him on the wing, she carries her body nearly upright in the air, and her tail, which is elevated or lengthened for the purpose, curved inward and upward. In this way she approaches the part where she designs to deposit the egg, and, suspending herself for a few seconds before it, suddenly darts upon it, and leaves the egg adhering to the hair. She hardly appears to settle, but merely touches the hair, with the egg held out on the projected point of the abdomen. The egg is made to adhere by means of a glutinous liquor secreted with it. She then leaves the horse at a small distance, and prepares a second egg, and, poising herself before the part, deposits it in the same way. The liquor dries, and the egg becomes firmly glued to the hair. This is repeated by these flies till four or five hundred eggs are sometimes placed on one horse.

The skin of the horse is usually thrown into a tremulous motion on the touch of this insect, which merely arises from the very great irritability of the skin and cutaneous muscles at this season of the year, occasioned by the heat and continual teasing of the flies, till, at length, these muscles appear to act involuntarily on the slightest touch of any body whatever.

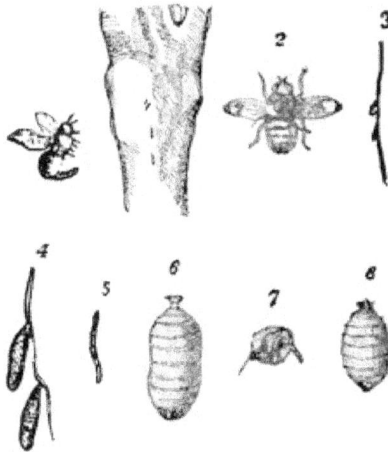

THE ŒSTRUS EQUI.

1. The female fly, about to deposit an egg.
2. The male fly.
3. The egg; its natural size.
4. The egg magnified.

5. The newly-hatched bot.
6. The bot full-grown.
7. The head of a bot magnified.
8. The chrysalis.

The inside of the knee is the part on which these flies are most fond of depositing their eggs, and next to this on the side and back part of the shoulder, and less frequently on the extreme ends of the hairs of the mane. But it is a fact worthy of attention that the fly does not place them promiscuously about the body, but constantly on those parts which are most liable to be licked with the tongue, and the ova, therefore, are always scrupulously placed within its reach.

The eggs thus deposited I at first supposed were loosened from the hairs by the moisture of the tongue, aided by its roughness, and were conveyed to the stomach, where they were hatched; but on more minute search, I do not find this to be the case, or, at least, only by accident; for when they have remained on the hairs four or five days, they become ripe, after which time the slightest application of warmth and moisture is sufficient to bring forth, in an instant, the latent larvæ. At this time, if the tongue of the horse touches the egg, its operculum is thrown open, and a small active worm is produced, which readily adheres to the moist surface of the tongue, and is thence conveyed with the food to the stom-

a. b. At the first hatching, as we observe, it is a small, active worm, long in proportion to its thickness; but as its growth advances, it becomes proportionably thicker and broader, and beset with bristles.

Bots are very frequent in horses that have been at grass, and are, in general, found adhering to the white insensible tissue or coat of the stomach. They usually hang in dense clusters to the white cuticular lining of the stomach, and maintain their hold by means of two dark brown hooks, between which a longitudinal slit is seen, which is the mouth of the larvæ. When removed from the stomach by the fingers, by a sudden jerk, so as not to injure them, they will, if fresh and healthy, attach themselves to any loose membrane, and even to the skin of the hand. For this purpose they sheath or draw back the hooks almost entirely within the skin, until the two points come close to each other. They then present them to the membrane, and, keeping them parallel till it is pierced through, they expand them in a lateral direction, and afterward, by bringing the points downward toward themselves, they include a sufficient piece of the membrane, to remain firmly fixed for any length of time, as if at anchor. These bots pass the autumn, winter, and spring months in the stomach, and arrive, about the commencement or middle of the summer, at their full growth, requiring a year to fully complete their structure.

The Œstrus Hemorrhoidalis, or Fundament Bot.—The parts chosen by this insect for this purpose is the lips of the horse, which is very distressing to the animal, from the excessive titillation it occasions; for he immediately rubs his mouth against the ground, his fore-legs, or sometimes against a tree, with great emotion, till the animal, at length, finding this mode of defense insufficient, quits the spot enraged, and endeavors to avoid it by galloping away to a distant part of the field; and if the fly still continues to follow and tease him, his last resource is in the water, where the œstrus never is observed to pursue him. These flies appear sometimes to hide themselves in the grass, and, as the horse stoops to graze, they dart on the mouth or lips, and are always observed to poise themselves, during a few seconds, in the air, while the egg is preparing on the extended point of the abdomen. When several of these flies are confined in a close place, they have a particularly strong, musty smell; and I have observed both sheep and horses, when teased by them, to look into the grass and smell it very anxiously; and if they, by these means, discover the fly, they immediately turn aside and hasten to a distant part of the field. I once saw, in a meadow or field, upon the cliffs of Margate, a fly of this sort teasing a horse that was confined to a small space by a spike stuck in the ground, to which a cord was tied. He could not get away from its attack, and became quite furious, for in kicking at the fly with his fore-foot, which he did vehemently, he often struck the bone of the lower jaw, creating excessive pain; for in that direction, while grazing, the fly comes to the beard of the lower lip.

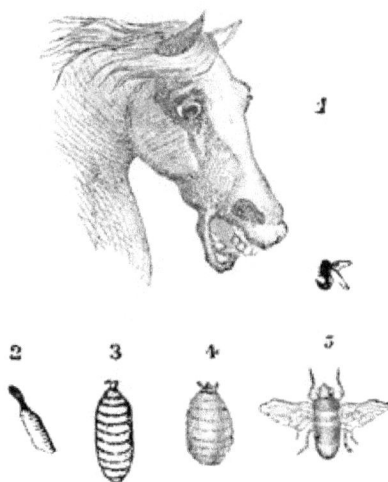

THE ŒSTRUS HÆMORRHOIDALIS.

1. The female about to deposit an egg.   3. The bot.
2. The egg magnified.   4. The chrysalis.
                  5. The male fly.

The eggs of this species are difficult to be seen upon the horse's skin or beard, owing to the agitation of the beast, and from the color of the egg being dark, like that of the skin of the horse. The animal has been generally too impatient, while undergoing this operation, to let me examine them very well. I ascertained, however, its form by pressing one of these eggs from the abdomen.

The larvæ, or grubs, of this species inhabit the stomach as the former, generally adhering to the white lining, and are disposed promiscuously in dense clusters, after the same manner. They may, however, be distinguished from them by being, in general, smaller and longer in proportion to their bulk.

The larva of this species may be obtained from almost any horse that has been much the preceding year at grass, and exposed to these flies, and will be found, during the summer months, sticking more or less within the verge or opening of the anus, adhering to its soft lining, and producing considerable irritation and uneasiness. Indeed, I once well remember being on a tour of pleasure in the Isle of Wight, and experiencing much annoyance from these larvæ. The little horse I had hired for the journey became so lazy and unwilling to go on, and moved so awkwardly, that I could not keep pace with my company, and I was at a loss how to proceed; but, on casually taking up the tail, I discovered three or four of these insects hanging to the rectum, and their removal instantly proved a cure."

It is very likely that the fundamental or hemorrhoidal bot does produce some irritation within the rectum, but not more so than the ascaris, or the filaria, which inhabit the intestines, including the rectum, of both man and horse. Taking a rational and scientific view of this subject, the author is inclined to believe that all parasites infesting the bodies (or living citadels) of the superior orders of creation are only, in their essence, so many modes of death, more or less sudden or gradual, as the case may be, or as the exigencies of Nature require. In some cases it is supposed to be a beneficent provision that some horses, as well as men, should have a gradual exit out of life, when, by virtue of their constitution, they inherited no long lease of it. The parasites here alluded to, and others not mentioned, may be classed as so many diseases, which never prove mortal unless the individual's lease of life has expired, and then, as before remarked, he has no tenure on longevity. In taking this view of the subject, the reader will, of course, infer that all extraneous or accidental causes of death are excluded.

Disease, which probably includes the presence of all parasites, in the bodies of men and animals, may be considered as the trials of pilgrimage life, and, therefore, may be classed as the checkmates to Nature. In order to play the game of life so as not to get checkmated, we must understand the law of rational physiology, and use our intelligence accordingly, to preserve health. In the ordinary acceptation of the term disease, it does not mean much of any thing; for it does not express the manner in which the animal is affected. It simply means that he is sick, from

some cause or other. A man or animal may be diseased without
our expression of the manner in which either are affected, just as
a flower may be said to be colored without designating its particular hue.   In each case we receive definite, although negative,
ideas—in the one, that the creature is not in health ; in the other,
that the flower is not colorless.   And so the same reasoning may
apply when bots become the subject of our consideration.   In a
given case, they may prove injurious (according to our ideas), and
in another they may remain, for a certain length of time, perfectly
harmless ; so that the animal infected or infested may or may not
suffer any infliction, but trot on through the journey of life just
as long as horse-life is worth wearing.

As regards the several modes of death to horse flesh, in which
bots are said to play an active part, they may thus be stated:
Bots sometimes congregate, in large numbers, at the superior or
inferior inlet or outlet of the stomach, and thus produce mechanical obstruction to the passage of both semi-fluids and solids ; and
the result is, equine death.   The author once held an autopsy,
which revealed the presence of a cluster of bots, numbering about
one hundred, located in the cardiac region of the œsophagus.
They completely obstructed the passage of food into the stomach.
The horse died suddenly, as if suffocated, from distension of the
œsophagus and pressure on the trachea.

The following case, furnished for the " Veterinarian " by Surgeon MATHER, illustrates another mode of death from the presence of bots :

" The inclosed tongue and larynx I took from a foal this evening, which I shall feel obliged to you for examining, as at the base
of the tongue you will find about a score of what appear to me to
be perfect specimens of a small bot.   I have only made a partial
examination, as I was anxious to send the specimen off to you
before it became decomposed, as the weather is very hot.   The
history of the case is as follows : Two days ago I was requested to
go to Lord Glasgow's breeding establishment, to see a foal that
was unwell.   I inquired of the stud-groom how long the animal
had been ill, and he informed me that three days ago it appeared
to have a difficulty in swallowing, and that he noticed some saliva
issuing from its mouth ; and that, thinking it had a sore throat,
he had rubbed some embrocation upon the parts.   On examining
the animal, I found it in a very debilitated state.   Pulse, very

weak, and numbering 80; ears and legs, extremely cold, associated with shivering fits of the entire body; mouth, filled with partly masticated hay and grass; tongue, furred. The animal had seemingly a desire to eat, but was unable to open its mouth more than about an inch. It was likewise unable to swallow any thing but liquids, and these only in very small quantities. In spite of all I could do, the power of swallowing got worse, and this morning it died, very much to their disappointment, as it was a very highbred animal.

Could this foal have taken in the ova of the gad-fly, and some of them become hatched, and, instead of passing into the stomach, have become adherent to the base of the tongue, and there set up such an amount of irritation as to cause the parts to swell so as to prevent the animal from swallowing?

Observations on the Case.—We are indebted to Assistant Professor Varnell for the following remarks upon this singular case:

'The morbid specimen forwarded by Mr. G. Mather, is of very rare occurrence. We have never met with such an one before, nor are we aware that a similar instance has ever been recorded. The parts came to hand in a good state of preservation, although the temperature at the time was very high, so that we were enabled to form a pretty correct opinion as to the nature of the changes that had taken place during the life of the animal. The dorsum of the tongue, its root, the whole of the surface of the fauces, the *velum palati*, and the margin of the glottal opening, as well as the lining membrane of the larynx and pharynx, were very much inflamed, and, in places, considerably thickened. These parts were likewise covered with a deposit seemingly consisting of a mixture of mucus, lymph, and pus.

The most remarkable feature, however, of the case was in the cause that had given rise to the inflammatory state of the throat; namely, the existence of a number of small bots—the *œstrus hemorrhoidalis* (the 'red-tailed horse bot' of Bracy Clark)—which were found to be firmly adhering, by their tentacula, to the mucous membrane. Besides the number which were still in *situ*, many others had apparently been attached to the parts, but which had, most likely, been lost in the handling they had undergone.

By a little reflection, we may probably form a correct opinion as to how these parasites became located in this unusual situation. We imagine that the foal, by licking, gathered the ova, in a state

12

of perfection, from the hair of its dam, on which they had been deposited, rather than from its own coat. The perfected ova, being thus taken into the mouth, became first entangled in its mucus, and finally lodged in those shallow recesses at the root of the tongue, where they became hatched. Some of these creatures measured as much as three-eighths of an inch in length, and rather less than one in diameter. They were firmly fixed, by their little hooklets, apparently to the inner surface of the mucous follicles. Their free extremities were of a deep red color, and their bodies of a yellowish brown.'"

Treatment.—The only remedy for ridding the system of bots is a run at grass, in the month of May.

GASTRITIS, OR INFLAMMATION OF THE STOMACH.

Acute inflammation of the stomach seldom, if ever, occurs among horses as a spontaneous affection. It is usually occasioned by irritating medicines or poisons, or food of an inferior quality. When an irritating poison has been received into the stomach, and excites inflammation there, or when acute inflammation arises from any cause, the symptoms which mark that inflammation are pain, distress, and restlessness; a loathing of food; for, if any thing be given by the mouth, it creates increased pain. The animal breaks out in a cold sweat, lies down and quickly rises again, becomes early and greatly prostrated in strength, and has a pulse usually quick and much oppressed. Sometimes purging sets in; at others, the bowels are constipated. The disease is very apt to run into gastro-enteritis—inflammation of the stomach and bowels.

Treatment.—The general treatment consists in the administration of slippery-elm mucilage, and half-ounce doses of chlorate of potass, two or three times per day. Should gas accumulate in the stomach or bowels, give one ounce of hyposulphite of soda, dissolved in a pint of warm water. The ordinary drink should consist of hay tea, to which a small quantity of molasses may be added.

RUPTURE OF THE STOMACH.

Rupture of the stomach is always a fatal malady, over which neither art nor nature has any control; yet, the accident being one of frequent occurrence, it becomes necessary for us to know something regarding its cause and symptoms, which are all detailed in the following case, from the author's note-book:

Death of a Horse from Ruptured Stomach.—The subject, a black gelding, nine years old, had been driven a distance of twelve miles into the city; on arriving at the "City Hotel stable," he first manifested signs of uneasiness, by pawing with his feet and turning his head toward the belly. He finally got down, rolled, tumbled about, and passed considerable flatus, etc. His owner, supposing him to be the subject of colic, administered a dose of "fire-water," in the form of gin and cayenne pepper, followed by three ounces of spirits of niter. He urinated freely, passed considerable flatus, yet grew no better very fast. Four hours from the commencement of the attack our attention was called to him. The case then appeared hopeless, consequently we declined prescribing. The following were the symptoms: Pulse, small and thready; in number, fifty-six; respirations, sixty—laborious, emitting, occasionally, a sort of grunt, indicative of pressure, from a distended stomach, on the diaphragm and lungs. The eyes were amaurotic (pupils dilated and loss of vision); sclerotica (white of the eye) slightly injected; conjunctival membrane, within the eyelids, surcharged with blood. The buccal membrane, within the mouth, intensely inflamed, induced, probably, by the fire-water drench. The tongue was of a leaden hue, and the nasal membranes were somewhat reddened. The body, or external surface, was the seat of profuse perspiration, which ran down the limbs like rain. They, as well as other parts of the body, remained comfortably warm until within a few minutes of death. There appeared to be considerable distension in the region of the stomach, and less at the flanks, from which circumstance we concluded that the patient was laboring under tympanitis (windy distension) of the stomach. At the moment of death, the reverse was the case, which circumstance seemed to confirm this opinion; for then, the stomach being ruptured, the gas would diffuse itself in the abdominal cavity and increase in bulk.

Aspect and motions of the body.—The head was kept within striking distance of the ground; the tail was elevated, and kept in a quivering condition (involuntary). The fore extremities were advanced, one leg frequently crossing the other; the hinder parts maintained a crouching attitude, and the penis hung pendulous. When not in this position, the patient would revolve in a circle;

then lie down flat on his side, turn his head toward the ribs, and, after a short time, raise the forward parts, and sit on his haunches like a dog. At last he fell over, gave two or three convulsive struggles, and died, just thirty-five minutes from the time we first saw him.

Next morning, in company with Dr. WOOD, we proceeded to Ward's establishment, but came very near being too late; for the metamorphosis of the horse into grease, food for swine, etc., had commenced. We however had an opportunity to make a casual examination of some of the organs, and found a rupture of the stomach of some six or eight inches. The contents, some of which were found in the abdominal cavity, were very imperfectly (crudely) masticated. It may be proper, therefore, to name the disease indigestion; its consequences, tympanitis of the stomach; the cause of death, rupture of the stomach. The loss of vision referred to is one of the symptoms which generally attends acute indigestion.

A person, who was present during the last moments of the animal's life, asked if we were not going to give or administer something. Our reply was, that we never tortured dying horses by forcing drugs down their throats. One of our duties, as a physician, was to know when to withhold medicine—when to do nothing; and another equally important one was to be able to discriminate between a hopeless and curable case. The one in question was beyond the aid of science, and, therefore, in view of "clearing our skirts" of being the cause of death by administering the last dose (a charge that is often ungenerously laid at the door of medical men), we preferred to watch the dying animal, and give directions calculated to lessen the pains of death.

Unfortunately for us and our patients, many employers form an estimate of medical qualification in proportion to the amount of medicine administered. They have an idea that the more medicine we give, the greater are our efforts to control the disease, and the more willing are they to pay for the same, when, in ninety-nine cases out of a hundred, the very reverse is the case; for many diseases to which horses and cattle are subject have a definite type and limited duration, and would terminate favorably with less medicine; provided, however, the patient have the advantages of pure air, suitable diet, and proper management. Many medical men are compelled to swerve from the line of their own convictions merely to satisfy their employers. This should not be

If a man employs a physician, (quacks are not here considered,) he should have unbounded confidence in the ability of the former to treat the case. If any thing occur to lessen that confidence, let the attendant be discharged immediately. The employer has no right to dictate how or in what way his horse shall be treated when he abandons him to scientific men. The science of medicine is never a "God-send." No man is thus favored. It is the result of diligent application and study, in and out of the dissecting-room, and the legitimate student of medicine is a person entitled to the full confidence of reasonable men for its legitimate practice. Therefore, we contend that the qualifications of a veterinary surgeon should never be estimated in ratio to the quantity of drugs administered, nor the length of his newly-invented syringe, nor by the number of instruments contained in his pocket-case; but by his works let him be judged.

There is one practice, prevalent among some horsemen, which we very much deplore; that is, the drenching of sick animals with gin, pepper, and molasses. This is a universal stable panacea in certain quarters, and is often the cause of much pain and annoyance to the animal, and more frequently of an aggravated form of disease. Gin, or the fire-water sold under that name, is not always to be procured pure. A great deal of it appears to us to be composed of spirits of turpentine, new rum, and spirits of juniper, which, of course, in certain cases, excites inflammatory action. We have seen a horse's mouth in a most shocking condition—the interior surfaces almost completely denuded—after a drench of the same; consequently, in cases of inflammation of the stomach or bowels, (which the stable oracles are apt to mistake for colic, or "bots," as they term it,) the remedy (poison) is certainly objectionable, because it not only operates unfavorably on the horse by aggravating the disease, and sometimes changing a simple into a grave one, but much valuable time is lost, in not sending for a competent person to treat the case. Besides, if a medical man be called on after the animal has been improperly treated, he then has two things to contend against: a medicinal disease and the original one; and the former is sometimes more difficult to control than the latter; and this is the reason why some medical men, ourselves included, refuse to take charge of what we term a second-hand case.

The above case illustrates our argument. The remedies were

not adapted to the emergency. Even the simple article molasses was objectionable, because it is an acid. The case required alkalies, in the form of ammonia or chloride of lime. Gin was objectionable, because it is an excitant of the brain and nervous system, when, instead of the brain, the stomach ought to have been stimulated. Niter, which was given, is an agent that increases urinary discharges only, and, therefore, was not needed. The trouble was in the stomach—to arouse its function and condense the gas which distended the same. Alkalies and tonics should have been administered, followed by such other treatment as the nature of the case demanded. We look upon this gin and molasses treatment as decidedly unprofitable to owners of live stock. It savors too much of our grandmothers' castor-oil practice, which often proves more heroic than salutary, does more harm than good, and, at the present day, is decidedly unpopular.

Our advice to horse-owners, therefore, is, if you have a sick animal, and can not comprehend the nature of the disease, never attempt to prescribe for him, but call in the aid of a professional man. We give this advice under the conviction that domestic animals are subject to many of the diseases that afflict our race; that the treatment should be somewhat analagous, and the tact and skill not inferior, but superior, to that of the human practitioner.

GORGED, OR OVER-DISTENDED STOMACH,
(COMMONLY CALLED "GRAIN FOUNDER.")

Gorged stomach is usually the result of overfeeding, by which means the stomach becomes overtaxed in its function, or overburdened in its capacity; so that, in the first place, digestion is partially or totally suspended; and, secondly, the stomach, being distended beyond its physiological capacity, can not perform the mechanical action, viz.: contraction and expansion, so necessary for the physiological process of digestion. The act of digestion is consummated through the aid of Nature's chemistry, by means of the salivial fluids, bile and pancreatic juice. These are the solvents of all the digestible food. This chemical action receives much aid in the mechanical action of the muscular fibers of the stomach, which, by alternate contractions and expansions, keep not only the stomach in motion, but also its contents, and thus the food is digested. As regards overfeeding, however, it may be proper to remark, lest the groom or feeder take offense, that an animal may be regularly fed from meal to meal (not carelessly overfed), yet the stomach, being disordered, does not digest the usual meal. A portion of it is "left over," as the saying is, and thus gradually accumulates in that organ, until over-distension, with its consequences, commands our attention. The reader must also bear in mind that horses sometimes free themselves from the halter at night, or when the groom is absent, and voluntarily gorge themselves; therefore the person who has charge of and feeds the subject is not always at fault. When engorgement of the stomach occurs, and is associated with flatulency, the sufferings are doubly severe, because then the intestines are also the seat of distension, and the pain from that alone is sometimes terrific.

The following case, which lately occurred in the author's practice, fully illustrates the condition just alluded to. The subject was a bay gelding, aged about seven years, the owner of which informed me that the animal had, in the early part of the morning, broken loose from his stall, and immediately commenced a desperate attack on a bag of oats, which happened to be within reach. The horse was known as a ravenous feeder, and the probability is that he did not spare the oats, but took a pretty large dose, so as to over-distend his stomach. The services of the animal being required, a few hours afterward he was hitched up. He soon, however, showed symptoms of distress, attempted to lie down, and finally got into a profuse perspiration. Shortly after this, my attention was called to him. I found the patient in great distress—his head pendulous; breathing, much embarrassed; pulse, very rapid; body, bedewed with a chilly perspiration; legs and ears, cold; continual eructations of gas; at the same time efforts to vomit, which brought away nothing more than salival secretion, which, however, was quite profuse. The patient rolled and tumbled about considerably, and would occasionally stand up for a few seconds at a time.

Treatment.—It appeared that the indications, in view of relief, were to arouse the action of the stomach, and, at the same time, to arrest fermentation. I was under the impression that the contents of the gorged stomach could only be got rid of in the ordinary way, viz.: by digestion. I am aware that it is customary to give drastic cathartics, in view of getting rid of the contents of the stomach; yet I would not advise the reader to do this, for the horse may die ere the medicine operates. The following drench was administered:

No. 20   Fluid extract of ginger.............. 1 oz.
         Fluid extract of golden seal........ 2 oz.
         Hyposulphite of soda.............. 2 drachms.
         Water......................... 4 oz.

After the exhibition of the above, I threw into the rectum a
couple of quarts of soap-suds, to which was added a handful of
salt.   In the course of about an hour the animal appeared to be
somewhat relieved, and passed a large quantity of oats, whole.
I repeated the dose; also the enema.   The patient very soon afterward passed a large quantity of excrement, mixed with oats, and
so rapidly improved that I left him, and did not see him until the
next morning, when he appeared to have entirely recovered.
With the exception of a bran-mash or two, the above comprises
the whole of the treatment.

It was lucky for the horse as well as the parties concerned that
the animal did not get corn instead of oats; for the same quantity
of corn would have surely caused death, from the fact that when this
article of fodder is submitted, within the stomach, to the action of
heat and moisture, it increases in bulk in a ratio of about five to
one, and the usual result is rupture of the stomach.   Still, should
any of our readers be called upon to treat a case of the latter kind,
there is no other plan of treatment with which I am acquainted
that will be likely to succeed in saving the animal than the one
here indicated.

Another case.—The patient, a gray gelding, aged about eight
years, the property of Messrs. Wright & Bros., of Chicago, had
performed an ordinary day's work without showing any symptoms
of ill-health.   About five o'clock in the evening he was unhar-
nessed and put up for the night.   In a few minutes it was noticed
that he began to bloat, and that the abdomen increased in size very
rapidly.   A messenger was immediately dispatched for me.   On
arriving at the stable, I found the animal in a very dangerous condition.   The abdomen was enormously distended, and an eructation
of gas from the stomach, by the mouth, was continually occurring.
The respirations were laborious and accelerated, and very much
quickened; pulse, very indistinct; extremities and surface of the
body, quite chilly; rectum, protruding; and the animal was very
uneasy, and appeared to suffer much pain.   Occasionally he would
get down and make desperate efforts to roll on his back, yet he
did not succeed, for he was round as a barrel; and when he could

scarcely stand alone, in fact, required the services of three men to steady him, he could walk, and seemed most at ease when kept in motion. I commenced immediately to inject warm soap-suds into the rectum; but so soon as the fluid entered the same it was immediately ejected, there being no room for it within, in consequence of the distension of the walls of the intestines, which pressed on the rectum. I next attempted to administer a colic drench, composed of ginger, golden seal, and hyposulphite of soda, but did not succeed in getting much of it down; for as soon as the fluid entered the æsophagus, it was met by a volume of eructating gas, which brought the fluid back by way of the mouth and nostrils. I learned, however, that, previous to my arrival, Mr. Smith, the foreman, had succeeded in drenching the patient with a full dose of the colic mixture. This probably had the effect of limiting the generation of gas; otherwise the animal might have died ere my arrival. The case now appeared almost hopeless; so I procured a trocar and canula, and plunged them through the flank into the large intestines. (See puncture of the intestines). On withdrawing the cutting instrument, a steady volume of gas issued through the canula for a period of ten minutes. The relief was immediate. I next gave a dose of colic drench, and left the animal in charge of the owner for the night. Early next morning I visited the animal, and found that he was all right. He was fed lightly for a few days, and, without any other medical treatment, was again put to work.

The dangers to be apprehended in cases of this character are either rupture of the intestines or diaphragm. In either case, death is sure and certain. Sometimes, however, neither of these ruptures occur; then the distended intestines exert such pressure on the diaphragm and organs of respiration that the animal dies of suffocation and loss of pulse. When, therefore, it becomes impossible for the patient to swallow medicine, in a case of this character, recourse must be had to the trocar and canula, in view of liberating the imprisoned gas. The lack of a little knowledge on this subject has been the cause of the loss of very many valuable horses.

METEORIZATION OR SUDDEN DISTENSION OF THE INTESTINES.

Meteorization is the name given to a disease known as bloat or hoven in cattle. When it appears in a horse, it receives the above name, or else is termed acute tympanitis. This form of windy or tympanitic distension is always sudden in its attack; animals apparently in the very best of health all at once show symptoms of enormous distension of the abdomen, and, in the course of a few moments, their lives may be said to be in peril. The distension of the stomach and intestines is due to the presence of an immense volume of gas—probably carbureted hydrogen—which augments either by spontaneous generation, or in consequence of the food running into fermentation instead of being digested.

Cause.—As regards the cause of this malady, we may, with safety, assert that it is the result of a temporary derangement of either the digestive function or the organs of digestion; and such condition is inaugurated by overtaxing that function, or those organs which carry it on, either by feeding indigestible food, or in feeding with too liberal a hand. I have noticed that many animals have periodical attacks of this malady. Such are said to be voracious feeders. Their abdomen is unusually large, and out of proportion to other parts of the body; hence, in so far as their conformation is concerned, they may be said to be predisposed; yet the disease sometimes occurs accidentally, and without the intervention of any hereditary tendency.

Physiologists contend that very many diseases which afflict horses are due to hereditary tendencies. For example, diarrhea and colic are, to a certain extent, hereditary, inasmuch as they are very prone to attack animals of particular form and constitution, as those with narrow loins, large sides, and of what is generally termed a "washy" appearance. If such animals be overworked, especially soon after being fed, if their food be suddenly changed, or if they be allowed an unusual quantity of water, they are almost certain to be attacked by purging or tympanitis. The tendency of these diseases appears, in such cases, to depend on a want of adjustment among the different organs of the body—a want of balance among the functions of digestion, circulation, and respiration.

The following case will give the reader some idea of the symptoms and treatment of this malady :

History of the Case.—The horse had performed his usual daily labor, without any fault-finding on the part of his driver, when, all at once, he commenced to bloat, appeared uneasy, and, by his actions, demonstrated very conclusively that he needed some assistance. On examination, I found that the abdomen was distended to its utmost capacity. The walls of the abdomen were rigid as a plank, and the respirations, in consequence of the great pressure on the diaphragm and lungs, were very laborious and much accelerated, amounting to about one hundred per minute. The pulse was thready and indistinct; the tongue, livid; eyes, glassy and protruding; surface of the body, bedewed with a cold, clammy perspiration; paralysis of the optic nerve had set in, and the animal was "blind as a bat;" in fact, he was dead to all intents and purposes. This was not the worst feature of the case. He was continually vomiting from both nostrils, and, as the story goes, "a horse that vomits surely dies." Occasionally the animal would fall on the ground with violence, and then assume a variety of positions, in view of finding some relief, but, alas! to no purpose. His agony was intense, and there seemed to be no relief for him, unless by a surgical operation. Medicine was out of the question; the animal had lost the power of swallowing, and an attempt to administer a drench would have choked him to death; so I selected the most salient or tympanitic spot on the left flank, and then sent a trocar and canula through the walls of the abdomen into the large intestine. So soon as I had withdrawn the trocar, the gas escaped very rapidly through the tube, making a noise like a steam-whistle. The animal obtained almost immediate relief, and the bad symptoms rapidly subsided, so that I was enabled to give a colic drench. This operated to prevent fermentation, and gave tone to the digestive organs. Two subsequent drenches, of four ounces each, were given, and at the end of twenty-four hours the animal was taken home. The slight wound made through the walls of the abdomen into the intestine healed readily, so that I had no trouble with that; and, to conclude, I would inform the reader that puncture of the intestine is the only rational plan of treatment in a case of this character.

Colic drench is made up of the following ingredients:

No. 21. Fluid extract of ginger...................... 2 oz.
        Fluid extract of golden seal................ 2 oz.
        Hyposulphite of soda....................... 1 oz.

Tympanitis (Belly filled with Gas).

Case in Illustration.—At midnight, June 16, 1865, I was called upon to visit a horse afflicted with this disease. On arriving at the stable, I found the animal on the floor, apparently in great agony. The abdomen was distended with gas to an immense capacity; the pulse was feeble; respirations accelerated, and very laborious; body was bedewed with a cold, clammy perspiration; tips of the ears, cold; extremities in a similar condition. Eructations of gas from the stomach, by the way of the mouth, were constantly occurring, indicating that the stomach, as well as the intestines, were occupied with gas. The animal had bruised himself very badly in struggling, and extensive abrasion of the skin had taken place in various parts of the body. I was informed that the horse had just returned from a very long journey, and had not tasted food for sixteen hours. On arriving at the stable, where I found him, a bountiful supply was placed before him. In his weak and exhausted condition, this was about the worst that could have been done; for the stomach and digestive organs, sharing, either by direct sympathy or otherwise, with other parts of the body, were not in a fit state, until a period of rest had occurred, to digest even a small quantity of food. The cravings of hunger, or a morbid appetite, induced the animal to devour most, if not all, of what was placed before him. The consequence was, the food, instead of undergoing digestion, ran into fermentation, and generated gas known as sulphureted hydrogen.

Treatment.—The animal was urged to rise. I then gave him a colic drench, composed of two ounces of fluid extract of golden seal, and one ounce of hyposulphite of soda. The surface of the body was then rubbed with wisps of straw, which produced some reaction, so that the surface of the patient became warmer. Occasionally the animal was led about for a short distance, and then was led back again to the stall, where he would get down, and roll and tumble about, as if in great agony. I administered enemas of soap-suds and salt, but did not succeed in bringing away either feces or gas, and all the gas which escaped from the alimentary cavity passed by the mouth. Two hours after the administration of the first dose, finding that the animal was still unrelieved, I repeated the dose of colic drench, and threw soap-suds into the rectum. Soon after some feces were voided, and with them quantities of gas. A slight improvement was soon perceptible, yet it appeared that the fermentation was still going on, as the abdomen did not decrease in volume; so I gave two ounces of hyposulphite of soda, in view of arresting the generation of gas. This had the desired effect; for, as the gas was liberated, the abdomen decreased in size. It was not, however, until a lapse of ten hours from the attack that the patient obtained complete relief from his sufferings. This case only differs from meteorization in the suddenness of the attack of the former.

Persons who attempt to treat cases of this character must exercise due patience. The animal must suffer a certain length of time, notwithstanding the best-selected remedies are exhibited. In other words, the reader is advised not to be in too much of a hurry in getting the animal on its legs again. Take good care of him; give him plenty of bedding to rest or roll on; in short, nurse him as if he were a child, and, above all, do not give him too much medicine.

FLATULENT COLIC.

This is a disease of very frequent occurrence among horses. It is known by a distension of the intestines and abdomen, with flatus or gas. In the early stage, no perceptible abdominal distension occurs, yet it very soon manifests itself. One way of satisfying ourselves of the presence of flatus is, to apply the ear to the abdominal region, within which a sort of active rumbling is heard, often accompanied by a tinkling or metallic sound. But while making our examination, we may, perhaps, perceive that the horse passes flatus by the anus, or eructates it from the stomach by the mouth. In either case, our doubts, if we have any, are set at rest. This feature of flatulency, accompanied by others which will be alluded to, complete the chain of evidence.

In some cases the gas accumulates in the large intestines until the abdomen resembles, in size, that of an ox when "hoven." It is very distressing to witness the sufferings of an animal in this condition; and when the intestines are so enormously distended as to threaten rupture, the only chance of relief is to send a trocar and canula through the walls of the abdomen, into some portion of the distended intestine, and thus liberate the gas.

Symptoms.--It may be well to bear in mind that flatulent colic is always sudden in its attack, and some horses are liable to it under every variety of circumstance—in the stable, on the road, or at grass. At the commencement of the attack the animal becomes uneasy from pain, and commences to paw with his fore-feet. He soon gets down, and, if space be sufficient, he commences to roll from side to side, often remaining for a few moments on his back, in which position he seems to obtain temporary relief. Sometimes, as quick as thought, he is on his legs again, gives the body a shake, and then anxiously regards his flanks, by turning his head toward one side or the other, as much as to say, " Here is the seat of my trouble." Soon he is down again on the floor, rolling and tumbling about. Now and then the animal remains quiet for a time, in a sort of crouching attitude, the limbs being gathered beneath the body, until the distension is so great, or the pains so severe, that he must shift his position, when again we find him rolling, or standing with his hind extremities stretched backward, and the fore ones advanced, thus representing the attitude of a horse in the act of urinating. Supposing, at this period, that there be no flatulency present, yet the respirations are hurried, the pulse wiry, the eyes glassy, and the patient excessively nervous and uncontrollable, the case is then of a spasmodic character (see Spasmodic Colic) ; but should the animal pass flatus, or the abdomen increase in volume, the case is unmistakable—it is flatulent colic.

Causes.—I shall now offer some brief remarks on the cause of colic. If we could only stretch the imagination, so as to take for granted all the popular causes assigned for the production of colic, then their name would be legion. Some writers inform us that a drink of cold water, when the animal is heated, is the most potent cause. Then we must infer that the horses ridden by the Russian Cossacs, in a country where cold water is abundant, should be the subjects of colic; but the very reverse is the case. The disease is almost unknown among them.

Previous to the introduction of Cochituate water into the city of Boston, colic occurring among horses was partially attributed to the cold well-water ther. used; but now they all drink the former, yet colic is just as prevalent, and, indeed, more so, than at the time referred to.

There are no people that pay more attention to the watering of horses than the English grooms, and it is customary, when a horse is taken from his home to a distant race-course, to remain away

only a few days, to have sufficient water transported with him
to last during his stay, such a dread have English grooms of
"strange water," as they term it. Now, it is a notorious fact that
English horses are more subject to bowel affections than those of
this country. From this circumstance, I infer that bowel affections are hereditary, or, in other words, a predisposition to such
is transmitted from parents to offspring.

I can well remember the time when physicians would not allow
a fever patient to taste water; but now they act more in accordance with reason, and allow the sick just what they crave for. It
is not good policy, however, to allow a horse, when performing a
journey, a bountiful supply of water, because active exercise somewhat suspends the digestive function, and, therefore, the water
may remain in a portion of the large intestines without undergoing the usual transformation. It then occupies space, and, being
weighty, may, in the rapid movements of the animal, operate unfavorably in various ways, more particularly on the gut itself, it
being pendulous from the spine.

A horse is often brought into the stable in a state of profuse perspiration, and, of course, is somewhat exhausted. Now, it would
be decidedly wrong to either feed or water him until he is rubbed
dry, and has rested awhile; for, at such times food is just as likely,
and, I think, more so than water, to operate unfavorably on an
exhausted animal. Many horses, however, even when in this
exhausted state, will fill their stomachs with food and water, and
yet enjoy immunity from colic. Therefore, should colic occur in a
horse after he has had a hard drive, he having partaken of a
draught of water, it is rather difficult to decide which of the two,
if either, was the cause of the malady. One fact is certain, and
that is, the stomach is deranged, and, therefore, fermentation pre-
cedes digestion; hence the gas.

Notwithstanding our best efforts to prevent it, colic will occasionally occur. Green grass, clover, carrots, and turnips are said
to occasion it. Then, again, it appears in stables, where nothing
but corn, oats, and hay are used. One horse is attacked immediately after a draught of cold water; another has the chill taken
off his, yet he is often found in the same predicament. Warm
water is the most insipid and nauseating drink that you can offer
a horse; and many would prefer to continue thirsty for some time
ere they would imbibe it.

It is well known to physiologists, and I presume the reader must have observed, that both men and animals inherit peculiar idiosyncrasies. Each are predisposed, either through parental defect, temperament, or conformation, to certain forms of disease. This peculiarity, or predisposition, is said to lurk in breed, and those conversant with the horse's structure and temperament can readily determine whether he be predisposed to certain forms of disease or not. For example, a horse predisposed to flatulent colic is often observed to have a capacious belly, voracious appetite, and does not properly masticate his food ; and he is not over-particular as to the kind of diet, for we often find him devouring, with apparent relish, the filthy straw that has served as bedding. Often he proves to be a "crib-biter." (Cribbing is a defect, either inherited or acquired.) Ordinarily the salivial fluid is augmented, yet it does not look healthy ; 't is more like soap-suds, and of a dirty color. The tongue is also foul, and the breath somewhat fetid. One or more of these peculiarities generally predominate in colicky horses. I contend, therefore, that some horses are predisposed to colic ; and this explains the reason why the ordinary exciting causes, such as cold water, exposure, fatigue, irritating food and medicine, are operative on the system of one horse and inoperative on that of another. Colic is the heritage of some of the best horses in the world.

Now, should the question be asked, What are the causes of colic? I answer, It is occasioned by predisposing, direct and indirect causes, operating conjointly or not. In all cases of colic, the function of the stomach is either disturbed or partly suspended.

Treatment.—The principal objects in the cure of colic are, to restore the tone of the stomach, by means of stimulants and tonics, and to arrest fermentation and absorb the gases of the alimentary canal. I shall not undertake to say that the remedies here alluded to are the very best ; but, as I have met with remarkable success in their use, I can safely recommend them. I have an impression that when a horse is the subject of colic, he requires to be treated just as any intelligent physician would treat one of us. There is no necessity to convert his stomach into a "slop-shop." The patient should be treated in a rational manner, by the same means and with the same skill as if one of our race were concerned.

I do not like to see an animal trotted up one street and down another, followed by a biped, whip in hand, and a crowd of idlers when he is the subject of excruciating abdominal pains, and the sweat is pouring off him like rain, from sheer agony. The custom is decidedly wrong. Our reasoning faculties confirm this opinion; and what reason teaches, man should endeavor to put in practice. No practitioner of human medicine dare advise such a course as compelling a man to hop off a bed of sickness, and start on a fast trot up and down stairs; for, should he so advise, he would very soon entitle himself to a ticket of leave. Should, however, the attack be slight, a little walking exercise does no harm; and if the patient appears disposed to roll, it were better to let him do so—on the ground, rather than in a narrow stall.

My usual directions are: Give the patient a wide stall and plenty of bedding. Let him lie down, rise, and tumble about just as much as he pleases, only watch and see that no accident happens to the animal. The colic drench used by the author, during the past ten years, is composed of the following:

No. 22. Fluid extract of Jamaica ginger........ 2 oz.
Fluid extract of golden seal........... 1 oz.
Powdered hyposulphite of soda......... 1 oz.
Water...................... .... 4 oz.

Dissolve the hyposulphite of soda in the water, then add the other ingredients to it. The dose may be repeated, if necessary.

A good wisp of straw, vigorously applied to the belly and flanks, and also to the limbs, may do some good, because we thus preserve the equilibrium of the circulation. Enemas of soap-suds should be administered often, and, provided the case be a curable one, the horse will soon recover.

SPASMODIC COLIC.

The usual causes of spasmodic colic are, mental emotions, external chilliness by exposure, cold water drank hastily, drastic cathartics, poisons, etc. Spasmodic colic is usually confined to the small intestines, and they become cramped and contorted. In this condition the patient is much tortured. The pain occasionally remits, and the patient will appear easy, but only for a short time. In some cases the patient will stamp and paw almost continually, and strike his belly. He throws himself violently on the floor, and occasionally rolls on his back. The abdomen is not distended, as in the case of flatulent colic, but is rather tucked up, and the

13

abdominal walls are rigidly contracted. In the commencement of the malady, the pulse is usually hard and wiry, but varies as the disease progresses; and the same is true of the respirations, viz.: they are accelerated and laborious. The patient occasionally suffers much pain in consequence of retention of urine and over-distension of the bladder, which is the result of sympathetic spasm in the region of the neck of the bladder. Such are the notable symptoms of spasmodic colic.

Treatment.—The disease being of a spasmodic character, requires the exhibition of medicines of an antispasmodic character, such as assafœtida, lobelia, sulphuric ether, and copious enemas of warm water, or an infusion of lobelia. The latter is the most valuable antispasmodic now in use. Should the case, however, be a desperate one, I advise the reader to follow the treatment indicated in the following case, taken from my note-book:

On making an examination of the animal, the following symptoms were observed: Pulse, very strong and wiry, averaging twenty beats above the natural standard; membranes of the mouth, nose, and eyes very vascular, or highly reddened; the surface of the body was bedewed, in patches, with perspiration; there appeared to be much rigidity of the muscles, especially in the abdominal region. All at once the animal would throw himself violently on the floor, and throw his limbs about in the most wild and reckless manner. He would assume all sorts of positions, yet seemed to obtain most ease when flat on his back; still, if closely approached or touched, he would kick and strike with fury, as if he were intent on mischief. In short, the horse cut up such antics that it was almost impossible to approach him; yet I finally succeeded in delivering an antispasmodic drench while the animal lay on his back, in which position I also gave an enema, composed of warm water and lobelia.

The animal kept growing more restive and uncontrollable, till, at last, it became very evident that convulsions had set in. The breathing had become fearfully laborious and rapid; his nostrils were dilated to their utmost capacity; the sight had become so affected, through temporary paralysis of the optic nerve, that temporary blindness had set in.

There seemed to be little hope for the animal, and, fearing that he might either kill himself by violence, or injure those in attendance, I concluded to chloroform him, and thus put a stop to his dangerous performances. He had had the best of us long enough, and now it was for me to show what science had in store for such otherwise unmanageable cases. I procured a mixture, composed of four ounces of chloroform and the same quantity of sulphuric ether; next, a sponge was tied to a broom-handle, and covered with a towel. Thus we were enabled to chloroform him at a safe distance.

The patient did not seem to relish this mode of practice. He fought for some time, tried to strike and kick me, but, being temporarily blind, I had the advantage of him. It was soon evident that the chloroform was beginning to do its work. The patient gradually settled himself on the floor, and was soon completely etherized. As it is dangerous to keep a horse under the full effects of chloroform any great length of time, I now removed the sponge, and only applied it occasionally, slightly saturated, so as to insure a sort of incomplete state of insensibility. It was an encouraging sight to behold the once powerful and furious animal now lying free from pain, and deprived of the power of injuring himself or those in attendance; and it is also gratifying to know that science ministers to the wants and necessities of the inferior as well as the superior orders of creation.

At the expiration of an hour, during which time the animal was more or less under the anæsthetic agent, he was allowed to rise. He gave himself a few shakes, and seemed very much relieved and more tranquil. I now gave a drench of spirits of niter and infusion of lobelia; administered, also, a lobelia enema, and left the patient to the care of his attendants. A few hours afterward I again visited the patient, and learned that he had a slight spasm occasionally, for which I prescribed two drachms of powdered assafetida. I then introduced the male catheter into the bladder, drew off a small quantity of urine, and this completed the whole of the treatment. The patient recovered.

Another case of Spasmodic Colic, treated by Inhalation.—The subject was a large and powerful stallion, aged about eight years, the property of a gentleman residing in Chicago. The horse was attacked in the street with symptoms of acute abdominal pain, and, in consequence, was hurried home to the stable. The moment he arrived there, he threw himself down and commenced rolling, and grew excessively uneasy and irritable. In the course of an hour our attention was called to him. He then exhibited symptoms of acute abdominal pain. The eyes appeared wild, and their membranes congested; the nostrils were dilated; pulse, about sixty; extremities cold; body, bedewed with perspiration; no tympanitis, yet the animal anxiously regarded his flanks. He would paw with his fore-feet, and strike forward in a most reckless manner, and fall down, as if in convulsions; then he would roll on his back, curve his head to one side, and, as quick as thought, jump up again.

He was now so spiteful and unmanageable that it was impossible to administer either drench or ball. We tried every means to get something down, but all to no purpose; he was determined to make us keep beyond the reach of his fore-feet and mouth. It now became evident to us that the patient must be mastered, and, in this view, not knowing at the time that the agent we were about to use would act beneficially on the disease, we procured three ounces of chloroform, and the same quantity of sulphuric ether. An old sheet and a sponge were used as an inhaler. Watching an opportunity, when the horse was down, we got two men to secure him there by the neck. We then saturated the sponge with the mixture, and applied the same to the nostrils, enveloping them with the folded sheet, leaving an orifice for the admission of air. The patient struggled violently at first, but soon became quieter, and, in the course of two minutes, trembled, breathed deep and loud. The pupils then became dilated, the breathing stertorous, and the animal was fully etherized. Having him completely under control, we were not disposed to let him up in a hurry. The sponge, however, was removed, and so soon as he showed signs of partial consciousness, it was again replenished with ether, and applied.

Supposing that the case might be running on to intussusception, or entanglement of the bowels, we thought there could be no harm in keeping the animal under the influence of an agent that seemed to act so well; consequently he was kept under its influence for fifty-five minutes. His head was then brought toward the door, and a few drops of cold water were sprinkled on him. Soon he raised his head, looked bewildered, yet showed no symptoms of pain nor uneasiness. Shortly afterward he got on his legs, but had to be supported by several men, who soon rubbed him dry. He was then drenched with laudanum, five drachms; sweet spirit of niter, one ounce. He was led to a stall, entirely free from pain partook of some gruel, and rested well during the night. Next morning he got five drachms of aloes and one of ginger, in solution ; the bowels responded to the same, and four days after he was at work. We would not, however, have the reader suppose that the same agent shall suit all cases of this character ; for another similar feat might prove a failure, and a third equally unmanageable. A case might yield to the ordinary sedatives, however, provided the patient could be induced to swallow them.

Inflammation of the Peritoneum (Peritonitis).

The peritoneum is a serous membrane, which not only gives a covering to the intestines, but also a lining to the abdominal cavity. Our pathologists teach that the peritoneum, like other serous membranes, is very ready to take on inflammatory action upon the operation of certain exciting causes. Watson informs us that "acute inflammation, beginning in one spot, is almost sure to transfer itself to any other spot that happens to be in contact with the first, and is very apt to extend itself rapidly to the whole membrane. The disease tends to the effusion of serum, and of coagulable lymph. It is of the adhesive kind, and its effects are those of distending the peritoneal cavity with fluid, or of gluing its opposite surfaces together, so as to obliterate that cavity, or of forming partial attachments. In all these respects the analogy between inflammation of the peritoneum and inflammation of the serous membranes of the chest and the covering of the heart is perfect."

Peritonitis is generally a manageable disease when treated according to the principles of rational medicine ; but when treated according to the dictum of Blain and others, who have never done any original thinking, in so far as veterinary science is concerned, it is decidedly the reverse. For example, here is a passage from Blain's works :

"The treatment of peritonitis is much the same as for enteritis. The only difference is, that the animal will lose more blood before he displays any symptoms of uneasiness. We may also be more bold with purgatives. The three ounces of solution of aloes may be increased to five, the pint of oil enlarged to a pint and a half ; and if this produces no signs of its having acted on the bowels, it may even be repeated when another three hours and a half have elapsed. All the other measures are alike, in both cases."

Those who follow such advice as this would be very apt to lose their patients, either through the direct influence of meddlesome medication, or by converting peritonitis into dropsy of the abdomen. But let us see what WATSON says on the subject of the treatment of peritonitis:

"Most writers whose worth I am acquainted with recommend purgatives as highly serviceable in peritonitis. I do not think the good which they are calculated to do, as antiphlogistic remedies, can at all be put in competition with the harm that I am persuaded they may produce, by increasing the peristaltic action of the intestines, and so causing additional friction and tension of the inflamed membrane. I believe that in all cases of well-marked and pure peritonitis, when the inflammation is limited to the serous membrane, it is far better and safer to restrain than to solicit the internal movements of the alimentary tube."

In human practice, physicians, instead of giving purgatives, resort to the use of narcotics, in view of tranquilizing the system, and for the purpose of mitigating the intense pain which exists during the acute attack.

Symptoms.—The most marked symptoms of peritonitis are as follows: In the early stage, the animal paws slightly, but not with that sort of viciousness which is characteristic of colic or strangulation of the intestines. Soon, he shows febrile symptoms, the pulse runs up to seventy or more, and a shivering fit sets in. When pressure is made on the abdomen, the animal evinces signs of pain. He will sometimes get down and lie on his back, but he soon finds out that the best way of lessening his pains and aches is to keep as quiet as possible. His tongue is usually coated, and the bowels are inactive

Treatment.—Give the patient mucilaginous drinks, as in the case of enteritis, and drench him occasionally with an infusion of poppy-heads, in pint doses. Empty the rectum by means of soap-suds enemas, and then throw in a quart of infusion of hops. Cloths, wet with warm water, should be applied to the abdomen, and the surface of the body should be kept warm by means of blankets. This kind of treatment, with good nursing, will probably restore the animal to health. Should symptoms of dropsy set in, give two ounces of fluid extract of buchu, daily.

ASCITES, OR ABDOMINAL DROPSY.

Ascites signifies a collection of serous fluid within the abdominal cavity. The subject of this disease is noticed as having a symmetrical and uniform enlargement of the abdomen; and, if the patient be a mare, some persons are likely to infer that she is pregnant, but examination, by pressure, will suffice to determine the conditions; for, if pressure be made with the fingers, in the abdominal region, a waving or fluctuating motion is discovered, similar to that which follows the forcible displacement of liquid. This fluctuation is the discriminating symptom between ascites and pregnancy, for in the case of pregnancy no such fluctuation can be elicited.

Ascites is usually the result of some chronic or acute disease of the peritoneum (lining membrane of the walls of the abdomen), or, rather, a loss of equilibrium between the local functions of secretion and absorption. This disease, like many others, may be rapid or insidious. If rapid or acute, it is apt to prove fatal in the course of a few days; if insidious or chronic, the animal may survive for a long time, as the following case will show :

Case of Dropsy of the Abdomen, ending in Rupture of the Stomach.—The subject of the above-named affections was a mare, owned by the North Chicago Railroad Company. She had been in their employ during a period of three years, and had always performed the work required of her without any manifestation of disease; in fact, had gone her usual trips up to within a few hours of her death. She had just partaken of a hearty breakfast, when, all at once, she began to falter, and the surface of the body was soon bedewed with a profuse perspiration ; she trembled, the head became pendulous, and the case assumed an alarming aspect.

This happened at about six o'clock in the morning. At eight, my attention was called to her, when the following symptoms were observed. Pulse, small and indistinct ; respirations, humid, and rather laborious; abdomen, quite tense from distension ; percussion over the region of the abdomen yielded a dull sound, yet fluctuating, showing that the abdominal cavity was occupied by some kind of fluid. From the peculiar conformation of the abdomen, I at first suspected that the animal was pregnant, which proved not to be the case.

On percussing over the region of the stomach, a tympanitic

sound was elicited, showing that the stomach contained a large quantity of gas, and was over-distended. Her extremities were cold, and other symptoms were present which indicate death. I suspected that the food was running into fermentation, and, in view of arresting it, I gave one ounce of the hyposulphite of soda, and the same quantity of fluid extract of ginger and golden seal. The body was bathed with new rum, and she was rubbed vigorously, by three men, for about half an hour, from which treatment she seemed to rally some. Several injections of salt and soap-suds were thrown into the rectum. They brought away a slimy substance, but very little feces. Finally nothing came away but that which was injected, and from this I was led to infer that, either from pressure or stricture, the small intestines were closed.

I now determined to send a trocar into the abdomen, for the purpose of discovering what it contained, and, accordingly, selected a point in the lower middle of the same. In withdrawing the cutting instrument, a fluid commenced to run, which, in color and consistence, resembled urine. For an hour and a quarter this fluid continued to run through the tube, and the amount obtained must have been over three buckets-ful. I was in hopes that the withdrawal of this large amount of fluid might remove the pressure on the intestinal tube, and thus the gas, which was pent up in the stomach, might find its way to the anal outlet. Not so, however; for, notwithstanding all our efforts and attention, the mare died, at noon, with rupture of the stomach.

The autopsy revealed a rent, or rupture, in the stomach, of about eight inches in length, through which a portion of the food and fluids of the stomach had escaped into the abdominal cavity. The small intestines were the seat of stricture in several places; their external surface, or, rather, peritoneal covering, was studded with albuminous and fibrous deposits, resulting from abdominal dropsy; otherwise, the contents of the chest and abdomen presented the appearance of perfect health. I should judge that the water had been accumulating in the abdominal cavity for a long time. The foreman at the stable observed that the animal always appeared very portly, so much so that the abdomen was square with the hips, which had led some persons to suppose that her hips were " knocked down."

Most authors contend that when a horse is the subject of rupture of the stomach or intestines, he squats on the haunches, like a dog;

out this animal did not exhibit any such symptoms, but stood up until within a few minutes of death.

Treatment.—A case of this kind requires the exhibition of tonics and diuretics, and the following is the best remedy which can be given :

No. 23　Fluid extract of buchu............ }
　　　　Fluid extract of snakehead......... } equal parts.
　　　　Mix.

Dose, two ounces per day.

INFLAMMATION OF THE INTESTINES (ENTERITIS).

This disease is located on the inner or mucous coat of the intestines, sometimes, however, involving the muscular portion of the same.　The causes are various: improper management, bad food, worms, irritating medicines, musty hay or clover, chilling the body by cold water, exposure to rain, etc.

Symptoms.—The disease often commences with a shivering fit, to which succeeds heat of the skin, restlessness, loss of appetite; the mouth hot and dry; the inner membrane of the eyelids and nostrils are redder than usual.　As the disease progresses, the pain increases, and the animal will lie down.　There is no intermission of pain; it is persistent.　The breathing is accelerated, and pressure in the abdominal region elicits symptoms of pain.　The bowels are usually constipated, and if any feces are voided, they consist of small, hardened balls.

An examination after death reveals a gangrenous state of the small and sometimes the large intestines.　Often extravasated blood is found within the intestinal tube.　So soon as the pain ceases, and the animal appears no better, we infer that gangrene has set in; cold sweats set in, and the animal shortly dies.

Treatment.—The treatment of enteritis does not, and should not, differ from that of any other acute affection.　Mucilaginous drinks, chlorate of potass, hyposulphite of soda, in doses as recommended for inflammation of the stomach.　Occasional enemas of soap-suds are to be thrown into the rectum, so long as the feces remain hard, or are voided with difficulty.　The objectionable method of back-raking—so highly recommended by some authors—should never be resorted to; it is a disgusting affair, and actually unnecessary; for, let the rectum be ever so impacted, it can be evacuated by

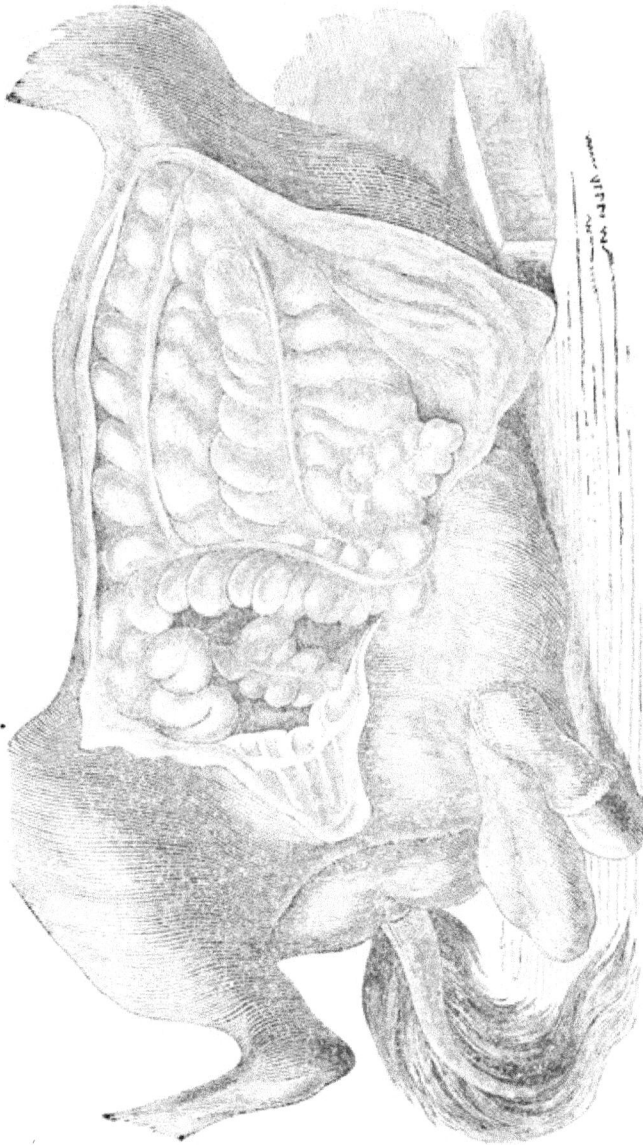

APPEARANCE OF THE INTESTINES ON OPENING THE ABDOMINAL CAVITY.

enemeta, composed of soap-suds, to which add a little ginger or lobelia. If the weather is not too cold, cloths wet with warm water may be applied to the abdomen, over which throw a blanket, so as to prevent evaporation and chill. The outrageous treatment recommended by some authors is a disgrace to the profession. They recommend bleeding until the horse faints or falls ; then to scorch the sides with ammonia and cantharides; and, lastly, to empoison what little blood he may have left in his system, so that death may occur *secundum artem*.

Should the subject of inflammation of the intestines suffer much pain, a drench of half a pint or more of infusion of hops, or poppies, may be given occasionally; or one ounce of the fluid extract of one of these agents may be substituted, to be given two or three times during twenty-four hours.

The treatment of inflammation of the intestines, as well as inflammation of any other part of the body, consists in using those means and agents which ward off or remove any agent which, in its effects, does, or tends to, deprive any of the organs or tissues of the living body of their vital action. The remedial means used to accomplish this object must be such, and such only, as sustain the weakened or increase the reduced vital action in any of the organs or tissues. This is the doctrine which the author attempts to teach in this work. And is the doctrine not plain? Could any thing accord more directly with reason, or commend itself more clearly to the common sense and unbiased judgment of intelligent people? If the vital action of any of the organs or tissues become weakened or enfeebled, should they not be strengthened? Can this be done by any other agent and means than those which increase and strengthen this very vital action? If this action, in any of the organs or tissues of the animal body, becomes reduced or impaired, must it not be increased by agents and means which coöperate with the vital power? Then, how can agents and means which always impair vital action (poisons) cure disease?

Different remedial means have different effects upon the organs and tissues of the system ; but all poisons and agencies which impair the integrity of the organism of a well animal must have a disastrous effect on one that is sick and suffering. Unhealthy and poisonous agents impair and reduce the vital action of certain organs and tissues according to the nature and tendency to affect various parts of the living structures.

Now, if it can be made to appear that disease consists, in its nature and effects, in increased vital action or increased health of any portion of the body diseased, then will it be proved that poisons are, and of necessity must be, medicines or the agents to cure disease. But what are we taught by our brethren of the heroic school upon this point? "In spontaneous disease," that is, disease which occurs without any apparent cause, HUNTER contends, "we know of no method that will entirely quiet or remove the inflammatory action or disposition. All we can do, probably, is to lessen every thing which has a tendency to keep it up. Inflammation is either an increase of life or an increase of a disposition to use it."

Again: "One of the means of curing this inflammation," says the same author, "is by producing weakness. The means of producing weakness of an absolute character are, bleeding, purging, and the exhibition of poisonous agents. The inconvenience, however, arising from this practice is, that the sound and healthy parts must suffer nearly in the same proportion with the diseased; for, by bringing the inflamed parts on a par with health, the sound parts must be brought much lower, so as to be too low. The first method, namely, bleeding, will have the greatest, the most permanent, the most lasting effect; because, if it have any effect at all, the diseased action can not be soon renewed; the second, purging and poisoning, will act as auxiliary, so that these processes go hand in hand; but neither these nor sickness can possibly lessen the original inflammatory disposition. Lessening the power of action belonging to the inflammatory disposition can only lessen or protect its effects, which, however, may be of service, as less mischief will be done, and this will often give the inflamed disposition time to wear itself out. But this practice must not be carried too far. It must be followed with great judgment. Nothing debilitates so much as purging, if carried beyond a certain point. Even one purging may kill when the constitution is much reduced. The internal medicines generally ordered for the cure of inflammations are such as have a similar effect to that of bleeding; namely, lowering the constitution or the action of the parts. As inflammation has too much action, which action gives the idea of strength, such applications as weaken are recommended."

Here, then, are the reasons for the use of poisons to cure diseases, as believed and taught by our brethren of the heroic school. Such agents and means we do not want, for the plain reason that disease can be cured quicker and better without than with them. Our system of practice, and that recommended in this work, tends to sustain the vital powers during the progress of disease; and this is accomplished by resorting to sanative medication. This sanative system of medication must, eventually, commend itself to the candid and to the unprejudiced reasoner and scientific inquirer after truth.

STRANGULATION OF THE INTESTINES.

When an animal becomes the subject of strangulation of the intestines, it is very difficult, at first, to diagnose the case, for the symptoms very much resemble those of spasmodic colic; yet as the treatment of the two diseases does not differ much, the mistake in itself is of little consequence.

Symptoms.—The difficulty is, at first, accompanied by uneasiness. The animal moves about in his stall, and commences stamping with one of the fore-feet. Frequently he gets down and rolls on his back, in which position he will remain sometimes for several minutes. So soon as he rises, he commences to paw and stamp with a vengeance. As the disease advances, he totters, and sometimes drops on the floor. The pulse is usually irregular, the mouth cold, and visible surfaces pallid: the tongue appears of a bluish cast; and the body is generally bedewed with sweat. Finally, the animal makes great efforts to evacuate the rectum, and in so doing strains violently, so as to cause blood to escape from the rectum; and sometimes the rectum itself protrudes. Should the symptoms not improve, the animal soon dies of strangulation and mortification of the intestines.

Treatment.—This must be of an anti-spasmodic character. An ounce or two of fluid extract of lobelia should be given, every two hours, in a pint of warm water; the body should be enveloped in bandages, wet with warm water, and enemas of assafetida or lobelia should be thrown into the rectum; but if the rectum protrude, enemas are inadmissible. On some occasions of the kind I etherize the patient, with marked relief.

Wounds penetrating the Abdominal Cavity.

The treatment of a penetrating wound into the abdomen will depend somewhat upon the nature of the same. Should the intestines protrude through the wound and exhibit a puncture, put on the twitch and Rarey-strap. Then return the intestines into the abdominal cavity, and proceed to suture or sew up the intestinal rent, by means of a small needle, armed with sewing silk—first, by means of sutures through the abdominal muscles, one end of the suture being left long enough to hang out of the wound; second, close the skin (the sutures should be distant from each other about one inch); next place a cold water pad on the parts, and encircle the body in the injured region with a cotton-roller, over which place a surcingle, and securely fasten it. In the event of the intestines being swollen and distended, it may be necessary to cast the patient; then roll him on his back, and thus the protrusion may be returned; but should they still prove too large for the orifice, it must be dilated by means of a probe-pointed knife.

Splenic Apoplexy (Carbuncular Fever).

Mr. Gamgee has made some translations from foreign authors on the subject of Splenic Apoplexy, and he tells us that " this is the disease described as carbuncular fever by Renault and Reynal. It is most common in hot and tempestuous seasons. In its carefully-observed symptoms, M. Anginiard is induced to regard the disease as dependent on miasmatic exhalation. Until recently, the author had studied the malady only in sheep, but has frequently observed it, of late, in the horse. It is not so suddenly fatal in the latter as in the former. The prostration and stupor are very striking. The horse can barely stand, and is very listless; the muscles of the limbs tremble, and partial sweats bedew the body. In some parts the skin is dry, and the coat staring; the breathing is slow and irregular; the conjunctiva, or lining of the eyelids, is white, and indicative of an anemic or bloodless state; the pulse is feeble, wiry, and accelerated, offering a remarkable contrast to the tumultuous action of the heart; the loins are very sensitive; and there are often indications of pain on pressure on the abdominal walls. Colicky symptoms are frequently noticed. At first, all appetite is lost, but, as the disease advances, the animal eats and drinks freely. In attempting to bleed an animal, great difficulty is experienced in filling the vein; and when the blood flows, the animal instantly manifests great weakness. The blood is black, and does not coagulate as in health. The symptoms rapidly grow worse, the breathing labored, the eyes sunken, the prostration extreme, and progression impossible, until the animal reels over and dies. The disease runs a rapid course, and, on an average, death occurs within fifteen or eighteen hours. There is an absence of the external swellings characteristic of ordinary carbuncular fever; and the internal distensions of organs, particularly of the spleen, so ably described by Renault and Reynal, are sufficient, according to M. Anginiard, to distinguish the two diseases. In the treatment of this disease, Anginiard trusts mainly to quinine. He administers a drachm and a half as soon as possible, either in an infusion of coffee or other aromatic mixture. If the symptoms do not subside two hours afterward, two drachms are given. Generally this is sufficient. If not, a third and fourth dose have to be exhibited. Bleeding is to be condemned. The body of the animal may be rubbed with a wisp, or a turpentine liniment may be employed as a derivative. Free ventilation, and leaving the animal uncovered, to permit free exhalation, are essential points in the management of these animals. M. Dumesnie has tried this treatment, and found it most successful. Anginiard believes in small doses, and not often enough repeated."

Apoplexy is a disease which, if not discovered early, is very apt to prove fatal. Whether discovered or not, in its early stage it generally baffles the skill of those who, in this country, have attempted to cure it. The incurability of such malady may, under the circumstances, be accounted for on the belief that it often rages as an enzoötic, affecting animals in isolated spots, without regard to locality, and attacking only those susceptible to its influence. What the conditions are which favor its propagation I am unable to determine, for I have known it to affect equally those which appeared healthy as well as others unhealthy in appearance; hence it is very difficult for any one to point out the direct causes of the affection. I have not seen many cases of either enlargement of the spleen, or apoplexy of the same, occurring in cattle, from the fact that my practice has been mostly confined to horses in cities, where cattle were not very numerous; but a friend of mine, located in a cattle-raising district in the State of Maine, informed me that this disease, accompanied by enlargement, often prevails there enzoötically, and that most of the animals thus affected die.   I have had considerable experience in the treatment of such affections in horses, but, as I have just observed, the result is very unsatisfactory.   I believe I have but one case of cure on record, and the medicines used on that occasion were stimulants and tonics, with forty grains of iodide of potass per day.   The spine was also rubbed, night and morning, with equal parts of cod-liver oil and spirits of camphor.

ENLARGEMENT OF THE SPLEEN.

Symptoms.—The symptoms of enlargement of the spleen in horses (they do not differ much in cattle) are as follows: Feeble pulse; respiration not much disturbed; the tongue is usually coated; mouth, hot, and the breath has a feted odor; the membranes lining the mouth and eyeballs have a slight yellow tinge; the head droops, and the tips of the ears and lower parts of the limbs are chilly; the patient is rather unwilling to move, and, when urged to do so, exhibits a staggering gait, and sometimes falls never to rise; the urine is scanty, and, in the last stages, blood oozes from the anus and nostrils; colicky pains attend the disease.

In a case that terminated fatally, I made the following record of the post-mortem examination: On opening the abdominal cavity, the spleen was found to occupy a very large space.   It was of immense proportions, and weighed nearly twelve pounds.   The ordinary weight is three pounds.   It presented the appearance of a spleen in the chronic stage of disease, being altered in structure, of a dark, pitchy color, and surcharged with dark, venous blood.   Before death, I percussed the left side, in the locality of the spleen, and the sound elicited was of a solid character, which indicated enlargement of the same.

The early symptoms of splenic apoplexy do not differ much from the above, only they are of a more acute character.   In the latter stages, the malady is complicated with a painful affection of the bowels.   The diagnostic symptom of splenic apoplexy, with enlargement of the same, is a notable enlargement on the left side of the abdomen, well up toward the ribs.   When standing behind an affected animal, and casting one's eyes carefully along the sides of the abdominal walls, a perceptible eminence will be seen, which can not be mistaken. The region of the spleen is between the stomach and the false ribs on the left side.

In view of rendering this subject more valuable and interesting, I now introduce the following, which was written for the "Edinburgh Review:"

"*Splenic Apoplexy.*—This disease broke out in the year 1858, on a farm characterized by its extreme richness of soil, in the north of Northumberland. In 1859 the malady reappeared, and did not cease until twenty-three head of cattle had been affected. Splenic apoplexy is a malady that has not hitherto been observed in the north of Britain, and its occurrence recently is evidently to be attributed to the special method of farming and feeding stock where it has, for the last two years, proved so destructive.

The farm is a peculiarly dry one. About three hundred head of cattle are annually fattened on it. One lot was made ready by Christmas, and sold as fat beef in Newcastle; a second was kept back until January, when they were allowed an abundance of turnips, meal, and the best oat straw. The cattle affected were three years old; and it would appear that those fed on Swedish turnips, especially from a certain field, suffered most; but some fed on yellow or white turnips have been affected, and we should regard the meal, and perhaps the oat straw, as more likely to produce this disorder. According to Delafoud's researches, leguminosæ, or forage very rich in nutritive principles and deficient in water, from artificial pastures, are frequent causes of splenic apoplexy. The farmer in Northumberland suspected that the artificial manures had produced the trouble, and the following season ne dressed his land with home manure; but the properties of the last crop have proved as deleterious as those of the preceding one. The fact is, the crop had nothing to do with the development of the disease.

The malady stopped suddenly, about the middle of February, and this was probably due to the cattle suffering only when the system was taxed by change from rather moderate to very high feeding. It is an interesting fact that in cows the disorder only affects those which are approaching the period when the secretion of milk is stopped, and when there is a tendency to lay on flesh, they then being liable to plethora. Cows fed on food capable of producing splenic apoplexy die of it.

Change of diet proved of no avail at the farm referred to (showing very conclusively that the food was not the special cause). Common salt was recommended as a preventive; but Delafoud has shown that salt is not a reliable remedy in such cases. The malady stopped spontaneously, and the point now to settle is the prevention of disease during the future seasons.

For the benefit of our readers who may not know the symptoms of splenic apoplexy, we may mention that often there are premonitory signs—the animals apparently enjoying the best of health early in the morning, may be dead some time before noon. Sometimes there are symptoms of excitement; the eyes are prominent, and the visible mucous membranes are injected or reddened. Symptoms of uneasiness suddenly manifest themselves, and colicky pains indicate abdominal disorder. The urine voided is high-colored and red, and there may also be blood in the feces. The back becomes arched, and the animal fixes himself, hanging on to any thing by which he is tied, or he will press his haunches on some resting-place in the corner of the stall. After this the animal soon drops, and is seized with convulsive pains and twitchings. In addition to the discharge of feces and urine, tinged with blood, there is a red, frothy liquid, which escapes from the nostrils; the animal bellows, moans, and soon dies."

Treatment.—Tonics, alteratives, and glandular stimulants are indicated. Take of

No. 24. Fluid extract of blood-root............ 4 oz.
   Fluid extract of golden seal.......... 6 oz.
   Chlorate of potass.................... 4 oz.

Dissolve the potass in one pint of boiling water; then add the extracts. Give the patient a wine-glassful every six hours, until amendment is perceptible, when the dose may be decreased to two ounces every twelve hours. Rub the body with a portion of the following:

No. 25. Proof spirits..................... 1 pint.
   Tincture of capsicum.............. 4 oz.

Two or three applications, in the course of twenty-four hours, will suffice. Should the animal suffer much from colicky pains, dissolve one ounce of hyposulphite of soda in four ounces of hot water, to which add half a table-spoonful of powdered ginger, and drench with the same. Pure air and a light diet will aid in restoring the animal, provided the case is curable.

FUNCTIONAL DISEASE OF THE LIVER (JAUNDICE OR YELLOWS).

The liver, like other parts of the digestive apparatus, is liable to become deranged, in either function or structure, by errors in diet, which is the most potent cause of both functional and organic disease of the same. In almost all cases of liver disease, we observe a yellow tinge of the visible surfaces—membranes of the eye and mouth. In order to account for the yellow tinge, we must bear in mind that bile is formed in the blood; that the office of the liver is to absorb or withdraw it from the circulation constantly, as fast as it is formed. Disease of the liver, simulating jaundice, manifests itself whenever the due separation of the bile from the blood is not accomplished; hence the accumulated bile seeks other outlets, namely, the kidneys and intestines, and excretory gland of the skin.

When animals become jaundiced suddenly, we may infer that the difficulty, whatever it may be, in the liver, is of an acute character; that is to say, it is recent in origin and active in form. Jaundice is a phenomenon of several diseases peculiar to horses. During the progress of the epizoötic known as influenza a jaundiced state of the visible surfaces is always present, which disappears when the patient convalesces, showing that the liver is only functionally deranged. When jaundice accompanies indigestion, and disappears under the action of alteratives or changes in diet, we may also infer that, if not absolutely functional, it is, at least, not purely inflammatory.

It is plain, therefore, that jaundice is of but little consequence in prognosis, since the symptoms with which it is often associated declare, much more positively than itself, the gravity of the affection in which it occurs as a symptom. It sometimes happens, however, that the function of the liver becomes suddenly interrupted, causing pain, after the fashion of what accompanies the bilious colic of man.

The following case will probably give the reader some idea of the nature of acute functional disorder of the liver. The patient, a gray gelding, aged nine years, was suddenly attacked with symptoms of colicky pains, for the relief of which the owner had given a colic drench, which seemed to afford temporary ease; but soon the symptoms reappeared, the animal got down, and seemed to be the subject of acute pain. Under these circumstances, the owner sent for me.

Symptoms.—On arriving at the stable, I found the patient standing, his head somewhat pendulous, appearing dull and sleepy. The
membranes within the eyelids and mouth were of a bright yellow
tinge; pulse, rather feeble; bowels, constipated; in short, the usual
appearances attending functional derangement of the liver were
present; yet the "key-note," or diagnostic symptom, was the bright
yellow tinge of the visible surfaces.

Treatment.—The following drench was given:

No. 26. Fluid extract of mandrake............ 3 oz.
         Fluid extract of golden seal.......... 1 oz.
         Carbonate of soda..................... 2 dr.
         Water................................. ½ pint.

An injection of soap-suds was thrown into the rectum, which
brought away a small quantity of hardened, dark-colored excrement.
Next day I found the patient in about the same condition, and repeated the dose of medicine and injection.   In the course of a few
hours the bowels responded, evacuating a large quantity of semi-
fluid, dark-colored feces, mixed with a substance resembling shreds
of albumen.   Next day the animal appeared much better.   I then
gave two ounces of fluid extract of golden seal, and one drachm
of hyposulphite of soda, and thus ended the treatment.

This case is an illustration of a simple mode of treatment far
superior to the old-fashioned system, which contemplates strong
doses of aloes and calomel, agents which usually intensify the
morbid phenomena, often producing a medicinal affection, which
requires much time for the animal to recover from.   It is the
author's opinion that a vast number of very fine animals in these
United States die from the heroic system of practice, which over-
looks the best means of aid presented by the resources of mild
and sanative medication.

We have long been an advocate of the physiological method
of treating disease, and we contend that is more philosophical,
safe, and the surest means of restoring an equilibrium of vital
action in the animal economy, in all forms of morbid conditions;
and we earnestly entreat our readers to exercise great faith in the
powers of Nature to aid and assist in the cure of all maladies,
and place less reliance in the power of art.

In view of strengthening the faith of young converts to sanative medication, we quote the language of Sir JOHN FORBES, a
man of undoubted authority.

"Such has ever been the want of trust in Nature, and the over-trust in art, prevalent among the members of the medical profession, that the field of natural observation has been, to a great extent, hidden from them—hidden either actually from their eyes or virtually from their apprehension. The constant interference of art, in the form of medical treatment, with the normal processes of disease, has not only had the frequent effect of disturbing them in reality, but, when it failed to do so, has created the belief that it did so, leading, in either case, to an inference equally wrong—the false picture, in the one instance, being supposed to be true; the true picture, in the other, being supposed to be false."

In all cases of liver disease, it is good policy to change the diet, and offer the animal green vegetables—sliced carrots, and grass if it can be had. In all cases of chronic disease of the liver, a run at grass is the most potent remedy for the cure of the same.

It was formerly supposed that the function of the liver was merely to eliminate bile, but modern physiologists have discovered that the blood itself is changed while circulating through that gland, which elaborates fibrine from albumen, and fat from sugar. The liver stores up fatty matter, so that, should the food be deficient in fat, the liver yields up what it contains. It is very curious to notice the peculiarity in the circulation of the liver. It is supplied, like other organs, with arterial blood for its own support; and it also receives an immense quantity of venous blood through veins commencing on the gastro-intestinal cavity, which terminate in a vessel named *vena porta*. This vessel, on entering the liver, ramifies in various directions, like an artery, and ultimately terminates in the veins peculiar to the liver; so that the blood, instead of flowing directly from the stomach and intestines, is made to circulate through the liver, and thus the various transformations are effected. The liver must, therefore, perform the double function of secretion and excretion. It secretes bile, and excretes carbon and hydrogen from the system. Should an animal labor under diseased lungs, he may, under proper medical treatment, recover, provided the liver be healthy. This organ can, for a time, eliminate carbon and hydrogen; but, instead of passing off as they do from the lungs, in the form of carbonic acid, they accumulate in the liver, in the form of fat, or else the liver increases to an extraordinary size In Strasburg they prepare a favorite dish from the livers of geese, artificially enlarged by the cruel process of depriving them of light and liberty, confining them in a warm room, and stuffing them with food. We find that, under such barbarous management, there is a disproportion between the oxygen respired and the carbon taken in the form of food. The respiration which active exercise increases and rest diminishes are, now that the animal is confined, slow and uniform. The lungs fail in eliminating carbonic acid, and hence carbon and hydrogen are driven back to the liver by means of venous circulation. An excess of carbonaceous material in the liver is liable to result in jaundice, which is a simple functional derangement of that organ. It occurs generally in the warm months, and is usually sudden in its attack.

Among the most prominent symptoms are high-colored urine, yellow tinge of the visible surfaces, languid pulse, and slow action. If the liver be the seat of inflammatory action, the pulse will be quick and bounding, respiration hurried, the patient feverish, and pressure over the region of the liver elicits symptoms of pain. If simple functional derangement exists unchecked for any length of time, it leads to organic lesions and structural disease. When the bile accumulates, it is very apt to thicken and produce gallstones or calculi; if these accumulate in the gall ducts, the subject, unless relieved, soon dies.

Diseases of the liver have hitherto been considered the bane of tropical climates, but they are equally prevalent in cold and moist regions. Horses and men are as frequently attacked with it in northern as in southern latitudes. The celebrated sheep-breeder, Bakewell, knew that early disturbance of the liver led to the accumulation of fat, and, in order to derange the liver, he was in the habit of folding his sheep in wet pastures. Now, the English agriculturists are well acquainted with the fact that water meadows have a tendency to produce that almost incurable disease termed rot. Rot originates from a diseased liver, and, in the early stages of it, the animal accumulates fat very fast; so that by rotting sheep he was able to bring fat ones early to market, and thus steal a march on his more conscientious neighbors. This state of the liver termed rot is associated with the existence of parasites termed *distoma hepaticum*, commonly denominated flukes, and these parasites are considered the cause of rot, when, in fact, they are the results of deranged functions of the liver. The rot, therefore, is not local. It can be produced in any country by exposing animals to the debilitating effects of moisture and neglect. It is not only prevalent in England, but also in Asia, Africa, and America. In Egypt, on the receding of the Nile, it spreads to a fearful extent. In warm climates the livers of men and animals become diseased, not so much from local causes as from stimulating drinks and diet. If a man attempts to consume the same amount of food in warm that he has been accustomed to in cold climates, he is apt to become bilious, and this is a premonitory indication of perverted function of the liver. A very important fact has lately been discovered in reference to the function of the liver. A French chemist has ascertained that one function of the liver was to produce sugar out of the animal organism. This is the secret of the formation of the fat. Chemically speaking, sugar and fat are nearly allied, the chief difference being that one contains a small portion of oxygen.

Many valuable animals are lost from organic disease of the liver, the origin of which may, in many cases that have come to our knowledge, be traced to simple functional derangement, which has existed for some time. Its symptoms are either unobserved or, if perceived, disregarded. Some of the subjects that have been put under our care for the treatment of diseased liver had a strong predisposition for the malady, their color being black and temperament bilious. Such horses are the ones that require special attention whenever they appear to be "ailing." It is remarkable to observe, however, that many horses of a temperament diverse from the bilious are often attacked with organic and functional derangement of the liver, and this peculiarity can only be accounted for on the ground that the horse has no gall-bladder—no receptacle, such as is found in cattle, for the accumulation of bile ; consequently, the horse's liver must be an active organ, especially when the animal is permitted to make but one meal per day, and that meal occupies a period from sunrise to sunset, and daily and weekly continues to occupy no less time. The consequence is, the organ is overworked, and is the seat of local exhaustion. The liver requires regular periods of rest to recover from fatigue or functional duty, yet how is it possible to secure the same when a horse is permitted to make a perpetual hay-rack and corn-bin of his stomach ? The herculean feat of converting a hay-stack and corn granary into bone, muscle, and nerve at the rate of 2.40 is beyond the physiological capacity of any vital organ ; therefore, in the language of the sailor, the vessel sinks from local inundation—the liver caves in from overwork, too much food, and too little discretion. A hard pull on an omnibus, or a similar feat in a marketer's wagon, would be the very best medicine in the world for a plethoric horse, or one predisposed to hepatic derangement.

THE PANCREAS AND ITS FUNCTION.

The pancreas is a glandular body lying across the spine, in the epigastric region, underneath the diaphragm, near the small curvature of the stomach. The body of this gland is pierced by the *vena porta* (known as the *gate rein*); has one attachment to the spleen, and another to the left kidney. It is divided into head, body, and tail. Structure, similar to salivary glands. Color, pale red, speckled. It is composed of many lobules. Every lobe has a distinct set of nerves, arteries, veins, and ducts. Every lobe is, therefore, considered as a distinct gland, and the same is true of the salivary glands. The duct pierces the duodenum (known as the second stomach), alongside of the hepatic duct.

The pancreatic arteries are derived mostly from the hepatic. Several, however, come from the splenic, in its course to the left side of the abdomen, and one or two from the gastric. The veins are tributary to the *vena porta*. The nerves come from the *cœliac plexus*. The fluid secreted by the pancreas is a colorless, limpid fluid. It forms an emulsion of fat; hence it dissolves the fatty matters of the food. It is an active agent in the preparation of chyle. That the pancreatic juice possesses the property of emulsifying fat is proved from the fact that when the pancreas is destroyed, and the animal fed on food containing fatty matter, the latter passes with the feces, just like fat in an unchanged state.

Mr. GAMGEE says: "The pancreatic juice has another important use which remains to be spoken of, and that is a peculiar action on nitrogenized substances. If raw meat be placed in some of the juice, the meat speedily softens and putrefies. The same occurs with albumen or caseine in the raw state; but if the azotized principles are previously boiled, or acted upon by the gastric juice, the pancreatic therefore serves many purposes, and acts on every kind of aliment.

Circumstances affecting the digestive function in general have a special influence on the pancreas, and modify its secretion. Such is not the case with the salivary glands and their products. The pancreatic juice may, like every other secretion, be found to contain adventitious substances that have accidentally entered the blood. Thus, iodide of potassium may be eliminated from the system by the pancreas. It is not so with every salt, the prussiate of potash, for example, never being seen in the pancreatic juice.

The pancreatic secretion is formed during infra-uterine life, but we are at a loss to account for its uses there. It is difficult to determine what nervous influences affects its production. A dose of ether excites it, and pressure on the abdominal viscera likewise tends to its increase ; the efforts of vomiting stop it."

WORMS OR PARASITES WHICH INFEST THE INTESTINAL CANAL.

There are various forms of parasites which infest the alimentary canal of horses and cattle which, no doubt, are the cause of some annoyance to the infested ; but really they are not, at all times, so injurious as some writers make them out to be. They are rarely if ever found in the intestines of healthy animals, and their presence is generally due to a deranged condition of the digestive organs. They very frequently originate spontaneously.

"Certain independent organisms, both vegetable and animal, are found in the body. The vegetable growths are all microscopic, and belong to the lowest order of plants, the algæ and fungi. They are never met with except upon cutaneous or mucous surfaces, nor while these surfaces remain healthy, usually. A secretion of fibrine or mucus, undergoing decomposition, forms the soil in which they grow. In some cases, they are believed to be the media of contagion.

Animal parasites are very numerous. Many of them are infusorial. Many belong to the class of insects and mites, as fleas, lice, bugs, and the acari, of which the most important one is the itch-mite. A class of higher consequence comprises several sorts of worms. Those which infest the intestinal canal are extremely common, and are the *oxyuris vermicularis,* or thread-worm, which inhabits the rectum ; the *trichocephalus dispar,* or long thread-worm, which is found in the large intestine, and especially in the cæcum ; the *oscares lumbricœdes,* or round worm, whose ordinary residence is the small intestine ; the tape-worm, or tænia, which also affects the same part. The kidney is occasionally the seat of a round worm, called the *strongylus gigas*, measuring from five inches to three feet in length, and from two to six lines in thickness.

The diseases with which even large intestinal worms are connected appear to be sometimes the cause and sometimes the effect of the presence of these parasites. Very often they exist in considerable numbers without producing the least disturbance of the economy, but in other cases they are unquestionably the cause of much suffering and ill-health. How far they are themselves the result of a morbid state of the organs in which they appear is still an undecided question.

The origin of parasites is extremely obscure, and has long been a mooted point among naturalists. It may not be inappropriate to present a summary of the opinions which are entertained respecting a subject of so much interest, but, in doing so, we shall confine our remarks to the parasitic animals which inhabit the interior of the body, or entozoa.

It is evident that these animals must originate in one of two ways; that they must be derived directly or indirectly from without, or be created out of materials existing within, and furnished by, the body. No other supposition is possible. If an entozoön is in any manner derived from without, it must be admitted that this takes place either through the reception of the animal itself or of its ova. If either opinion be assumed, it follows that the parent animal must exist somewhere external to the body. But the parasites in question have never, in any case whatever, been detected except within the organism. If it is objected that many of these animals are so minute that they might easily elude discovery in the elements around us, the argument fails when applied to the giant strongylus, the stout lumbricoid worm, and the tænia, measuring many yards in length. Besides, even admitting for a moment the possibility of the parasites which inhabit the intestine, and other mucous cavities, having once existed externally, the insuperable difficulty still remains of explaining the entrance of entozoa into shut cavities and parenchymatous structures—into the eye, or the muscles, for example, and their presence in the unborn child, and even in the bodies of larger entozoa of a different species. On the other hand, if it is maintained that the ova are alone received, it must still be shown that the ova exist external to the body, which has never been done. Nor would the admission of this explanation be sufficient; for many of the entozoa are not propagated by eggs, but belong to the viviparous class, so that in regard to them the difficulty remains undiminished. But granting the existence of ova without, and their reception into the body, it is still impossible to explain the development from them of the animals found in the parenchyma, in the embryo, etc., without, at the same time, admitting that the ova are not only carried to these localities through the blood-vessels, but actually pass through the walls of the capillaries. Such an admission would be a physiological absurdity; for the extreme vessels will allow of the passage of a single blood-globule at a time, and no more, and will not permit any denser fluid than the plasma of the blood to permeate their walls. How, then, could they afford a passage in any manner to ova, the least of which is ten times as large as a blood globule?

If the hypothesis now presented is untenable, it only remains to adopt the alternative one, to-wit: that entozoa are generated or created anew out of the materials or the products of the living organism. It may be urged affirmatively, in support of this doctrine, that each organ possesses its own entozoa—the kidney, a species different from those of the intestine, which are, again, unlike the parasites of the liver. Even more: the several parts of the same organ generate dissimilar animals. The small intestine produces the round and the tape-worms; the large intestine, the two species of thread-worms. These facts seem to show that some extremely local concurrence of circumstances is essential to the production of the several entozoa. It may also be argued, and we think the argument unanswerable, that if spermatic animalcules, which exist in the testicle, are there spontaneously generated, no violence is done to probability in supposing parasitic animals to be produced in the same manner. It will hardly be denied that spermatozoa are literally evolved from the constituents of the semen; but it is objected to the doctrine of spontaneous generation that it is against analogy, which every-where supports the famous dogma, *omne vivum ex ovo*. This objection is a mere begging of the question. The decision of the case in hand involves the truth of the theory just quoted, and, as we believe, must be allowed to show that this theory is not absolutely universal in its application. Other facts, also, among which are the following, tend to invalidate it. Nothing can be more certain than that all organized beings were, at some time or other, created. Geology proves that successive genera and species have been thus created, at long intervals apart; and the history of disease renders it probable that one affection, at least (syphilis), which is only propagated by direct descent, *ex ovo*, as it were, is really of comparatively recent origin.

In conclusion, after the review of the preceding outline of an argument upon the generation of parasitic entozoa, we feel obliged to admit that the weight of facts and probabilities is wholly on the side of the doctrine of spontaneous generation. At the same time, we can not but look with interest to the results of future observations in this field, nor altogether suppress the hope that the simple law of Nature, *omne vivum ex ovo*, may even yet be found to embrace the classes which now appear to form so striking an exception to its provisions."*

Symptoms.—The usual symptoms indicating the presence of worms in the intestinal canal are, sometimes, a scurfy, yellow mucous accumulation around the margin of the anus; the appetite is not uniform; the bowels are irregular, sometimes loose, and sometimes constipated; and the feces are often shrouded with shreds of mucus, of a yellow color. When ascarides are present (located in the rectum), the horse is much inclined to rub his tail against any post or fence he can get at, and he employs himself at such feats against the sides of the stall, in view, probably, of allaying the intolerable itching which usually attends the presence of these parasites.

The lumbrici, or round worms, sometimes exist in the intestines in great numbers. They are then apt to be hurtful. In such cases, the coat looks unthrifty, and the hair has lost its natural glossiness, and appears rough to the sense of touch. It has been noticed, also, that when horses are afflicted with intestinal parasites, their breath gives out a fetid odor, and they have a sort of dry, hacking cough. It is well known, however, that some horses, enjoying, apparently, the very best kind of health, are often the subjects of worms.

Treatment.—Most of the remedies recommended in the ancient works on farriery are just about as likely to kill the horse as the parasites, and consequently are, to say the least, unsafe. The true theory is that worms are the result of indigestion; hence our

_____

* Stillé's "Elements of Pathology."

object, by way of prevention, is to give tone to the digestive organs; and soon after the worms have congregated in the intestinal canal, bitter tonics and alteratives are obnoxious to them. They then die, are digested, and pass from the anus as fecal matter.

Suppose we give the infested horse powerful purges, tartar emetic, calomel, turpentine, etc., as the books recommend. We may succeed in their expulsion, but that does not mend the matter for the agents just named are all more or less prostrating, and create a worse state of digestive derangement than that which previously existed, and soon, by virtue of spontaneous generation, a new crop of parasites make their appearance.

When a horse shows any of the symptoms here alluded to, indicating indigestion, or the actual presence of worms, I recommend that an occasional dose of the following be given:

No. 27. Powdered poplar bark (populus tremuloides). ⎫
      Powdered sulphur....................... ⎬ equal parts.
      Powdered salt .... ⎭
      Mix.

Dose, one table-spoonful, to be mixed with bran or oats.

EXAMPLES OF VERMIFUGES.

No. 1.—Fluid extract of wormwood, four drachms; to be given in the morning, before feeding time, for several successive days.

No. 2.—Give one ounce of fluid extract of pink-root, every morning, before feeding, for a week.

A GOOD REMEDY FOR THE TAPE-WORM.

Give four drachms of the solid extract of male fern (aspidium felix mas) every other morning, during a period of about ten days. This is said to be a *specific* for tape-worm.

Finally, as regards the treatment of worms, Nature has endowed these parasites with such tenacity of life, that no matters known to us will effect their destruction, though a few may answer the purpose of their expulsion. Bots are so hardy as apparently to survive immersion in oil, in alcohol, spirits of turpentine, and even powerful solutions of mineral acids. The continued use of salt, mixed with the food, appears, however, obnoxious to them; for sometimes, under its use, their hold gives way, and they are ejected. Bitters, purgatives, and the mechanical irritation of pointed bodies, as pewter, tin, etc., have no effect whatever upon bots; but with regard to the other parasites, rather more success may be expected from medical aid, in the form of vermifuges It has been attempted to effect the removal of worms mechanically, by dissolving the mucus they are supposed to be imbedded in, for which purpose lime-water, oil, solutions of aloes, etc., have been injected by clyster up the rectum, and which practice is most to be depended on for the ejection of ascarides when in the rectum. This practice of washing away the mucus of the intestine, and thus to deprive the intestine of the secretion given for its protection, is not to be recommended, though oil, for this end, would be harmless. Strong purges are given with the same intent, which may remove them also from the whole alimentary track. Remedies have likewise been exhibited to destroy them within the body, by the mechanical irritation of their spiculi, under which view tin, brass, iron, and pewter are thought remedial. The Indian caustic barley and Indian pink are reputed vermifuges against the teres and ascaris. The oil of turpentine has also been strongly recommended as an excellent general vermifuge; but, except for the destruction of the tænia, or tape-worm, it certainly does not appear to deserve that character."*

---

* "Blaine's Outlines"

SECTION VIII.

DISEASES OF THE URINARY ORGANS.

INFLAMMATION OF THE KIDNEYS (NEPHRITIS).

A NEPHRITIC affection is often mistaken for what some persons term "sprain across the kidneys" (lumbar sprain). It is a mistake of some importance, from the fact that when sprain or strain is suspected, people are apt to resort to the use of irritating embrocations or liniments, which may do much harm, as their action is to augment heat and pain, which, in case of nephritis, is to be avoided.

Symptoms.—Hard and accelerated pulse; quickened respiration, indicative of pain; back, arched; legs, straddling; the head is often turned toward the loins, or region of pain; the animal is unwilling to describe a circle with its body, and, while the acute stage lasts, scarcely if ever gets down on the floor; the urine is reddened and scanty; finally, the animal crouches when pressure is made over the region of the loins, and, as is the case in all acute affections, thirst and loss of appetite are observable.

Treatment.—The treatment of nephritis, in the acute or inflammatory stage, is just such as would be proper supposing the case to be one of enteritis, or peritonitis. Twenty or forty drops of fluid extract of gelseminum may be placed on the tongue two or three times, at intervals of four hours; fomentations of hops or poppy-heads (warm) should be applied to the loins, and occasional enemas of warm water may be thrown into the rectum. The drink should consist of what is known as flaxseed or slippery-

(224)

elm tea. Soon the inflammatory symptoms will subside, the patient will manifest some relief from pain, and the color of the urine will change, become lighter and thicker in consistence. We then discontinue the above treatment, and administer one ounce of fluid extract of buchu, morning and evening. This treatment, aided by rest and good nursing, usually completes the cure.

Should it be suspected that the animal has a fit of pain, caused by the "gravel," or passage or presence of urinary calculi, then two drachms of muriatic acid should be mixed in the ordinary drink, every time the animal is watered. He should also have a small quantity of powdered slippery-elm or flaxseed mixed with the food. Horses the subjects of urinary calculi pass urine which, on being caught in an earthen vessel, deposits phosphates and other earthy matter. When this occurs, and the animal has a fit of pain or gravel, we may infer, in the absence of more positive proof, that urinary calculi are present in some portion of the urinary apparatus.

INFLAMMATION OF THE BLADDER (CYSTITIS).

The principal symptom of inflammation of the bladder is frequent urination, accompanied by straining and pain. Sometimes the urine dribbles away, involuntarily or not, as the case may be. It appears that the least distension of the bladder causes pain; hence the effort to keep it empty. The urine is usually high-colored, or, rather, of a dull red color. The animal stands with his hind limbs widely separated. The treatment is precisely the same as that just recommended for inflammation of the kidneys.

STONE IN THE BLADDER.

One very remarkable symptom attending the presence of calculus, or stone in the bladder, is, that after the urine is voided there comes a painful sensation, which causes the horse to groan. This is caused by the walls of the bladder coming forcibly in contact with the calculus, which is now a foreign body. Occasionally the horse is urinating a full stream, when, all at once, the stream is suddenly arrested, the animal still straining until urination again commences. This is a pretty sure sign of stone in the bladder. Stones that have been taken from the bladder, after death, present a highly-polished surface; hence they do not produce that amount of irritation which proceeds from intestinal calculi, the surfaces of which are always rough.

Mr. YOUATT says that "the symptoms of stone in the bladder much resemble those of spasmodic colic, except that, on careful inquiry, it will be found that there has been much irregularity in the discharge of urine, and occasional suppression of it. When fits of apparent colic frequently return, and are accompanied by any peculiarity in the appearance or the discharge of the urine, the horse should be carefully examined. For this purpose, he must be thrown. If there is stone in the bladder, it will, while the horse lies on his back, press on the rectum. Several cases have lately occurred of successful extraction of the calculus, but to effect this it is always necessary to have recourse to the aid of a veterinary practitioner."

It is not necessary to cast the horse in view of ascertaining the presence of calculus. The hand can be introduced into the rectum while the horse is standing. The rectum should, however, first be evacuated by enemas of warm water. But we have a better plan than this. A sound or jointed catheter has been invented, which, when passed into the bladder, determines, by percussion, the presence of calculi. The operation of lithotomy can only be performed by one skilled in the science; yet the reader may desire to know something about the mode of performing it; therefore I introduce a case of the kind, which occurred in the practice of Mr. FIELD. The following is the case:

"The animal was cast, in the usual manner, and both hind legs were drawn to the shoulders, as if for castration. Read's new flexible catheter being passed into the bladder, a quantity of warm water was injected sufficient to distend that organ and the urethra moderately. The catheter being withdrawn, and holding the penis with the left hand, a slightly-curved grooved staff, two feet long, was introduced, so as for the curved part to come into the subanal portion of the urethra, above the posterior edge of the ischium, extending toward the sphincter ani. An assistant, kneeling on the left side of the horse, drew the penis forward with his left hand, and gently pushed the staff backward with the right, at the same time keeping the groove exactly beneath the raphe. This elevated the portion of the urethra to be incised. I then made an incision, a line from and on the right side of the raphe, through the sk... and fascia, extending the length of from three to four

15

inches; and, pushing the penis a little on one side, I gradually divided the muscular and spongy portion, and exposed the mucous membrane of the urethra, when the finger readily detected the groove of the staff, into which a small incision was made sufficient to admit the *bistoire cachée*, following which with the index finger of the left hand, the membrane was divided to the rectum. Very little blood flowed, and the water of the urethra gushed out. The staff being removed, I easily introduced the small forceps through the urethra into the bladder, and grasped the stone, a portion of which flaked off. The large forceps were then employed, and, my brother holding the handles, I directed the blades upon the stone, my left hand being in the rectum. Having placed the stone in a proper position, I grasped it with the forceps, and, with both hands, gave it a half-turn, so as to place its widest axis between the pubis and rectum; and thus, with a moderate force, I gradually and evenly drew it out, the neck of the bladder readily dilating. Two stitches were inserted in that part of the incision nearest the anus, the lower part being left to itself."

Suppression of Urine.

Suppression of urine signifies that condition in which no urine is either secreted or voided. The affection is due to either functional or organic disease of the kidneys, or it is an accompaniment of various forms of disease, and, in such cases, is termed functional. For example, I lately treated a horse the subject of jaundice; he did not pass a drop of urine for two days. I inferred that the kidneys were inactive, and did not secrete urine, therefore he had none to pass; but after this period, a slight improvement in the disease (jaundice) having taken place, he urinated, yet at first only in small quantities.

When the affection comes on periodically, it indicates organic disease of the kidneys, and, finally, is apt to prove fatal. Watson says "that if no urine be separated from the blood, coma soon supervenes, and death. It is believed that these consequences result from the detention of urea in the system. Urea is a mere excrement, which, in health, is removed from the blood by the kidneys as fast as it enters that fluid. When it is not so carried off, it accumulates in the blood, circulates with it to every part of the body, and acts as a poison, especially upon the brain.

To render it thus poisonous, however, its decomposition in the blood appears to be requisite.

This is one of the numerous instances showing that the carrying fluid of the body may become the vehicle of disease and death, if it be not duly purged of deleterious matters which pertain to the unceasing processes of organic life. If carbonic acid be not extricated by the lungs, the animal functions are as certainly and almost as speedily extinguished by that gas as the flame of a taper might be, and we know that when the outlet of the liver is shut up, when the blood is not pumped from the excrementitious bile, the powers of animal life are weakened and sometimes utterly and rapidly destroyed."

Frequent complaints are made to me, by horsemen, that their horses do not urinate sufficiently, and I am requested to furnish diuretics. I usually advise the former not to feel alarmed, for Nature regulates these matters, and, in time, all will be well; provided, however, good food is furnished, and proper care is taken of the animal. The subject being one of importance, both as regards man and horse, I here introduce the following article from WATSON's "Lectures:"

"Suppression of urine, for a considerable time, is not, however, necessarily and universally fatal. Patients laboring under the epidemic cholera would secrete not a drop of water for some days, and yet recover. It was remarkable how entirely free such patients were from any approach toward coma. Was the urea here drained off from the blood in the enormous and unnatural flux from the stomach and bowels? I think it probably was; but chemical search has not detected that substance in the fluids so effused. Schmidt thinks, indeed, that it would be found but for its rapid decomposition into carbonate of ammonia. There are, however, some very singular instances on record of persons who have passed days and even weeks without secreting urine, and without showing any other indication of impaired health. What degree of credit such narratives deserve I do not know; but assuming that there was neither fraud nor mistake, it may be suspected that either the natural secretion was compensated by some vicarious or supplemental discharge, or that a small quantity of urine was actually separated by the kidneys. 'If any water, however small the quantity,' remarks Sir Henry Halford, 'had been made in these cases, I should have thought it possible that the patient might have recovered; for it has often surprised me to observe how small has been the measure of that excrementitious fluid which the frame has sometimes thrown off, and yet preserved itself harmless. But the cessation of the excretion altogether is universally a fatal symptom in my experience, being followed by oppression on the brain.' The same eminent physician states that in three of his five cases there was observed a remarkably strong urinous smell in the perspiration for twenty-four hours before death. This I believe is of common occurrence in such cases. Other patients have vomited, or passed by the bowels, watery matters possessing some of the sensible qualities of urine; and a urinous fluid is said to have been found in the ventricles of the brain in some of the fatal cases.

I have spoken of suppression of urine as a malady, though it probably is never any thing more than a symptom; yet it is one of those symptoms which, from our uncertainty respecting their origin and determining cause, we are obliged to treat and to study as if they were substantive diseases. In the only well-marked instance that I have seen of suppression of urine coming on in an apparently healthy person, some blood had appeared in the urine for a day or two before the secretion was totally suspended, and the kidneys were found gorged with blood. Extreme congestion or inflammation of the substance of the gland is probably at the bottom of many of these cases. The same train of symptoms supervenes not unfrequently upon organic renal disease. They happen, too, sometimes, when the ureters become impervious from disease, or from impacted gravel. In this condition urine continues to be secreted, for a time at least, and distends the ureter behind the seat of the obstruction. The apoplectic state which ensues may arise from a reabsorption of the secreted fluid; or, in consequence of the obstacle, the secretion itself, after going to a certain point, may stop, and then the case becomes a case of suppression."

Treatment.—Persons desirous of administering medicine for the treatment of this affection, are advised to give half an ounce of powdered chlorate of potass, dissolved in the drink, every night, and half an ounce of fluid extract of buchu every morning.

RETENTION OF URINE.

Retention and suppression of urine are terms very often confounded with each other. We have just shown what is meant by suppression, and now it remains for us to understand the true nature of retention. Retention signifies an undue delay of natural discharges. The secretion of urine may be active as ever, yet the animal has not the power to void it. The bladder soon becomes distended beyond its physiological capacity. Its muscular fibers are overstretched, and thus have not the power to contract again until the fluid is evacuated by means of the catheter; hence the treatment of such a case as this pertains more to the art of surgery than that of medicine.

Treatment.—In most of these cases we find the neck of the bladder spasmodically contracted. The spasm may be either primary or it may accompany spasmodic colic, which is often the case. In fact some animals, when suffering from retention of urine, act just as if they had colic; hence it is, in such cases, highly necessary that the bladder be examined by introducing a hand into the rectum. By this means, should the bladder be distended, it can easily be discovered. The catheter must then be introduced, or the animal will die from rupture of the bladder. Provided no catheter should be at hand, I should throw into the rectum copious enemas of warm water, and administer one or two ounces of tincture of assafetida as an antispasmodic, which may possibly have the desired effect.

DIABETES, OR PROFUSE STALLING.

Many veterinary writers contend that diabetes is caused by the administration of diuretics, or else in consequence of improper food. In some instances this may be true. Errors of this kind may produce an excessive flow of urine, but this does not constitute diabetes.

Symptoms.—The characteristic symptom of diabetes is a most remarkable change in the quality of the urine. It becomes loaded with sugar. It has been found that in true diabetes the urine is never without sugar. The most ready test for sugar in urine is as follows: Invert a test-tube, filled with urine, to which a small quantity of yeast has been added, into a saucer, also containing urine; set the mixture in a warm place, and, if sugar be present, fermentation will soon commence, and carbonic acid rising in the tube will depress the upper surface of the urine.

Very few experiments have ever been made on the diabetic urine of horses; but large quantities of saccharine matters have been obtained from the diabetic urine of man, by a process of evaporation. WATSON tells us, in his "Lectures," that he has seen large flat cakes of beautifully crystallized diabetic sugar. "It differs somewhat from common sugar, the produce of the sugar-cane, and approaches more nearly to the sugar of grapes. This kind of sugar, which may also be produced artificialy from starch, chemists have named glucose. By rapid evaporation of the water, a thick syrup is produced, resembling treacle; but Dr. Macintyre, who has presented to our hospital museum some very fine specimens of this sugar, prepared by Dr. Blandford, informs me that to get it well crystallized, the evaporation in a steam-bath should be stopped while the urine is of thin consistence. It may be quickly reduced to one-half, perhaps, of its original quantity; then it should be set aside, in shallow plates, and in the course of ten days or a fortnight the sugar will be deposited."

An animal the subject of diabetes is usually very thirsty. The urine is light-colored, almost transparent. It has not the ordinary odor of common urine, but something like musty hay.

Treatment.—The indications in the treatment of this affection are, to give tone to the system and sustain the general health. A drachm or two of the sulphate of iron may be occasionally mixed with the oats (the latter must be of the best quality), and one ounce of the fluid extract of buchu may be given every night.* Good wholesome food and an occasional drink of slippery-elm tea are also indicated. Should the disease not yield to such treatment as this, the case may be considered incurable.

Cause.—In regard to the cause of diabetes, very little is known; but, to set the matter right in the minds of some who believe that nothing but diuretics and inferior provender excite it, I offer the following quotation from the pen of the author just named. I think, however, that bad food is most likely to produce diabetes.

* The buchu is not a direct diuretic, like resin or niter; therefore it may be given with safety. It acts as a tonic and sudorific, and operates physiologically on the kidneys.

"It is probable enough that the exciting cause of diabetes may sometimes lie in the digestive organs, as Mr. McGregor supposed The results of his experiments do not conflict with M. Bernard's. Mr. McGregor, you may remember, found sugar in the partly-digested food brought up from the stomach of a diabetic patient. He detected it also in the saliva; and in the feces, which, when allowed to dry spontaneously, became covered, after the lapse of some time, with distinct crystals of sugar. And yeast having been administered to two diabetic patients, in ounce doses, after each meal, had soon to be discontinued, because the patients, to use their own expressions, felt as if they 'were on the eve of being blown up.' There being sugar in the blood, we need not be surprised that he met with it in the gastric and intestinal secretions from the blood. Dr. Harley has observed that by injecting irritating matters into the portal vein, (ammonia, ether, chloroform, alcohol,) a saccharine condition of the urine may be artificially produced. It is conjectured that these substances act upon fibers of the pneumogastric nerve, whence an impression is transmitted to the nervous centers, and thence is again reflected upon the liver through the splanchnic nerves. It is not difficult to imagine that irritating substances may find their way into the portal blood through a faulty digestion, or through the use of certain kinds of food or of medicine. Again: since contrived irritation of the brain at the origin of the pneumogastric nerves will make the urine saccharine, the cause of diabetes in the human subject may reasonably be placed, in some instances, within the skull; and we may understand how injuries or diseases of the brain, or even mental disquiet and dejection, operating through the brain, may produce it. Some striking cases have been published by Dr. Goolden, in which head symptoms were accompanied by saccharine urine, and in which the diabetic symptoms were checked or removed by remedies addressed to the head affection—by blisters especially, and by purgatives. Nay, we may ask whether there may not, in fact, be two varieties of diabetes mellitus, in one of which the animal and in the other the vegetable form of sugar may be present in the urine, and whether the one of these varieties may not be more hopeful of cure or recovery than the other. Bearing in mind the name and the distribution of the pneumogastric nerve, may we not indulge the conjecture that disease or injury of the brain near the origin of that nerve may directly affect the functions of the stomach, and thus prevent its digestive power, or the functions of the lungs, and thus interfere with the chemical destruction of sugar in those organs? This last notion might seem to receive support from the frequent association of pulmonary disease or disorder with saccharine urine."

RUPTURE OF THE BLADDER.

It is well known that rupture of the bladder may arise from over-distension with urine; yet it may occur from other causes, as the following article, contributed by Dr. C. M. WOOD, will show:

"*September* 1.—My subject was a bay horse, seven years old, in high condition; that is, fat. On inquiry, I was informed, by the man who had the care of him, that when he entered the stable, at four o'clock, A. M., he found the horse rolling and tumbling, and in great pain. I asked, What has the horse been doing? and was answered that, having fallen into other hands, he had done little more than merely exercise for the past four months. On examination, the following symptoms were observed: The animal was standing, with the near fore extremity raised and extended, and in the act of pawing, but he did not disturb his bedding. He would suddenly place one foot down and raise the opposite one, when he would continue to paw, in like manner, for two or three minutes. He would then cease pawing, and extend himself in his stall, with the head elevated, and a fixed stare, as if in the act of urinating. He would then perhaps lie down at full length in his stall, or throw his head back upon his side, and remain in this position a few minutes; would again rise and commence pawing, as before. When standing, his hind extremities were wide apart. His pulse was forty-two; mouth, hot, but moist; the visible mucous surfaces slightly injected; his respiration little disturbed; in fact, his symptoms were not violent. I gave an antispasmodic drench, and left him. At eight, A. M., again visited my patient. Found him standing, apparently relieved, yet he occasionally manifested slight symptoms of a kind of dull abdominal pain. His bowels being costive, I gave him one ounce of cathartic mass, one drachm of calomel, and half a drachm of pulverized opium, in a ball. At noon he was still the same. Ordered him a bran-mash, which he did not eat; nor did he drink. Gave an enema, etc.

ulated the abdomen, and left him. Six, P. M.—He has laid down
quietly for two hours; stands easy, but neither eats nor drinks.
Nine, P. M.—Still the same; lies down occasionally. When
standing, paws considerably. Has had no evacuation of the bowels or of the urinary passages since the man discovered that the
animal was sick. Examined the rectum, but found no feces therein.
Gave half a drachm each of Venice turpentine and copaiba, in an
infusion of linseed; gave also an enema, and left him for the night,
leaving a pail of water in the manger before him.

September 2, six o'clock, A. M.—On entering the stable, my
patient appeared much the same as when I left him the previous
evening. I inquired how he had been during the night. The
groom informed me that 'he had remained easy until about one
o'clock, when he commenced pawing, as before, but without touching his bedding; that he would lie down for a few minutes quite
easy, when he would throw himself at full length in his stall, and
remain in that position a short time; then rise up and extend
himself, as if to relieve the urinary passages, but only a few drops
dribbled away, as usual.' As usual? said I. Have you never
observed any difficulty in his making water before? He answered,
'I have taken care of this horse for over two years, and have
never seen him make more than a wine-glassful of water at any
one time. He passed a little every ten or fifteen minutes. His
stall was constantly wet, but I have never seen any trouble on
account of it.' This statement was fully corroborated by several
other persons. I now examined my patient per rectum, and found
the bladder empty. Upon introducing the catheter not a drop of
urine was obtained, nor even could the odor of urine be detected
on the catheter. I now concluded that he had a ruptured bladder,
and gave up all hopes of his recovery. However, from the pain
at intervals, and apparent relief of the animal occasionally, I directed my treatment to the kidneys and bladder, by giving oily
laxatives, emollient enemas, and counter-irritation, in the hope,
not of cure, but of palliation, and each, in turn, gave relief for a
short time. Nine, P. M.—Have seen my patient several times
during the day. His symptoms increasing in frequency and violence. Pulse, sixty-six; respiration, hurried; body and extremities, still warm. He paws, rolls, and tumbles violently. Gave
an enema, which had the effect to remove the contents of the rectum, consisting of four or five small pellets of feces, covered with

234   DADD'S VETERINARY MEDICINE AND SURGERY.

mucus. He now, for the first time, drank six or seven swallows
of water, but had no desire for food. I ordered a pail of water
to be placed before him, and that a man should watch him during
the night.

September 3, six, A. M.—Has drank the water. Still paws,
lies down, suddenly rises, paws again, and thus continues. The
body is still warm; he is not at all violent, his pain being, as it
were, dull; the abdomen rapidly increasing in size. Nine, P. M.—
Have seen my patient several times since morning. See no possible chance for recovery. His respiration is quick and laborious;
pulse, seventy-four; visible mucous surfaces highly injected; head
and neck bedewed with a cold, clammy perspiration; tremor of
the fore extremities; still paws, but does not lie down. He puts
his nose in the bucket, plays with the water in it, but does not
drink. Abdomen appears to increase in size. Notwithstanding
every means was used for his relief, the symptoms continued to
increase in severity, and he died at one o'clock, A. M., on the
fourth.

Made an examination thirty-six hours after death, the knacker
neglecting to remove the horse before. On opening the abdomen the following appearances were visible: A slight blush of
inflammation pervaded the whole of the intestinal canal. Extreme distention of the cæcum and colon, but their contents were
pultacious. The bladder quite empty, and contracted at its cervix
into firm rugæ; in substance its walls thickened to half an inch;
its apex bore marks of ulceration, with a rupture of one and a half
inches through its parieties. Had it not been ruptured, I think
the bladder could not have been made to contain more than a
quart; and we think this condition of the organ is sufficient to
account for the constant dribbling spoken of by the groom. There
was an accumulation of several gallons of fluid in the abdominal
cavity, and the peritoneum bore evident marks of inflammation,
which, doubtless, had been rendered less solvent by the medicines
administered. The lining membrane of the bladder was clothed
with a kind of mucus, thick, muddy, and of a dark color. The
kidneys were almost destitute of the investing membrane. What
remained rubbed off with the slightest touch. In attempting to
remove them from their connections, they were torn with very
little force. They were of a brick red color, extremely soft, so
that they could be squeezed between the fingers like a paste: or,

when laid upon a board, they flattened like soft dough. The liver was softened, or, as it is generally called, rotten. It appeared pale, was of a clay color externally, and could be broken down with the slightest pressure. The stomach was healthy, and contained about four quarts of fluid. The thoracic viscera healthy; heart, large and firm; lungs, sound, but discolored, doubtless from lying so long after death.

An idea suggests itself in regard to the case now under consideration: that the liver did not properly prepare the blood for the secretion of the urine; hence the morbid condition of the kidneys and bladder. But, although these organs are shown to have been exercising a diseased secretory action, yet doubtless the primary seat of the affection was in the digestive apparatus, the stomach, and more particularly the liver. This question is not only a very important one in a pathological point of view, but it is also one of peculiar interest to the practitioner in his treatment of disease. It directs him to the proper remedies to employ, and the action of these remedies on the organs affected. In the treatment of all diseases, it is not only necessary to understand the cause, but to know the organ or viscus affected."

ALBUMINOUS URINE ("THICK WATER").

This is rather a rare disease among horses; yet, as it does occasionally occur, it may be proper to take some notice of it. Those cases which have come under the author's notice have occurred in animals of the scrofulous diathesis, most of which animals had a scurfy skin, and were otherwise unthrifty, just the subjects for disease of the kidneys; for it is well known that when the excrementitious function of the skin is interrupted, the kidneys have to perform double labor. Their function is then overtaxed; hence, disease.

Albuminous urine is often associated with various forms of disease, or, rather, is the result of disease, or may follow the use of improper medicines or bad food. Such are said to be the cause of this malady (functional) in the human; and, reasoning from analogy, we infer that the same causes operate on the horse, for all the functions of his body are carried on after the same general plan that obtains in the body of man.

WATSON teaches us "that some articles of food, and some medicines, have the effect, in some cases, of rendering the urine for a time albuminous. Perhaps it would be more correct to say that certain forms of indigestion may cause this change. Albumen has, also, been detected after a blister upon the skin, or under that general state of irritation of the skin called eczema rubrum, which is produced by the use of mercury. In the crisis of some febrile disorders, in some cases of pregnancy, of heart disease, and in epidemic cholera, the same phenomena has been observed. Whenever blood, proceeding from the long track of mucous membrane which lines the urinary organs, mingles with the urine, that fluid, of necessity, contains albumen, and coagulates, if tested by heat or by nitric acid." There is no albumen in healthy urine; neither can we recognize its presence by mere inspection. Horses often pass urine of a thick and ropy character, but that does not prove that it is albuminous. It may be loaded with morbid or excrementitious matter, yet contain not a particle of albumen. Healthy urine, when recently discharged, possesses the ordinary temperature of the body, is transparent, of a straw color, and exhales a peculiar ammoniacal odor, which it loses in cooling. About ninety-three parts in one hundred of healthy urine is water; the remainder are made up chiefly of urea, saline and organic matters. Albumen, being similar to the white of eggs, passes from the fluid to the solid state by boiling; therefore, in order to detect albumen in urine, it is only necessary to heat the suspected urine to the boiling point, when the albuminous opacity becomes visible.

A horse the subject of albuminous urine usually has a straddling gait; will stretch backward his hind legs; is stiff, and makes short turns with difficulty. He is usually thirsty and feverish, has a quick pulse, and the feces are hard and dark-colored. When the disease is of long standing, and the animal loses flesh, and the mucous membranes of the mouth appear pale, the probability is that the disease is organic, and the case may be considered incurable. In such cases, the morbid appearances after death denote degeneration, or structural change in the secreting surface of the kidneys, and in the glands also.

Treatment.—The best remedy for the treatment of this affection is fluid extract of buchu, two ounces per day. Green vegetables and carrots are also indicated.

HÆMATURIA (BLOODY URINE).

Hæmaturia consists of extravasation of blood from the urinary organs, which is mixed with the urine, giving to the latter a blood-red tinge. Small quantities of blood is very often excreted from the kidneys, in consequence of congestion of the renal blood-vessels. The affection often follows injury or sprain in the lumbar region, and may occur in consequence of active disease, scattered in either the kidneys or the bladder. Strong diuretics, such as turpentine, juniper berries, etc., are operative in producing hæmaturia. But many cases are very obscure. Blood is sometimes voided with the urine without any assignable cause. I have known mares, when menstruating, to pass considerable blood with the urine, which ceased on the termination of menstruation.

Treatment.—The proper method of treating this affection is to apply warm-water dressings to the loins, and drench the patient with one ounce of fluid extract of matico, night and morning. Mucilaginous drinks are also needed, which may be composed of, slippery-elm, flaxseed, or gum arabic.

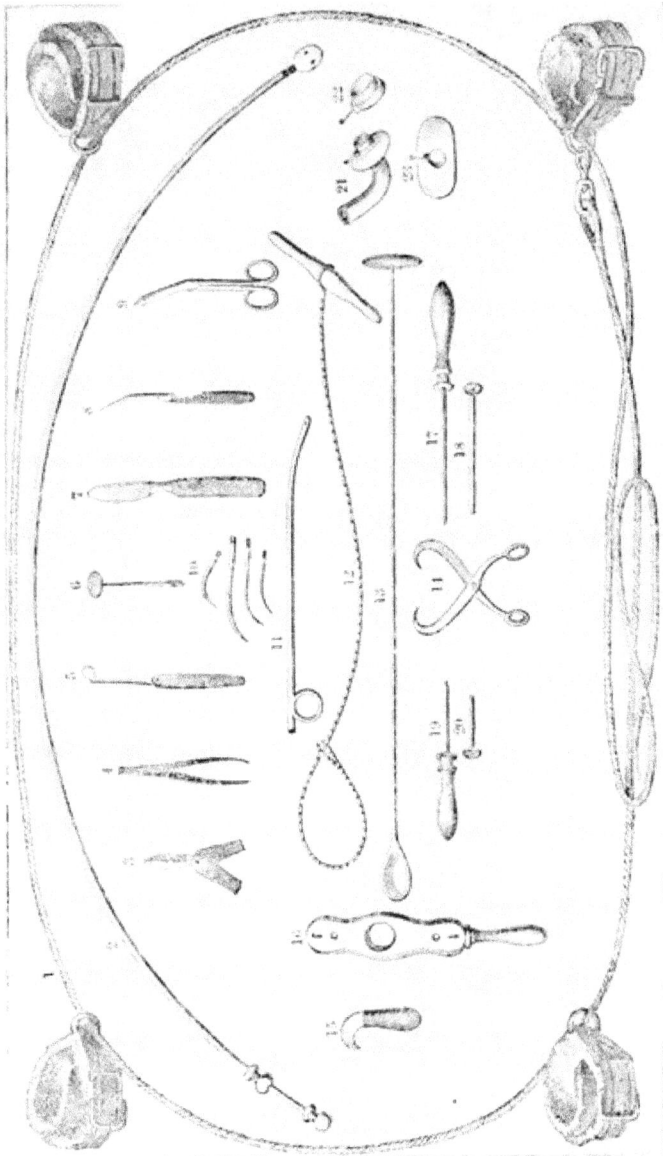

SURGICAL INSTRUMENTS USED IN VETERINARY PRACTICE.

EXPLANATION.—Fig. 1, Hobbles for casting; 2, Probang, for the removal of obstruction within the œsophagus; 3, Thumb lancet; 4, Artery forceps; 5, Gum scarificator; 6, Tube used for obstruction in the teats; 7, Castrating knife; 8, Beak-pointed scalpel; 9, Curved scissors; 10, Suture-needles; 11, Female catheter; 12, Parturient slip-noose; 13, Parturient lever; 14, Embryotomy forceps, or clasp-hooks; 15, Embryotomy knife; 16, Mouth-gag; 17, Trocar; 18, Canula or cover of the trocar; 19, 20, Trocar and canula for puncturing the chest; 21, 22, 23, Views of the several parts of the tracheotomy tube.

SECTION IX.

SURGICAL OPERATIONS.

Of Surgical Operations and the various Restraints it is sometimes necessary to place the Horse under for their performance—Casting—Slinging—Castration—French Method of Castration—Castration by Ligature—Lithotomy—Tracheotomy—Œsophagotomy—Neurotomy—Mode of Performing Neurotomy—Periosteotomy—Division of the Flexor Tendons—Amputations—Amputation of the Penis—Amputation of the Tail—Nicking—Firing—Blistering—Ammoniacal Blister—Rowelling—Setons—Abstraction of Blood, or Bleeding—Division of the Temporal Artery—Bleeding by the Palate—Bleeding by the Toe—Phlebotomy—Bleeding by the Jugular Vein.

Of Surgical Operations and the various restraints it is sometimes necessary to place the horse under for their performance.

" WHEN it is necessary to perform any painful operation on so powerful an animal as the horse, it is of consequence to subject him to a restraint equal to the occasion. Horses are very dissimilar in their tempers, and bear pain very differently; but it is always prudent to prepare for the worst, and few important operations should be attempted without casting. Humanity should be the fundamental principle of every proceeding, and we ought always to subject this noble animal to pain with reluctance; but when circumstances absolutely call for it, we should joyfully close our hearts to all necessary suffering. The resistance of the horse is terrible, and it is but common prudence to guard against the effects of it. The lesser restraints are various. Among them may be first noticed the TWITCH. The twitch is a very necessary instrument in a stable, though, when frequently and officiously used, it may have the ill-effect of rendering some horses violent to resist its application. In many instances blindfolding will do

more than the twitch; and some horses may be quieted, when the
pain is not excessive, by holding the ear in one hand, and rubbing
the point of the nose with the other. A soothing manner will
often engage the attention and prevent violence; but it is seldom
that either threats or punishment render an unruly horse more
calm. Inexperienced persons guard themselves only against the
hind legs; but they should be aware that some horses strike terribly with their fore-feet. It is prudent, therefore, in all operations, to blindfold the animal, as, by this, he becomes particularly
intimidated, and if he strikes he can not aim. When one of the
fore extremities requires a very minute examination, it is prudent
to have the opposite leg held up (it may, in some cases, be tied);
and when one of the hinder feet is the object of attention, the fore
one of the same side should be held up, as, by this means, the animal is commonly prevented from striking. If this precaution be
not taken, still observe to keep one hand on the hock, while the
other is employed in what is necessary, by which means, if the
foot become elevated to kick, sufficient warning is given, and the
very action of the horse throws the operator away from the stroke.
Without the use of these arts, the practitioner will expose himself
to much risk. The TREVIS is the very utmost limit of restraint,
and is seldom used save by smiths, to shoe very violent and powerful horses. Whenever recourse is had to it, the greatest caution is
necessary to bed and bolster all the parts that are likely to come
in contact with the body. On the Continent we have seen horses
shod in this machine, and apparently put into it from no necessity
greater than to prevent the clothes of the smith from being dirtied.
Horses have been destroyed by the trevis, as well as by casting;
or their aversion to the restraint has been such, they have died
from the consequences of their own resistance. The SIDE-LINE is
now very generally used, not only in minor operations, but also in
those more important. Many veterinarians do not use any other
restraint than this, in which they consider there is safety both to
the horse and to the operator. It is applicable to such horses as
are disposed to strike behind, and consists in placing a hobble-strap
around the pastern of one hind leg, and then carrying from a web
collar, passed over the head, the end of a rope through the D of the
hobble, and back again under the webbing round the neck. A
man is then set to pull at the free end of the rope, by which the
hinder leg is drawn forward without elevating it from the ground.

By this displacement of one leg the horse is effectually secured from kicking with either. Occasionally it is thus applied: hobbles are put on both hind legs, and the rope is passed through each of the rings. According to this last method, the horse is actually cast, as he must fall when the ropes are pulled. Take a long rope, and tie a loop in the middle, which is to be of such a size as it may serve for a collar; pass the loop over the head, letting the knot rest upon the withers; then take the free ends, pass them through the hobbles, and bring it under the loop. Let two men pull at the ropes, and the hind legs will be drawn forward.

CASTING.

The objections to this practice arise from the dangers incurred by forcing the horse to the ground. Mr. Bracy Clark simplified casting by inventing some patent hobbles, having a running chain instead of rope, and which, by a shifting D, made the loosening of all the hobbles, for the purpose of getting at a particular leg, unnecessary. These were still further improved by Mr. Budd, so as to render a release from all the hobbles at once practicable. Hobble leathers and ropes should be kept supple and pliant with oil, and ought to be always examined previous to using; nor should the D or ring of the strap be of any other metal than iron. Brass, however thick, is brittle, and not to be depended on. To the D ring, or ring of one pastern hobble, a chain of about four feet long is attached; to this a strong rope is well fastened, and, according to the way the horse is to be thrown, this hobble is to be fixed on the fore-foot of the contrary side. The rope is then passed from the hobble on the fore-foot to the D of the hind foot of that side, then to the other hind foot, and, lastly, through the D of the other fore-foot. After this, much of the ease and safety of the throw depends on bringing the legs as near together as possible. This should be done by gradually moving them nearer to each other, without alarming the horse, which will very much facilitate the business, and is really of more moment than is generally imagined. A space sufficiently large should be chosen for the purpose of casting, as some horses struggle much, and throw themselves with great violence a considerable way to one side or the other; and they are able to do this if the feet have not been brought near together previous to attempting the cast. The place should be also very well

16

littered down. The legs having been brought together, the assistants must act in concert. One, particularly, should be at the head,
which must be carefully held throughout by means of a strong
snaffle-bridle; another should be at the hind part, to direct the
fall, and to force the body of the horse to the side which is requisite. Pursuing these instructions, the animal may be at once rather
let down than thrown, by a dexterous and quick drawing of the
rope, the whole assistants acting in concert. The moment the
horse is down, the person at the head must throw himself upon
that member, and keep it secure; for all the efforts of the animal
to disengage himself are begun by elevating the head and fore-parts.
The rope is tightened. The chain is fixed by inserting a hook
through one of the links, of sufficient size not to pass the hobbles.
When the operation is over, the screw which fastens the chain to
the hobble, first put upon one fore-leg, is withdrawn. The chain
then flies through the D's of the other hobbles, and all the legs
are free, save the fore-leg first alluded to; the strap of this has to
be afterward unbuckled. There are also other apparatus used in
casting, as a strong leathern case to pass over the head, serving as
a blind when the animal is being thrown, and as a protection
against his rubbing the skin off his eyes when down. Then a surcingle is also used. This is fastened round the horse's body, and
from the back hangs a broad strap and a rope. The strap is fastened to the fore-leg of that side which it is desired should be
uppermost. The line is given to a man who stands on the opposite side to the generality of the pullers. On the signal being
given, the men having hold of the hobble-rope pull the legs one
way, while he who has hold of the rope attached to the surcingle
pulls the back in a contrary direction, and the horse is immediately cast.

SLINGING.

This is a restraint which horses submit to with great impatience,
and not without much inconvenience, from the violent excoriations
occasioned by the friction and pressure of the bandaging around
his body. Graver evils are also brought about by the abdominal
pressure. Some horses stale and dung with difficulty when suspended, and inflammation of the bowels has not unfrequently
come on during slinging. The slings are, however, forced on us
in some cases, as in fractured bones, the treatment of open joints,

and some other wounds where motion would be most unfavorable to the curative treatment. Suspension may be partial or complete. Suspension of any kind will require the application of pulleys and ropes affixed to the beams, that the whole body of the horse may be supported. A sling may be formed of a piece of strong sacking, which is to pass under the belly, the two ends being fastened firmly to pieces of wood, each of about three feet long, and which are to reach a little higher than the horse's back. To the pieces of wood cords and pulleys are to be firmly attached, by which means the sacking can be lowered or raised at pleasure. To the sacking, also, are to be sewn strong straps, both before and behind, to prevent the horse sliding in either direction, without carrying the sacking with him. Upon this so-formed cradle he is to recline. If horses when they are fresh should be placed in this machine, most of them would either injure themselves or break through all restraint. However, by tying up their heads for three or four nights their spirit is destroyed. The slings may then be applied without the fear of resistance. It is the best method not to pull the canvas firm up, but to leave about an inch between the horse's belly and the cloth, so that the animal may stand free, or throw his weight into the slings when he pleases. In this fashion a horse may remain for months in the slings, and at the end of the time display none of the wear and tear so feelingly described by old authors.

CASTRATION.

This practice is of very ancient origin, and is as extensive as ancient. It is founded on the superior placidity of temper it gives. The castrated horse no longer evinces the superiorities of his masculine character, but approaches the softer form and milder character of the mare. Losing his ungovernable desires, he submits to discipline and confinement without resistance; and, if he be less worthy of the painter's delineation and the poet's song, he is valuable to his possessor in a tenfold degree. In England, where length in the arms and of the wide-spread angles of the limbs is absolutely necessary in the horse to accomplish the rapid traveling so much in vogue among us, the exchange of the lofty carriage and high action of the stallion is absolutely necessary; and when we have added the lessened tendency of the gelding to some diseases, as hernia, founder, cutaneous affections, etc., we may be contei : to leave the sexual type with the racer for his breed; also with the dray-horse for his weight, and the fancy of his owner.

Supposing it, therefore, eligible to castrate our horses, what is the proper age for the operation? What are the relative advantages and disadvantages of the different methods of performing it? The proper age to castrate the young horse must depend on circumstances, as on his present appearance, his growth, and the future purposes we intend him for, observing, generally, that the more early it is done, the safer is the operation; for, until these organs begin to secrete, they are purely structural parts, and, as such, are not so intimately connected with the sympathies of the constitution. Some breeders of horses castrate at twelve months, others object to this period, because they think the animal has not sufficiently recovered the check experienced from weaning before this new shock to the system occurs. In the more common sort of horses, used for agricultural purposes, it is probably indifferent at what time the operation is performed, this consideration being kept in view: that the earlier it is done, the lighter will the horse be in his forehand; and the longer it is protracted, the heavier will be his crest, and the greater his weight before, which, in heavy draught-work, is desirable. For carriage-horses it would be less so, and the period of two years is not a bad one for their castration. The better sort of saddle-horses should be well examined every three or four months, particularly at the ages of twelve, eighteen, and twenty-four months, at either of which times, according to circumstances or to fancy, provided the forehand be sufficiently developed, it may be proceeded with. Waiting longer may make the horse heavy; but, if his neck appear too long and thin, and his shoulders spare, he will assuredly be improved by being allowed to remain entire for six or eight months later. Many of the Yorkshire breeders never cut till two years, and think their horses stronger and handsomer for it. Some wait even longer; but the fear in this case is, that the stallion form will be too predominant, and a heavy crest and weighty forehand be the consequence. Perhaps, also, the temper may suffer. Young colts require little preparation, provided they are healthy and not too full from high living. If so, they must be kept somewhat short for a few days; and, in all, the choice of a mild season and moderate temperature is proper.

When a full-grown horse is operated on, some further preparation is necessary. He should not be in a state of debility, and, certainly, not in one of plethora. In the latter case, lower his diet; and it would be prudent to give him a purgative. It is also advisable that it be done when no influenza or strangles rage, as we have found the effects of castration render a horse very obnoxious to any prevalent disease. The advanced spring season—previous, however, to the flies becoming troublesome—is the proper time for the performance of the operation upon all valuable horses; and be careful that it be not done until after the winter coat has been shed, which will have a favorable effect on the future coating of the horse, independent of the circumstance that at a period of change the constitution is not favorable to unusual excitement.

Castration is performed in various ways, but in all it expresses the removal of the testicles. There are methods of rendering the animal impotent without the actual destruction of these organs; for if, by any other method, the secretion of the spermatic glands is prevented, our end is answered.

Castration by cauterization is the method which has been principally practiced among us; but this by no means proves it the best. On the contrary, many of our most expert veterinarians do not castrate by this method. Mr. Goodwin, and many other practitioners of eminence, never castrate by cautery.

A preliminary observation should be made previously to casting, to see that the horse is not suffering from a rupture. Such cases have happened; and as in our method we open a direct communication with the abdomen, when the horse rises it is not improbable that his bowels protrude until they trail on the ground. Hernia, as a consequence of castration, may easily occur by the uncovered operation; for, as already observed, it makes the scrotal sac and abdominal cavity one continuous opening. It is not to be wondered at, therefore, if the violent struggles of the animal should force a quantity of intestine through the rings into the scrotal bag. Should we be called on to operate on a horse which already had hernia, it is evident we ought not to proceed with it unless the owner be apprised of the risk, and willing to abide by it. In such cases, we would recommend that the method of Girard be practiced; that is, to inclose the tunica vaginalis within the clams, (sufficiently tight to retain them, but not to produce death in the part,) pushed high up against the abdominal ring, and then to remove the testicle, being very careful to avoid injuring any portion of intestine in the operation. When a discovery is made cf the existence of hernia after an opening has been already made for the common purpose of castration, should the operator continue his process, and castrate? We should say, By no means; but, on the contrary, we would greatly prefer the method recommended by Mr. Percivall: firmly to unite the lips of the external wound by suture, allowing the testicle itself to assist in blocking up the passage, with a hope, also, that the inflammation caused by the incision might altogether stop up the scrotal communication with the abdomen. But, in the appalling case of immense protrusion of intestine, what is to be done? Mr. Coleman, in such a case, proposes to make an opening near the umbilicus large enough to introduce the hand, and thus draw in the bowels. Mr. Percivall would prefer dilating the external ring. But the testicle must be very firmly retained, and even permanently fixed against the dilated ring, or the bowels would again descend. The intestines probably would become inflated in any such case.

As unbroken young horses are the most usual subjects of this operation, and as such often have not yet been bridled, if a colt can not be enticed with oats, etc., he must be driven into a corner, between two steady horses, where, if a halter can not be put on, at least a running hempen noose can be got round his neck; but whichever is used, it should be flat, or the struggles, which are often long and violent, may bruise the neck, and produce abscess or injury. When his exertions have tired him, he may then be led to the operating spot. Here his attention should be engaged while the hobbles are put on, if possible; if not, a long and strong cart-rope, having its middle portion formed into a noose sufficiently large to take in the head and neck, is to be slipped on, with the knotted part applied to the counter or breast. The long pendent ends are passed backward between the fore-legs; then carried round the hind fetlocks, brought forward again on the outside, run under the collar-rope; a second time carried backward on the outer side of all, and extended to the full length in a direct line behind the animal. Thus fettered, Mr. Percivall says his hind feet may be drawn under him toward the elbows. It has been, however, often found that, at the moment the rope touches the legs, the colt either kicks and displaces the rope, or altogether displaces himself; but his attention can generally be engaged by one fore-leg being held up, or by having his ear or muzzle rubbed, or even by the twitch; if not, the rope may be carried actually round each fetlock, which then acts like a hobble, and this rope may be gradually tightened. This last, however, is a very questionable method, and the others, therefore, ought to be long tried before it is resorted to. In this way people have succeeded with very refractory colts; but it requires very able assistants, and, if possible, the man who has been used to the individual colt should be present. In either way, as soon as the rope is fixed, with a man at each end of it, behind the colt, let them, by a sudden and forcible effort in concert, approximate his hind legs to his fore, and thus throw him. Before the colt is cast, however, it should be endeavored to ascertain that he is free from strangles and hernia.

Being satisfied that no hernia exists on either side, proceed to cast the colt, turning him, not directly on the left side, but principally inclining that way; and, if possible, let the croup be very slightly elevated. It is usual to place him directly flat on the left side, but the above is more convenient. Next, secure the near hind leg with a piece of hempen tackle, having a running noose; or, in default of this not being at hand, make use of the flat part of a hempen halter, which should, for safety, be put on before the hobble of that leg is removed—as may be readily done, if the hobbles having shifting or screw D's (as described in casting) are made use of. Every requisite being at hand, the operator, having his scalpel ready, should place himself behind the horse, as the most convenient way to perform his manipulations; and, firmly grasping the left testicle with his left hand, and drawing it out so as to render the scrotum tense, he should make an incision lengthways, from the anterior to the posterior part of the bag. The resistance of the cremaster muscle has to be overcome before the testicle can be forced to the bottom of the scrotum, and this is the more readily accomplished if the animal's attention be engaged. The incision may be carried at once through the integuments, the thin dartos expansion, and the vaginal coat of the testicles with a sweep of the scalpel; but with one less dextrous at the operation, it will be more prudent to make the first incision through the scrotum and dartos only, to the required extent, and then to do the same by the vaginal coat, thus avoiding to wound the testicle which would produce violent resistance, and give unnecessary pain.

We, however, take this opportunity of noting that cases have
occurred when the tunica vaginalis was divided no testicle followed, firm adhesions between this tunic and the tunica albuginea
having retained it fast. In such cases the scalpel must be employed to free the testicle, by dissecting it away from the vaginal
sac.   When no such obstruction occurs, the testicle, if the opening
be sufficiently large, will slip out; but the operator must be prepared, at the moment of so doing, to expect some violent struggles,
more particularly if he attempt to restrain the contractions of the
cremaster, and, by main force, to draw out the testicle.   Prepara-
tory to this, therefore, the twitch should be tightened; the attendants, especially the man at the head, must be on the alert; and
the testicle itself, at the time of this violent retraction of the cremaster, should be merely held, but not dragged in opposition to
the contraction.   If the clams have been put on over the whole,
according to Mr. Percivall's method, they will assist in retaining
the retracting parts; but they must not be used with too much
pressure.   The resistance having subsided, the clams must now
be removed; or, if they have not been previously in use, they
must now be taken in hand, and, having been prepared by some
tow being wound around them, should be placed easily on the cord,
while time is found to free from the grip of the pincers the vas deferens, or spermatic tube, which is seen continued from the epididymis.   The Russians, Mr. Goodwin informs us, cut it through where
they operate.   Humanity is much concerned in its removal from
pressure, because of the excess of pain felt when it is included.

It is necessary, before the final fixing of the clams, to determine on the part where the division of the cord is to take place.
To use Mr. Percivall's words, 'if it be left too long, it is apt to
hang out of the wound afterward, and retard the process of union.'
On the other hand, if it be cut very short, and the arteries happen to bleed afresh after it has been released from the clams, the
operator will find it no easy task to recover it.   The natural
length of the cord, which will mainly depend on the degree of
the descent of the gland, will be our best guide in this particular.
The place of section determined on and marked, close the clams
sufficiently tight to retain firm hold of the cord, and to effectually
stop the circulation within it.   There are now two modes of
making the division: the one is to sever it with a scalpel, and then
to sufficiently sear the end of it as to prevent a flow of blood: the

other, and in some respects the preferable method, is to employ a blunt-edged iron, which is to divide by little crucial sawings, so that, when the cord is separated, it shall not present a uniform surface, but ragged edges, which will perfectly close the mouths of the vessels. This done, loosen the clams sufficiently to observe whether there be any flow of blood. Gently wipe the end of the cord, also, with the finger, as sometimes an accidental small plug gets within the vessel; this had better be removed at the time. Retain a hold on the clams a few minutes longer; and, while loosening them gradually, observe to have an iron in readiness again to touch the end of the cord, if any blood makes its appearance. Satisfied on this point, sponge the parts with cold water. No sort of external application is necessary, still less any resin seared on the end of the cord, which can only irritate, and will never adhere.

On the after-treatment much difference of opinion has existed, and even yet exists. The powerful evidence of accumulated facts has now convinced us of the necessity and propriety of some motion for the newly castrated horse, as a preventive of local congestion. Such practice is common in most countries, and seems salutary in all. Hurtrel d'Arboval, thus impressed, recommends the horse, immediately after the operation, to be led out to walk for an hour; and it is a general plan in France to walk such horses in hand an hour night and morning. Mr. Goodwin, in proof of its not being hurtful, informs us that whole studs of horses, brought to St. Petersburg to be operated on, are immediately traveled back a certain portion of the distance, night and morning, until they arrive at home. We have, therefore, no hesitation in recommending a moderate degree of motion in preference to absolute rest.

The French method of castration is advocated by Mr. Goodwin, and it is sufficient that it receives his recommendation to entitle it to attention. It is rendered the more so, as he observes, on the method in general use among us, ' that the operation performed by the actual cautery always induces, more or less, symptoms that often become alarming, and that it can not be performed on the adult without incurring more swelling and severer consequences than attend other methods of operation. If I ever use the actual cautery, it is for the sake of expedition, and then only on a yearling or a two-year old; but I am resolved never to employ it again on an adult.' These observations, as emanating from such a source, must be deemed important.

Mr. Goodwin then offers the description of the French method of operating, from Hurtrel d'Arboval: 'Castration by means of the clams is the method in general use, if not the only one now employed. It is the most ancient, since it was recommended by Hieroclius among the Greeks. It is performed in two ways, the testicle being covered or uncovered. In the former, the exterior of the scrotum, formed by the skin and dartos muscle, is cut through, and the testicle is brought out by dissecting away the laminated tissue, the gland being covered by the tunica vaginalis. The clam is then placed above the epididymis, outside the external peritoneal covering of the cord. In the uncovered operation, the incision is made through the servus capsule of the testicle. The tunica vaginalis being divided, the testicle presents itself, and the clam is placed well above the epididymis, on the cord. The operation, performed in either way, requires us to provide ourselves with a scalpel, a pair of clams, a pair of long pincers, made purposely to bring the ends of the clams together, and some waxed string. The clams may be formed of different kinds of wood, but the alder is considered the best, and generally made use of. To make a clam, we procure a branch of old and dry alder, whose diameter should be about an inch, and whose length should be from five to six inches; of course, the dimensions must, at all times, be proportioned to the size of the cord we have to operate on. At the distance of half an inch from each end, a small nick, sufficiently deep to hold the string, must be made, and then the wood should be sawed through the middle lengthways. Each divided surface should be planed, so as to facilitate the opening of the clams, either when about to place them on or take them off. The pith of the wood is then to be taken out, and the hollow should be filled with corrosive sublimate and flour, mixed with sufficient water to form it into a paste. Some persons are not in the habit of using any caustic whatever; then, of course, scooping out of the inside of the clam is not necessary. Notwithstanding, the caustic, inasmuch as it produces a speedier dissolution of the parts, must be useful, and ought not to be neglected.' The addition of the caustic, however, Mr. Goodwin objects to, with great reason, remarking that, unless it be a very strong one, and therefore dangerous to employ, it can not be of any use to parts compressed and deprived of circulation and life. He further informs us that he has operated in six cases in succession with the same effect, without any escharotic matter whatever. An experimental case of Mr. Percivall's terminated fatally. By the use of caustic the cord was greatly inflamed, as high as the ring, and which, unquestionably, produced the unfortunate result.

'The covered operation,' continues Mr. Goodwin, 'is the one that I am about to advocate, and which differs only insomuch that the scrotum and dartos muscle must be cautiously cut through, without dividing the tunica vaginalis. It was Monsieur Berger who was accidentally at my house when I was about to castrate a horse, and who, on my saying that I should probably do it with the cautery, expressed his surprise that I should perform the operation in any other way than on the plan generally approved of in France. Being a stranger to it, he kindly consented to preside at the operation, and, after seeing him perform on the near testicle, I did the same on the right, but, of course, not with the same facility. After opening the scrotum, and dissecting through the dartos, which is very readily done by passing the knife lightly over its fibers, the testicle and its covering, the tunica vaginalis, must be taken in the right hand, while the left should be employed in pushing back the scrotum from its attachments; and, having your assistant ready, as before, with the clam, it must be placed well above the epididymis, and greater pressure is, of course, necessary, as the vaginal covering is included in the clam.'

Mr. Goodwin further observes that in Russia he has seen hundreds of horses operated on, even after the human fashion, with safety; and, he remarks, it certainly produces less pain, the animal loses less flesh and condition, and is sooner recovered than when operated on by the actual cautery.

Castration by ligature is a painful, barbarous, and very dangerous practice, and consists in inclosing the testicles and scrotum within ligatures, until mortification occurs, and they drop off. It is practiced by some breeders on their young colts, but it is always hazardous and disgracefully cruel. The substance of the testicle, in some countries, is also broken down, either by rubbing or otherwise by pressure between two hard bodies. This is practiced in Algiers, instead of excision, and tetanus is a frequent consequence of it. In Portugal they twist round the testicle, and thus stop the circulation of the gland. Division of the vas deferens has been performed, it is said, with success, on many animals, and is proposed as a safe and less painful process than the emasculation of the horse. It consists in a longitudinal section through the scrotum, dartos, and vaginal sheath, so as to expose the cord, from which the vas deferens is to be separated and severed from the artery and vein. There is a certain consent of parts, by which the sympathy of an organ remains after its functional offices are apparently destroyed. There can be little doubt but the nervous excitement would continue, the vein and artery remaining entire. There are certain nice conditions of the organ necessary for propagation. Thus, the horse who retains his testicles within his abdomen, possesses all the roguish qualities of him with one perfectly evolved; he is lustful, and can cover, but is seldom fruitful.

Of the morbid consequences of castration we have little to say. By early evacuations, green food, a loose box, a cool air, moderate clothing, but, particularly, by walking exercise, swellings of the parts may be prevented; if not, bleed and foment. Should suppuration follow, and sinuses form, treat as directed under those heads; and if tetanic symptoms start up, refer to that article. There has been lately practiced in India a novel mode of castration, which is said to be the invention of a Boer, settled at the Cape of Good Hope. The cord is exposed in the usual manner. From the cord the artery is singled out. This vessel is scraped through with a coarse-edged, blunt knife, when the other constituents of the cord are cut away, and the operation is finished. This method is much praised by those who have adopted it, and is said to be always attended with success.

LITHOTOMY.

Hurtrel d'Arboval's account of the progress of lithotomy in veterinary practice commences in 1774; the second case was successfully operated on in 1794; and at later periods other veterinary surgeons have also performed it. In monodactyles there are two methods of operating for the stone—one through the rectum, the other through the bladder. The first, which consists in laying open the bladder by a longitudinal incision made through the parietes of the part of the rectum adherent to it, by means of a straight bistoury, is easily practiced, but in its consequences is dangerous in the extreme; in fact, it is an operation never to be adopted but in a case where the magnitude of the stone precludes its extraction through the neck of the bladder. In all other cases, lithotomy by the urethra is to be pursued. For its performance are required a straight probe-pointed bistoury, a whalebone fluted staff, and a pair of forceps, curved at the extremities. The animal should, if practicable, be maintained in the erect posture. The tail plaited and carried round on the right quarter, the operator feels for the end of the staff introduced up the urethra, and makes an incision directly upon it, from above downward, an inch and a half or two inches in length. Next, he introduces the sound, and passes it onward into the bladder. Now, placing the back of the bistoury within the groove of the sound, by gliding the knife forward, the pelvic portion of the urethra, and also the neck of the bladder, becomes slit open—the latter in two places, in consequence of a second cut being made in withdrawing the bistoury. The opening made being considered of sufficient dimensions, the operator introduces the forceps into the bladder, and seizes the calculus, one hand being up the rectum, to aid him in so doing. The forceps, clasping the stone, are now to be withdrawn, but with gentleness, and with a vacillating sort of movement of the hand from side to side, in order more easily to surmount any difficulties in the passage, and the more effectually to avoid confusion or laceration. M. Girard tells us 'that the cut through the pelvic portion of the urethra ought always to be made obliquely to one side. The operator should hold his bistoury in such a direction that its cutting edge be turned toward the angle of the thigh. By this procedure we shall gain easier access to the bladder, and not only avoid wounding the rectum, but also the artery of the bulb, as well as the bulb itself, and suspensory ligaments of the penis.

The parts cut through in the operation are, 1st, the fine thin skin of the perineum, smooth externally, and marked with a raphe; densely cellular internally; 2d, adhering to the tissue, the fascial covering, derived from the fascia superficialis abdominis, which has here become fibrous, it forms the common envelope to the parts underneath, and is closely connected with the corpus musculosum urethræ; 3d, the corpus musculosum urethræ, that penniform band of fleshy fibers which springs by two branches from the ischiatic tuberosities embracing the sphincter ani, and concealing the arteries of the bulb, whence they unite, and proceed to envelop the urethra; 4th, the corpus spongiosum urethræ, the part immediately covered by the muscular envelope, and which here is bulbous (it is more particularly worthy our remark, from two arteries penetrating the bulb, which come from without the pelvis, ascending obliquely outward to reach the part) ; 5th, the suspensory ligaments of the penis, pursuing the course of, and adhering to, the tendinous union of the erectores.   An attention to the relative position of these parts will demonstrate the advantages of the lateral oblique incision over one made directly along the raphe.   By pursuing the latter, we necessarily cut through the suspensory ligaments and into the bulb, wounding thereby the arteries ; whereas, by the former, all this danger is avoided, besides that it renders the operation more simple and facile.

TRACHEOTOMY (CUTTING INTO THE WINDPIPE).

Cases occur when this operation is required, as in strangles, when the tumors threaten suffocation, or when any substance has remained unswallowed in the œsophagus, the pressure of which obstructs respiration.   In a distressing case of gunpowder bursting immediately under a horse's nose, the effects of which tumefied his mouth and nostrils, so as to prevent free inspiration, the animal owed his life entirely to our excising a portion from the tracheal rings, about ten inches below the angle of the throat. The operation is a very simple one, and may consist either in a longitudinal section made through two or three of the rings, or a portion, occupying about an inch round, may be excised from the anterior cartilaginous substance.   The proper mode, when it can be done, however, is to make a circular opening with a very narrow knife, removing a portion of two cartilages, or taking a semicircular piece from each ; and this last, although it is seldom performed, is by far the best method.   The integuments should be first divided in the exact center of the neck, three or four inches below the obstruction ; then the skin and tissues should be sufficiently separated to allow a tube adapted to the size of the trachea to be introduced, the tube having an acute turn and a rim, which must be furnished with holes for the adaptation of tapes, to secure it around the neck.   There are several instruments of this sort in use, of which that adopted by the French, or the one invented by Mr. Gowing, of Camden Town, is to be preferred.   The operation has been also performed in cases of roaring, under an idea of dividing the stricture which impeded respiration; but, unless the exact situation of this were discovered, it would be but an experimental attempt.

ŒSOPHAGOTOMY (OPENING THE GULLET).

It was long thought that a wound in the œsophagus must be necessarily fatal, but we have now sufficient proofs to the contrary on record, so that we are not deterred from cutting into the œsophageal tube when it is necessary; but it is an operation requiring skill and anatomical knowledge, and its future results are sometimes very serious. The cases that call for œsophagotomy are the lodgment of accidental substances within the tube. An apple once so lodged was removed by incision by a veterinary surgeon at Windsor. Carrots, parsnips, beets, etc., are liable to produce such obstruction when not sliced. Too large a medicinal mass, also, has lodged there; and a voracious eater has, by attempting to swallow too large a quantity of not salivated bran or chaff, produced an obstruction, which pressed on the trachea and threatened suffocation. In all cases of obstruction of this kind, we will suppose that a probang, well oiled, has been previously attempted to be passed, and has completely failed. The probang for the horse, however, differs materially from that used for the cow. It is formed after the fashion of the one adopted by the human practitioner, consisting of a pliable piece of whalebone, having a sponge tied to one end. The operation being determined on, may be practiced standing. If the swelling be large, no fear need be entertained about cutting important organs, as the enlargement will push them on one side. Cut down, therefore, directly upon the center of the impacted substance. If the horse be cast, which is quite unnecessary, have him, of course, thrown with his left side uppermost. It will also be necessary to command a good light. The part of the neck chosen for the opening must, of course, be governed by the obstructing mass. A section should be made through the integuments and cellular tissue beneath them, right into the œsophagus, if possible, with one cut, and into the center of the pipe. If this be not done at once, and it requires some dexterity so as to effect it, mind to make all future incisions in a line with the first opening, as it is important that the cellular tissue should be little interfered with. The œsophagus, fairly cut into the impactment, should jump forth. Should it not do so, do not manipulate, or attempt to force it out, but enlarge the opening, and the substance will come through when that is long enough; but no fingering could compel its exit while the opening is too small. The end gained for which the incision was made in the œsophagus, the wound may be then closed by the interrupted sutures, each holding a small piece of tow above the orifice, and having their ends hanging out of the external opening, which should also be brought together by sutures. The after-treatment should be to interdict all dry food; the animal ought to subsist on very thick gruel for three, four, or five days. If the condition appears to suffer much, allow malt mashes, and when so doing watch the wound; and if the matters taken in are seen to ooze out, wash them away frequently with warm water, to prevent lodgment, which might encourage sinuses to form; and after each washing, syringe with some very mild stimulant, as a very weak solution of sulphate of zinc (white vitriol), etc.

NEUROTOMY (DIVISION OF THE SENTIENT NERVES OF THE FOOT).

Neurotomy has now stood the test of very extensive application. Our writers offer innumerable proofs of its restoring almost useless animals to a state of much utility; and if there are chances that it may occasion such injury as to hasten the end of some horses, it is usually in such as the disease would have done the same for at no distant period. Having stated thus much in its favor, it must not be supposed that we recommend it as an unqualified benefit, even where it succeeds best. No neurotomized horse ever after goes with the same freedom, nor with equal safety, as he did before the operation was performed. Indifference to the nature of the ground gone over is said to have fractured legs; it is quite common to batter the feet to pieces; and, although horses have hunted afterward, and hackneys have carried their riders long distances, yet it is more calculated to prove beneficial to carriage than to saddle-horses. This we believe to be a just statement of its merits; but there are benefits which it offers to the animal of a more extensive and constitutional kind. Those gained by the bodily system generally have been, in some cases, very marked. Thus, an aged and crippled stallion, from the irritation constantly kept up, became so emaciated as to be unable to fecundate; but, being relieved from a constant state of suffering by neurotomy, improved in health and condition, and was again used to cover. It happened, also, that a mare, similarly circumstanced, ceased to feel œstrum; but after neurotomy it again returned, and she resumed her character of a brood-mare. It appears to act with most certainty when a portion of the irritated nerve is excised. One case has actually occurred where the tetanus, occasioned by a wound in the foot, was arrested and removed by neurotomy. It also promises much in the painful state of some cankers, where the irritation has rendered the application of dressings almost impossible. Here, by depriving the foot of sensibility, we deprive the horse of that which is injurious to him. The sore itself is often amended by it; but in every instance the dressings can be effectively applied, and the healthy processes can not be at all suspended.

With respect to whether the lower or upper incision ought to have the preference, the decision should be guided by the circumstances, as regards the intensity and the seat of the disease. The operation commonly leaves, for a considerable time, some enlargement around the spot, the effects of the adhesive matter interposed between the severed portions of the nerve, and which can be remedied by no application of bandages. This bulging remains so long as life continues; and, however cunningly the incision be concealed, this can be felt with ease, and tells the truth ever after the operation has been performed. Such a circumstance has, however, led some practitioners, when it has been wished to make the upper section, and yet to avoid the chance of detection, to operate on the metacarpal nerve on the outside, and on the pastern or plantar nerve on the inside.

Mode of performing the operation.—The situation of the section through the skin being determined on, a guide to which may be gained from the perforatus tendon, and having firmly secured the leg to be first operated on, cut the hair from the part. This being done, and the exact course of the artery being ascertained by its pulsation, make a section close to the edge of the flexor tendon. Let the cut be near, but rather behind, the artery, if below the fetlock joint. The cellular substance being cleared away will bring the vessels into view, and the nerve will be readily distinguished from them by its whiteness. Elevating it from the vessels and its membranous attachments, by means of a crooked needle, armed

17

with thread, pass a bistoury under it, as near to the upper angle
of the section as possible. The violent spasm the division of the
nerve produces may be somewhat lessened by pressing the nerve
between the finger and the thumb, when an opportunity may be
taken, either with the scalpel or scissors, of dividing it; then,
taking hold of the lowermost portion between a pair of forceps,
excise about three-fourths of an inch of its trunk. Having fin·
ished, if both feet are affected, proceed to operate on the contrary
side of the other leg; after which turn the horse, and repeat the
operations on the like parts of each leg as they come in succession.
The integuments may be now drawn neatly together, and secured
by a twisted suture, the whole being properly covered by a light
compress. Tie up the head for a day or two, after which put on
a cradle. Keep the horse very quiet and low; give mashes, to
open the bowels; but we should avoid physicking, from the fear
that griping might occur, which would make him restless, or probably require exercise.

PERIOSTEOTOMY (REMOVING A TUMOR FROM SURFACE OF BONE).

This operation consists in having the horse thrown upon his
side, and the leg to be operated upon released from the hobble,
and extended upon a sack, filled with refuse hay or straw. This
is done by means of a piece of webbing passed round the hoof, and
the end given to a man to hold, who pulls rather violently at the
member. The operator then kneels down and feels for the exostosis he intends to perform periosteotomy upon. This may be a
splint or a node, and commonly exists upon the metacarpal portion
of the fore-limb. The operator having found the excrescence, snips
just below it with a pair of rowelling scissors. He then takes a
blunt seton-needle and drives it through the cellular tissue, and
immediately over the enlargement. Next, another slit in the
skin, above the exostosis, is made with the rowelling scissors, and
through this last opening the point of the seton-needle is forced
and then withdrawn. Into the free space thus made a curved
knife is introduced. The point of this knife is blunt, and the
blade curves upward, the cutting part being below. Some persons use a very diminutive blade, but the editor prefers a rather
large instrument, as being more under the command of the hand.
Having introduced this knife, he turns the cutting edge downward,

and with it incises the enlargement, sending the blade right through the periosteum, and also through the substance of the exostosis, if it be not too solid for the knife to penetrate. This latter fact is only to be ascertained by actual experience, and no opinion formed after an external examination can be of any value, such being much more the guess of a pretender than the judgment of a surgeon. The age of the animal may be some guide, but even this it is better not to depend upon too entirely. It is true that young horses freely cast forth exostoses, which aged animals mostly absorb; but this rule, though very general, has exceptions, and by no means is to be absolutely depended upon.

The enlargement being cut through, next take a seton-needle, armed with a tape, and draw it through the channel already made. Tie a knot at either end of the tape, large enough to prevent its being pulled through the opening at either end, and the business is over. The affair is very simple, and the horse may be at once let up. It is, however, in some cases, and only in some, of so much benefit that the horse, being thrown 'dead lame,' gets up and trots off quite sound. However, ere you adopt the operation, apprise the owner of the risk incurred, and that it is by no means a certain cure. Leave the choice with him, but be sure and tell him the opening made for the entrance and exit of the seton commonly leave a blemish behind them; and where the seton traveled, often there remains a thickening, which it may require months to obliterate.

The after part of the treatment consists in merely having the seton daily moved to and fro, though some persons apply an active blister all over the parts immediately in the neighborhood of the seton, under the idea that the vesicatory renders the operation of greater efficacy, which, however, is very questionable. When periosteotomy acts at all, it mostly does so at once; and when its benefits are not immediate, it is better to withdraw the seton, to prevent after-blemish, rather than hazard further and useless treatment by the application of a stimulant to skin already deprived of any connection with the deeper-seated structures.

DIVISION OF THE FLEXOR TENDON.

There are so many accidents and diseases that may produce contraction of the flexor tendons, that we only wonder we do not

more often meet with them, which we should certainly do, but
that the attendants, despairing of being able to afford relief, advise their destruction.   The operation consists in making a longitudinal incision, of about three inches in length, along the inner
lateral edge of the tendon, dissecting each portion from its cellular
attachments, so as to expose the nerve, artery, and tendons. This
opening will allow the perforans to be freed from the perforatus,
when a division should be made by a scalpel applied to its surface.
It is evident that this should take place below any thickening, or
adhesions, which may have permanently connected the tendon with
the neighboring parts.   Any lesser attachments will be broken
through, by forcing back the foot to the just position.   By Mr.
Dick this was done 'by placing his knee against the front or projecting part of the pastern, at the same time laying hold of the
foot with one hand and the upper part of the leg with the other,
and using considerable force; and this appears to be necessary, in
order to break any adhesions that may have formed.'   The limb
should now be placed in a poultice; and if any fear of future contraction should arise during the cure, lengthen the toe of the shoe
proper to the foot operated on.   Some slow exercise, after the
first week, may be allowed, but previously to that the horse should
be confined to a stall, during which the bowels must be kept open
with mashes.

AMPUTATIONS.

These have been hitherto confined principally to the tail, the
ears, and other parts of minor importance in the animal frame;
but veterinary surgery now takes a wider field, and the extremities are amputated with a certainty of making horned cattle still
serviceable for the purpose of yielding milk ; and, without doubt,
the same might be done with the brood-mare or stallion, particularly in fractures of the fore extremities.

Professor Dick, of the Edinburgh Veterinary College, furnishes
a case, sent to him by one of his pupils, to the following effect :
' I performed amputation upon a cow on the 7th of July. After
having properly secured the animal, and applied a torniquet above
the carpus, I made a circular incision through the integuments
round the leg, a little below the carpus; and, having separated
the skin so as to allow of its being pushed up a little, I cut through
the sinews, and lastly sawed off the stump. The parts are now

completely whole, although she has been going at grass all the time, and, now that she has got the advantage of a cork stump, makes a wonderful shift for herself, and yields a good supply of milk to her owner.'

Mr. Dick also notices another case of amputation of the fore-leg of a two-year old heifer; and of a third, where the hind leg was removed above the tarsus. Such operations have occasionally occurred from time immemorial, with a few enterprising characters. We have heard of them, but they were mostly regarded as mere matters of curiosity or wonder, and, therefore, were not imitated. We shall, however, probably, erelong have them more common, in cows, at least; for, occurring below the carpus and tarsus, they are as easily performed as nicking or docking; and there is no doubt but, were a hollow padded stump applied, such low operations might be prudent in many cases. Fractures, with great comminution of bone, considerable ravages of disease within the foot, or extensive gangrene, are the cases which might call for amputation.

Of the method of amputation little need be added to the above. The principal practical points are, the fixing of a torniquet of sufficient force, which should be padded to make its principal pressure on the leading arterial trunks, while its general circumference will act on the smaller vessels. A ring should be cut lightly below the intended place of operation, only through the integuments, which, when separated from their cellular adhesions for about six inches, should be turned back; and a circular incision may then be made through the muscles, etc., taking up by ligature such vessels, both venous and arterial, as display a disposition for much hemorrhage. The section thus made, free the bone from the soft parts by the scalpel, where the adherences are very intimate, and, having, by means of a crucial bandage, retracted the soft parts altogether above the bone, saw it through. Finally effectually secure the principal vessels, when, bringing the soft parts and skin over the bone, retain them there by proper bandaging, which suffer to remain without disturbance until the third or fourth day.

Amputation of the Penis.—Amputation of the penis is not unknown among us. It has been performed several times, and it is found that no canula is necessary to keep open the urethra, the force of the urinary flow breaking down any incipient cicatrization of its orifice. The sheath is first forced back, and the penis brought forward to its greatest possible extent. Whatever portion it is intended to remove is now cut through, by means of an amputating knife, when the remainder is retracted within the sheath, and little hemorrhage has afterward occurred, except at the time of passing the urine; but there appears to have been no alarming quantity of blood lost.

Amputation of the Tail, or Docking.—We are most happy to state this filthy and unnecessary operation is now discarded. It never consisted of more than the cutting off a portion of the stump with brute force, and the cruel application afterward of a hot iron to the small artery of the tail.

NICKING.

We should be grateful that this barbarous and dangerous process is no longer numbered among the necessary operations. It is so beset with accidents, which no skill or prudence can prevent, that no one who has a free will ought to mutilate a horse by nicking.

FIRING.

The practice of firing was not always confined to quadrupeds. On the contrary, it probably was first used on man; and to this day, in many countries, it is a very popular remedy among human surgeons. In India it is applied over the abdomen for the cure of scirrhosity of the liver.

Firing, in veterinary practice, has, by Mr. Coleman's pupils, been justified as only men will justify a favorite operation, the virtues of which have been impressed upon their minds by an eloquent teacher. When Coleman was the chief of the veterinary profession, firing, under his rule, was used for any and every occasion. It was ridiculously supposed to act as a permanent bandage, as if a few strokes with a heated iron could destroy the elastic property inherent in the skin. It was the favorite styptic of these practitioners, and was applied to arteries (as of the tail) as though it possessed within itself some medicinal virtue. It was used to promote absorption, as in callus; and was likewise resorted to for checking absorption, as in ulceration. It was called into action to promote granulation, in broken knees; and was also a favorite agent to check granulations, when they were too luxuriant. In short, there was no folly which a hot iron did not cover. It has now, happily, fallen into disuse. Most modern practitioners will now confess that their chief reason for exercising the iron is to satisfy the proprietor, not to benefit the animal. After such an acknowledgment, who would submit to have his patient servant's skin scored and burnt with red-hot metal?

The mode of cauterization differs according to circumstances. As a general rule, it ought, of course, to be applied in the direction of the hair, by which the blemish is lessened; but this rule can not be arbitrarily followed, although it ought to do away with all the false pride of displaying the taste in the figures scored upon a prostrate beast. The Veterinary College recommends that the limbs be always fired in perpendicular lines; others advocate all manner of fanciful marks. Some cast the horse; many surgeons perform standing. The irons used are of various shapes and dimensions. Some recommend the firing, of all things, to be very light; others persist there is no virtue in hot iron unless it burns very deep. The operation consists in having irons of some substance made red-hot, and then drawing them mechanically along, or twisting them about upon the skin. The figures are various; so is the depth of the incision. Both must be decided by the taste, judgment, or heartlessness of the operator.

BLISTERING.

This is an operation of very great utility, and is, perhaps, compared with its benefits and importance, the safest that is performed. When a vesicatory becomes absorbed through the pores of the skin, it inflames the sensible cutis underneath, the consequence of which is, an infusion of serum through the part, which, in the human subject, elevates the cuticle into a bladder equal to the surface inflamed, but in the horse, from the greater tenacity of the cuticular connections, it becomes separated in the form of small distinct vesicles only. If the irritating cause be quickly removed, the serum may be reabsorbed, and the surface restored by a slight effort of adhesive inflammation. If the irritant act in a still minor degree, it simply irritates the vessels of the cutis to an infiltration of fluid through the sensible pores, but produces no desquamation of cuticle. Such has been called a sweating blister.

But when, by continued irritation, the cutis is exposed, suppuration succeeds, and the part is fully blistered.

The salutary action of blisters is exerted in several ways—in promoting absorption, in combating deep-seated inflammations, and in aiding others. As a stimulus to the absorbents, they act beneficially in the removal of injurious deposits, as the coagula remaining after inflammatory lesions; but it is to be remarked that when any existing deposit is of long continuance, or is osseous, it requires that the action of the vesicatory be kept up. Mercurial blisters have been thought to have a superior influence in accelerating absorption. Mercurials, rubbed in some weeks or days previously to blistering, are certainly great assistants, and should always be employed in the treatment of obstinate osseous or ligamentary enlargements. Blisters are very important aids in inflammatory affections, as counter-irritants, derived from a law in the animal economy, that two inflammations seldom exist in the vicinity of each other; therefore, when such an affection has taken place in any part, and we wish to remove it, we attempt to raise an artificial inflammation in the neighborhood by means of blisters, which, if persevered in, destroy, or at least lessen, the original one. Occasionally, also, we blister the immediate inflamed part, with an intention to hasten the suppurative process by increasing the activity of the vessels, as in deep-seated abscesses, and also in those which attack glandular parts. We therefore employ blisters to hasten the maturation of the tumors in strangles. When the flagging powers vascillate between resolution and suppuration, as they often do in the phlegmonous inflammations of glandular or of deep-seated parts, blisters may either hasten the resolution, or they may add their influence to the attempted suppuration, and thus bring it to maturity. But we carefully avoid, in other cases, applying a vesicant to a part immediately in a state of active inflammation. Particularly we should avoid what is too often done, that of blistering over the tendons, ligaments, and articulatory surfaces of a tumid limb, laboring under a congested state of the parts from excess of vascular action. Here we should do great injury were we to blister, by causing a greater deposit of lymph, and by hastening its organization into an injurious bond of union between the inflamed parts.

The vesicatory, or blister, for general use in veterinary medicine, as a simple stimulant, should be principally composed of

Spanish flies. Cheaper substitutes are used, but they irritate violently. In extensive inflammatory affections, they are, on this account, perfectly inadmissible; and wherever a case requires any thing more, it will be noted. The mode of blistering with the Spanish fly is sufficiently known. The hair should be cut or short as close as possible from around the part; the blistering matter should then be well rubbed in for ten or fifteen minutes. If the pasterns and fetlocks are the parts to be blistered, previous to rubbing in the ointment, smear some lard, tallow, or melted suet over the heels, and within the hollow at the back of the small pastern. This will often prevent some troublesome sores forming, from the blistering ointment falling on these sensitive parts. While a blister is acting, the litter should be removed from under the feet, or it will tickle the legs and irritate; but, above all, the head ought to be most carefully secured, for two days and nights, to oppose lying down, more especially to prevent the horse biting the blistered part. On the third evening he may be permitted to repose; but a prevention should even then be continued, by means of what is called a cradle. This apparatus may be bought at every turning shop, or may be made of eight or ten pieces of round wood, an inch and a half in diameter, and two feet long. These are strung at each end on a rope, and fastened around the neck. When it is intended to blister repeatedly, the effects of the first should have subsided before the second is applied, the scurf and scabs also be cleared away, and the part well washed with soap and water. In all cases, the third or fourth day after the application the part should be thoroughly painted over, by means of a long-haired brush (such as are in use with pastry-cooks to glaze their crusts), with lead liniment, which should be repeated every day. And when it is proposed to turn a horse out, it should never be done until the whole blistered surface be quite healed; otherwise dirt, flies, etc., may prove hurtful.

It remains to observe that, instead of repeated active blistering, it is, in some cases, preferable to keep up a continual slight irritation on the original blister by means of stimulants, as iodine ointment, mild blistering applications, etc. Caution is, however, necessary, to avoid forming an eschar, and thereby a permanent blemish; but when a blemish is not of consequence, this plan will be found often more efficacious than firing, as in spiints, spavins, etc. Some practitioners blister mildly one day, and on the next wash off the blistering matter, thereby saving loss of hair; but there is more of apparent than real good in this plan. If a blister be necessary, it requires all its activity.

Ammoniacal blister.—Spanish flies are only efficacious when the animal can afford to wait their action, which is rather slow. In most of the acute diseases, the horse would perish before the blister began to rise, wherefore resort has been had to boiling water and red-hot iron. The action of these last coarse and brutal measures was alone controlled by the violence of the internal inflammation, and, if the practitioner was mistaken in his estimate of the immediate danger, extensive and lasting blemish was the consequence. We have in the liquor ammonia an agent quite as formidable as boiling water or heated iron, but it is rather longer in displaying its force; wherefore, it allows time for watching its action, and of checking it the instant it has sufficiently blistered the skin. It is true the liquor ammonia upon the skin can not be removed, neither need it be counteracted. Ammonia is like steam, only powerful when confined. The ordinary soap liniment, if covered over, would, because of the ammonia it contains, produce a lasting blemish; but every veterinary surgeon knows how very harmless a preparation that is when simply rubbed upon the surface. So, when we desire the active effects of liquor ammonia, we double a blanket or rug four or five times and hold it over the liquid. It takes from ten to twenty minutes to raise a blister, and it consequently can, from time to time, be observed; and when its action has reached the wished-for point, all we have to do, effectually to stop it, is to take away the rug or blanket. That removed, the free surface and the heat of the body occasions the ammoniacal vapor to be dispersed, and the animal is safe.

ROWELLING.

Rowels acts as foreign substances within the body. They cause irritation and suppuration, whereby more deep-seated inflammations are supposed to be removed. They are, however, often very convenient, because they stand as sign-boards to show the proprietor that something has been done. The common mode of making a rowel is after the following manner: A slit is first made by means of the rowel scissors, on any part of the integuments, held between the finger and thumb. With the handle of the scissors separate from its cellular connections a circle of two or three inches in diameter, into which introduce something to prevent the reunion of the skin. A piece of circular leather, tolerably stiff, with a central hole, is a very common substance used, but is objected to by some on account of the difficulty of changing it without injuring the skin. Tow, as more pliant, is frequently introduced into this cavity. If the rowel runs freely, it should be dressed every day, by changing the plug, if of tow, and by cleansing it, if of leather. No rowel should go undressed beyond the second day, for the comfort of the horse. They are very favorite applications with farriers, and frequently abused by being employed indiscriminately, but are falling into disuse, setons having almost superseded them.

Setons, in their action, resemble a very mild form of rowel, but are more convenient in their application. There is hardly a part of the body where a seton may not be conveniently placed. They have been put around the eye; they have also been entered at the withers, and brought out between the humerus and the sternum, so extensive or so diminutive can they be made. In sinuous ulcers of the withers and of the neck they may be placed; through the heels, in foot diseases, they have been inserted. In cavernous sores, they are entered at the superior part, and are brought out at an inferior, so as to form a depending orifice. The formation of a seton is very simple. A skein of thread, or a piece of tape, of a convenient size, may be used. At the one end place a large knot; arm the eye of a corresponding-sized seton-needle with the free end of the tape; introduce this into any proposed part, and, bringing it out at some other, either make a second knot, or tie the two ends of the tape together; which last method of fastening is, however, often objected to, from the danger of its catching in something and being torn out, to the disfigurement of the horse. When a seton is placed in a sinuous track, for the purpose of inflaming, it is moved twice a day, frequently, and moistened each time with some stimulant, as oil of turpentine, tincture of aloes and of benjamin. All setons require daily cleaning and moving. When they are required to act more quickly, the tape is infused in terebinthinate of cantharides, or small pieces of black hellebore are sewn within it. An old material, composed of woollen, flax, or cotton and hair, is also used instead of tape. Setons, however, are of small service in acute cases. They are chiefly in use for disturbances of a chronic description.

ABSTRACTION OF BLOOD, OR BLEEDING (ARTERIOTOMY.)

Blood is abstracted by opening the conducting vessels, arterial and venous. When taken from arteries, the process is called arteriotomy; when by the latter, phlebotomy. Some bleedings include both these operations, as general scarifications of the soft parts, bleeding at the toe point, divisions of the vessels of the cornea, etc. Blood-letting is called local when it is practiced on or very near the affected part; and it is supposed to act more immediately than general bleeding because it produces more effect with the loss of less blood. Local bleeding is, therefore, usually practiced on the minor branches of the arteries and veins, as on the temporal artery, the plate vein, the vena saphena, etc. Leeches are a means of local bleeding not often used by us in veterinary practice; but there is no reason whatever why they should not be employed. When applied to the eye, and occasionally to other parts, also, they adhere readily, abstracting blood rapidly, and, therefore, might be valuable aids in violent local inflammation. Cupping is also practiced, in France and other parts of the Continent, with very large glasses, and it is there supposed to act remedially in many local inflammations. By general bleeding we understand the depletion of the system at large, and this we practice in extensive inflammations.

Division of the temporal artery.—The proper spot for either its puncture or division is directly where the vessel leaves the parotid gland, to curve upward and forward around the jaw, a little below its condyle. When it is punctured, it usually affords much blood; and in such case, enough having been obtained, divide the trunk, when, the receding portions becoming pressed by the integuments, and lessening by their own contractility, the hemorrhage is stopped. It should be punctured by a lancet; a fleam may fix itself in the bone. Its division can be readily made, also, either by a lancet or scalpel.

Bleeding by the palate is also a species of arterio-phlebotomy, and is a very favorite spot for abstracting blood with most ignorant persons, who vehemently recommend it in spasmodic colic or gripes, and in megrims. In such cases, however, a want of knowledge of the anatomy of the parts has occasioned a serious hemorrhage to occur; it may prove a fatal one if the artery proper to the part be divided incompletely. The palatine artery and nerve run near each other, on each side of the roof of the mouth, so as to divide the inner surface of the hard palate into three nearly equal portions. No other than a direct division of the vein should ever be made; therefore, when bleeding is determined on at this place, do it by plunging a lancet or even a penknife in a direct line across the rugæ, one inch within the mouth, exactly between the middle and second nippers. There these vessels form a curve, which curve will be divided, and will then yield three or four pints of blood. If the instrument enter too much on one side, as about the middle of the second nipper, then a partial and longitudinal division of the artery may be made, and an alarming hemorrhage may follow. In this case, the section must be enlarged and deepened inwardly—that is, away from the teeth—which completely severs the vessel, and its retraction will stop the hemorrhage. A moderate or slight flow of blood from the palate may be obtained by light scarifications of the rugæ. But all bleedings here, except under circumstances of the most urgent necessity, had better be avoided.

Bleeding by the toe is also arterio-phlebotomy. By no means cut out a portion of the sole at the point of the frog, which frequently occasions abscess; but, with a very fine drawing-knife, cut down exactly in the line of union between the crust and the sole; then, by puncturing the part with a lancet, a vast flow of blood may be obtained, the benefits of which, in some cases, are very marked, particularly in acute founder. If the blood should not flow with sufficient freedom, place the foot in warm water. The bleeding finished, cover the puncture with some tow and a little tar, and lightly tack on the shoe. There are, however, other methods of bleeding from the toe. Mr. Maver uses a drawing-knife with a long curve, so that one sweep of the blade may cut a piece out of the foot. This appears to us bad practice, as it leaves Nature a space to fil! up, instead of a simple incised wound to heal. Others take away none of the horn, but merely make a slit through the outer covering on to the vascular portion of the foot. The flap of horn they hold up, so long as they desire blood, by the insertion of a piece of wood; and when they have obtained blood enough, they take out the wood, so as to let the horny flap down. This last method, of all others, appears to us the easiest and the best.

Sometimes the plantar vein is opened as a substitute. Scarifications are, also, occasionally practiced, which, of course, divide both venous and arterial branches. In France extensive scarifications used to be made into indurations before the suppurative process had commenced, which, in some cases, prevented that from going on, and the remedial wounds made were healed by adhesive inflammation, or by healthy granulation. The same method has also been occasionally practiced here, but it is not now often attempted.

Phlebotomy, or the puncture of a venous branch, is the most usual mode of drawing blood in veterinary practice, and may be employed on any point of the body; but some vessels are much more frequently opened than others, and most of all the jugular vein.

Bleeding by the thigh vein.—The saphena is a prominent vein continued from the inner part of the hock, and may be opened by the fleam, but with much greater safety and propriety by a lancet. The opposite leg being held up, the operator, placing himself in front of the thigh, and steadying himself and the horse by placing one hand on the hock, may fix the vein with the little finger of the other, while the lancet, held between the thumb and fore-finger, punctures it. This vein should never be opened save upon absolute necessity, as it is often troublesome to pin up. A horse has been cast for the trivial matter of stopping the hemorrhage.

Bleeding from the plate vein.—This vein is frequently opened to abstract blood after injuries of the fore extremities.

The superficial brachial vein is a continuation of the superficial division of the metacarpal veins, and in the passage upward receives more than one branch. Its principal trunk ascends along the inner side of the radius. It may also be well to remark that, when taking blood from the superficial veins of the arm or fore-arm, if any difficulty is experienced in obtaining a sufficient flow, the lifting up of the other leg, by throwing the muscles of the punctured one into action, will force the blood from the inner to the outer set, and an increased quantity may be obtained. The plate vein, or external thoracic, is often opened, as it emerges from behind the arm, and is pinned up without any difficulty.

Bleeding by the jugular vein.—The situation of this important vessel is well known, but its internal connections are not so familiar, though such knowledge is essential to the uniform safety of the operation. The horse has only external jugular veins, a right and a left one. As each emerges from the chest, it is found deep-seated, and approaching the trachea. It then passes forward, in company with the external carotid artery. Toward the middle of the neck it becomes more superficial, and is now distinctly seen progressing rather above and without the carotid artery and trachea, or windpipe. The carotid, therefore, in the future course of the jugular, is situated a little below and more deep-seated than the vein. The jugular is also separated from the carotid by a slight muscular band, derived from the levator humeri. Its further track is marked in the hollow formed by the inferior edge of the levator humeri, where it is covered by the panniculus carnosus and integuments only, when, having nearly reached the jaw, it makes its well-known division into two portions. Bleeding by the jugular is usually practiced with a lancet or with a fleam. The proper spot for the puncture may be found anywhere between two inches and six from the division of the vein. This latitude is here mentioned, because it is prudent to avoid puncturing directly over a former bleeding-place, known by the scar and enlargement. It should also be avoided where a little knot in the course of the vein will sometimes denote the existence of one of the venous valves. But in all ordinary cases, where these hindrances do not appear, operate at two or three inches from the division of the vein, which will be sufficiently evident when it is pressed on below the place punctured. Avoid operating low down in the neck, as there the vessel is deeper-seated, and near to important parts.

First moisten the hair and smooth it down; then, steadying and enlarging the vessel with one hand, with the other plunge the point of the lancet into the integuments, so as just to puncture them and the vein; then, by a slight turn of the wrist, carry the instrument obliquely forward to finish the cut. For opening the smaller veins, the lancet should always be used. In all but the practiced hand the fleam is the safest for bleeding from the jugular It is always prudent to have the eye of the horse covered. Unless the eye be covered, the horse will be likely to flinch at the moment of the stroke, and the puncture may be made in any place but where we wish. The hair being first wetted and smoothed, and the fleam being retained in the left hand, the unemployed fingers pressing on the vein so as to fix and swell the vessel, let the point rest exactly in the middle of the swelling. Strike the fleam sufficiently hard to penetrate the skin and vein. A blood-

stick is preferable for the purpose of striking the fleam.   There is
a vibration between two hard bodies when they meet, which, in
this instance, is favorable to a quick and moderate puncture of the
vein.   After the vein has been opened, moderate pressure with the
edge of the can which catches the blood is sufficient to keep up
the flow.   It may also be encouraged by putting a finger within
the horse's mouth.   The requisite quantity of blood being drawn,
remove the can.   The remaining process of securing the vessel
is of equal importance.   The sides of the orifice are first to be
brought in opposition, without pinching them, and without drawing them from the vein.   The same cautions should also be ob·
served when the pin is introduced.   Let it be small, with an
irregular point, and, when inserted, wrap around it a few hairs
or a little tow.

Common, however, as this operation is, and qualified as every
one thinks himself to perform it, yet there are very serious accidents which do arise occasionally.   It has occurred that the carotid
artery has become penetrated.   When the puncture has been made
through the vein, the accident is known immediately by the forcible and pulsatory gush of florid arterial and dark venous blood
together.   In one instance of this kind, which occurred to a French
practitioner, he immediately thrust his finger into the opening
through the vein, and thus plugged up the artery intending to
wait for assistance.   In this state he remained, we believe, an
hour or more, when, removing his finger, to his surprise, he found
the hemorrhage had ceased, and did not again return.   In another
case, where an English practitioner accidentally opened the carotid,
he placed a compress on the orifice, and had relays of men to hold
it there for forty-eight hours, when it was found the bleeding had
stopped.

The admission of air is also another serious accident that now
and then attends bleeding.   It sometimes happens from the sudden
removal of the fingers or blood-can, or whatever was used to distend the vessel by obstructing the return of the blood.   This, being
suddenly taken away, allows the escape of the blood toward the
heart, and occasions a momentary vacuum, the air being heard to
rush with a gurgling noise into the vein through the orifice; it
then mixes with the blood, and occasions, in some instances, almost
immediate death.   The animal begins to tremble; he next staggers,
and finally falls in a state of convulsion.   If the quantity of air

taken in has been considerable, death ensues. The remedy must, therefore, be instantaneous, and consists in again opening the orifice, or making a new one, to gain an immediate renewed flow of blood, which will, in most cases, renovate the horse, who has been found afterward to be tormented with an intolerable itching." *

* Blaine's "Outlines."

18

SECTION X.

DISEASES OF THE SKIN AND SUB-TISSUES.

ITCH AND MANGE—LICE ON ANIMALS—GREASE AND SCRATCHES—SURFEIT—DROP-SICAL LIMBS — SCARLETINA — RINGWORM — PRURIGO — POLE-EVIL — FISTULOUS WITHERS—WARTS ON THE SKIN—PURPURA HEMORRHAGICA—OUT OF CONDI-TION—HIDE-BOUND—HERPES.

ITCH AND MANGE.

ITCH, mange, and scabies are essentially local affections of the skin, and are occasioned by the presence of parasites known as "sarcoptes-equi." The eruption ensuing on the skin of a horse,

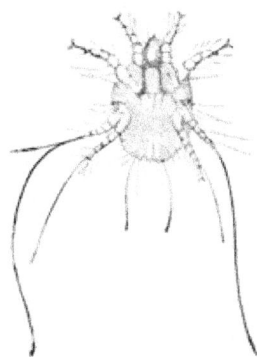

ACARUS, OR SARCOPTES-EQUI, THE ITCH PARASITE (MAGNIFIED).

when subject to this affection, is very similar to the appearance of itch in man, and probably is just as tormenting to the animal as in the case of his master. Numerous cases are recorded of trans-mission of itch from horse to man, and, when so acquired, it is impossible to dis-tinguish it from the human itch. Some persons suffer severely when attending mangy horses, but a clean person or clean horse are not apt to take the disease. It is well known that a healthy and clean horse may stand for weeks near a mangy one without taking the disease, showing, very conclusively, that the best preventive measures are those which promote health and cleanliness. As the English groom says, "plenty of elbow-grease, by means of brush and curry-comb, on the skin of the horse; regular daily exercise, pure air in the stable, and a sufficiency of good digestible food are sovereign remedies against itch and mange." This, we presume, is all the reader wants to know with reference to the cause of itch and mange.

Symptoms.—When a horse is affected with this malady, in either of the forms here named, the owner's attention will be aroused by the animal rubbing himself, whenever he can get a fair chance to do so; and, should the limbs or heels be affected, he will rub the fore-legs one against the other, and, at night, will constantly strike the floor with his hind extremities. The hair looks unthrifty, has lost its gloss, and does not lay smooth on the body. It falls off in patches, and an eruption is observed similar to the appearance of itch on man, yet not so distinctly recognized. A few pimples, scabs, and a slight reddening of the parts is all that may, perhaps, be seen. The skin of some horses, however, is more vascular and more delicately organized than others. In the latter case the disease is more marked, creates greater irritation, and the nervous system suffers more from sympathetic irritability than in the case of an animal having a coarse and comparatively insensible "hide." In addition to these symptoms, the affected animal will be observed to be more or less fractious and unmanageable.

Treatment.—The disease being of a local character, requires topical (local) remedies, and the best medicine is that which most quickly destroys the life of the itch parasite. It has been found that preparations of creosote kill the parasite in a few seconds, and GERLACH recommends the following preparation:

No. 28.	Creosote............................	2½	oz.
	Spirits of wine.....................	15	oz.
	Water...............................	40	oz.

Previous to the application of this, the same authority advises that the horse be smeared all over the body with soft soap, washing it off some time afterward with warm water, and having the animal well brushed; or a wash may be afterward used, consisting of one part of caustic potass to fifty parts of water; or, lastly, greasing the animal with linseed oil, train oil, etc., and, one or two days afterward, washing with soap and water or potash ley then apply the preparation of creosote.

The cheapest, and probably the safest, plan of treatment, and that which has proved most successful in the practice of the author, is as follows:

No. 29. Unslaked lime .... 1 lb.
       Flour of sulphur..................... 2 lbs.
       Water................................ 12 pts.

Put these ingredients in a stone jar, set it on the stove, or in a regular water-bath, until it boils. During the interval, the mixture must be constantly stirred, both in view of insuring a complete mixture, and of keeping the lime and sulphur from depositing in the bottom of the jar; for, should this occur, the jar will crack. The mixture must be stirred with a wooden spatula or glass rod. Keep the mixture boiling for about ten or fifteen minutes; then set it aside for twelve hours, at the end of which time pour off the clear liquor, bottle tightly, and cork the bottles. These should be put away in a dark closet, as the mixture soon decomposes when exposed to the sun's rays. This remedy will cure the itch on horses and cattle. It is, also, a sure cure when used on man. It is not adapted to the cure of the malady in sheep, as the sulphur is supposed to be injurious to wool. The only objection to its use is, that it has a very unpleasant odor— smells like sulphureted hydrogen. Before applying this mixture, the animal should be thoroughly washed with warm water and common brown soap; then wipe the animal dry, and apply the sulphur mixture, by means of a sponge. One application, if properly applied, will usually suffice to kill the parasites. Should it fail in the first instance, a reapplication will do no harm, as the preparation is perfectly inocuous.

Scabies is to be treated in the same manner; in fact, all diseases of the skin, supposed to be dependent on the presence of parasites, are to be treated after this fashion. GERLACH, of the Royal Veterinary School of Berlin, has made a series of experiments with various remedies for the destruction of the itch parasite, and the following table shows the result of his labors:

AGENT.	Dead in		OBSERVATIONS.
	Hours.	Minutes.	
Creosote, pure..	....	¼–1½	
Creosote, 1 part } Spirit, 10 parts } Water, 30 parts }	....	½–1½	
Creosote 1 part } Water, 80 parts }	....	2–2½	
Creosote with fat { 1.20 .... { 1.21	....	½–1½ 3½–5	
Creosote with oil { 1.21 .... { 1.40	....	3½–7 5½–9	
Iodine, tincture of, pure..........................	....	1–2	
Iodine, tincture of, with water, 1.4....	....	4–6	} Mathieu.
Iodide of potassium with water, 1.2....	....	9	
Iodide of potassium with water, 1.4....	....	20–26	
Caustic potash, 1 part; 24 do.	....	2–2¾	
Caustic potash, 1 part } Water, 10 parts } Tar, 2 " } Hartshorn oil, 2 " }	....	1¾	
Caustic potash, 1 part } Water, 16 parts } Tar, 2 " } Hartshorn oil, 2 " }	....	7–12	
Sulphuret of { with water, 1.10............. potassium { with oil, 1.10.............	 10–20	15–30	
Chloride of lime with water, 1.20......	....	15–30	
Hartshorn oil, pure.................................	....	3–4	
Hartshorn oil, with water, 1.10.............	....	30	
Oil of turpentine.....................................	....	5–9	
Barbadoes tar...	....	5–9	
Tar (Pix liquida)....................................	....	8–15	
Photogen { pure...................... { with oil { 1.5 .... { 1.10	1	7	{ Were living 7 hours after { immersion.
Concentrated vinegar, pure..................	....	1–1¾	
Concentrated vinegar, with water 1.1	....	2–3½	
Acetic acid with water, 1.1..................	....	2	
Concentrated sulphuric acid { 1.21.. with water { 1.48......	....	7–8 32–35	
Decoction of tobacco { 1.5 .... { 1.10 .... { 1.50	.... 2–5 4–10	10–20	{ Mathieu. { White arsenic, 1 part. { Sulphate of iron, 10 parts. { Water, 100 parts.
Tessier's arsenical bath........................	....	7–25	
Mathieu's arsenical bath.......................	....	15–16	{ Mathieu. { White arsenic, 1 part. { Alum, 10 parts. { Water, 100 parts.
Over-saturated solution of arsenic in water 1.6...	2–3	....	Hertwig.
Green soap...	½–1	....	{ A dilute solution does { not kill.
Liquor ammonia......................................	¼–1	....	Hertwig.
Solution of corrosive sublimate, 10 } grains, one ounce }	....	15–45	Hertwig.
Infusions of henbane, belladonna, } and Persian insect powder, 1.16 }	12–16	....	
Decoction of black and white helle- } bore, 1.16 }	6–36	....	
Infusion of digitalis, 1.16.....................	24–36	....	
Walz's ley, containing caustic pot- } ash, hartshorn oil, and tar }	6–48	....	

The Skin Diseases of Domestic Animals.

A celebrated writer on the diseases of the skin of domesticated animals, thus discourses on the subject:

"Few topics have engaged the attention of the successful breeder of our improved domesticated animals more than skin diseases produced by insects, in consequence of the manner in which both the breeding and feeding qualities of stock are affected by them. This arises from the little progress scientific inquiry has experimentally made in the physiology of these tiny parasites by means of the microscope, and the consequent paucity of reliable scientific works on the subject. Hitherto half the conclusions of our veterinary surgeons relative to their natural history have been drawn from the traditionary philosophy of our forefathers, and not from actual examination, either with the eye of the modern physiologist or morbid anatomist; while our entomologists have had too much to do to overtake the branches of zoölogical science, to fill their cabinets with specimens of the class in question—a class far more varied than the kingdom itself, every different race of animals being not only infested with a different brood of insects (mites), but almost every organic substance, animal and vegetable.

A very superficial acquaintance with these facts must convince the reader of the importance of the lecture on this subject, delivered by Professor Simonds, before the council of the Royal Agricultural Society of England, accompanied with drawings, exhibited on the wall, and specimens under his microscope. The former, drawn to a greatly magnified scale, illustrated, in a very conspicuous manner, the different stages of vitality, from the ovum to the insect, in all the vigor of matured life. But to us, and several other members who examined them, the latter appeared the most interesting, as nothing can exceed the fidelity with which the microscope exemplifies Nature, though all but invisible to the naked eye, confirming, in the most satisfactory manner, the soundness of the conclusions at which the Professor arrived, differing, as they did, in many respects, from those hitherto drawn, both by veterinary surgeons and entomologists.

Referring to the report of Mr. Simonds' first lecture, the task which devolves upon us is to apply it to the daily practice of the farmer, so as to profit by its deductions. These were principally confined, it will be seen, to scabies and ætrus in sheep and other

animals, with a few observations on the dog-flea (pulex canis).
Scabies is a pestiferous disease, whether it affects the horse, the
ox, the sheep, swine, or poultry, inflicting a loss not easily estimated; hence the maxim of every intelligent farmer is, to avoid.
With him "prevention is better than cure," and, therefore, his
grand desideratum is to guard against contagion. Sheep are, perhaps, more subject to it than any of the other animals, arising as
much from the nature of their skins and coats as from the fecun-
dity of the acarus ovis, and the greater vicissitudes of the weather
to which they (the sheep) are exposed. Certain parts of the body
are more liable to be affected than others; and so is an unhealthy
skin than a healthy one. Indeed, it has been said that an unhealthy skin will itself produce scabies (?), but this conclusion does
not appear to be well founded; for a disease dependent upon the
presence of living parasites can never arise spontaneously, but
must be effected by contagion, either by means of their eggs, or
the insect in some other stage of its existence.

Now, from what has just been said, it will appear obvious that
cleanliness, a healthy skin and state of the body, and a separation
from foul animals and ground, are the means necessary to avoid
contagion. The truth of this will, perhaps, be better understood
if we first review the important distinctions which Mr. Simonds
makes between the habits of the acarus scabiei of the human body
and the acari of our domestic animals, the former burrowing in
the skin, but the latter living on its surface, clinging to the skin,
hair, or wool with their trumpet-shaped, vesicular-cushioned feet,
to prevent their being thrown off by the animal when shaking or
nibbling itself. Hitherto distinctions of this kind have been overlooked, writers generally concluding that the acari of quadrupeds
burrowed in the skin like those of man, thus proving the little
use which had been made of the microscope in examining the
former, as it shows them to be incapable of living in the skin,
from the configuration of their bodies. Indeed, to have made
similar acari for naked skins as for those covered with hair, wool,
or feathers would have been an oversight on the part of Nature;
while the fact that the acarus scabiei will not live on the horse,
nor acarus equi on man, or acarus ovis on the ox, or acarus bovis
on the sheep, and so on, proves that greater differences than the
mere configuration of the animal structure exist, all pointing to
the above means as necessary, in every case, to avoid so great a

pest—one which appears to form an integral part of that curse inflicted on the whole creation by the fall of man.

The importance of cleanliness, and its concomitant health, to prevent contagion, may be further illustrated. We have just seen above that an acarus, although a loathsome pest, is yet very nice in its taste, and particular about a nidus in which to deposit and hatch its eggs. It enjoys the highest degree of prosperity on the unhealthy skin, multiplying there fastest; so that if it creeps from it to the opposite—the sleek, healthy one of the horse or ox, or dry wool of the sheep—it feels itself from home, and, before it even reaches the skin, may be bruised or shaken off. If, however, it creeps upon the unhealthy animal with its staring coat, it soon reaches the skin, and commences its direful work, every thing there being congenial to his happiness; hence the incredible speed at which it propagates its species, until it either consumes its victim alive, or is arrested at its fatal work by the timely unguent of the veterinary surgeon.

Again: when a dirty animal shakes itself, as it invariably does after rubbing itself against any thing, less or more scurf, dandruff, and dust is thrown into the atmosphere, and carried to a distance by high wind. Now, under such circumstances, when affected with scabies, it is manifest that the smallest of these puny insects, as well as their eggs, will be blown from one pasture to another; that the latter will lodge in the dirty staring coat of the unhealthy animal, when they will be blown off that of the clean sleek one, or be brushed off before they reach the skin, or any nidus capable of hatching them. In this manner we can trace contagion from one animal to another, and thus account, in harmony with entomological science, for what has hitherto been termed spontaneous cases of scabies in some of our domesticated animals, while others have escaped the disease, though all herding together in one field. We can also account for the fact why the disease is more liable to break out among sheep than horses and cattle, without coming in contact with strange flocks, because the coats of the latter are more likely to be impregnated with eggs than those of the former, while they afford a better nidus for hatching them.

With regard to health, it has even been said that the blood of scabbed animals is diseased: nay, that the blood of all animals is loaded, more or less with the eggs of acari, and that they are hatched under certain cutaneous affections, thus accounting for spontaneous cases differently from the atmospheric distribution of eggs. Hence the reason why sulphur and mercury are taken, internally as well as externally, to get rid of the disease. But this sanguiferous doctrine of distribution is more difficult to reconcile with entomological science than the atmospheric one.

Contagion by direct contact of clean animals with diseased, er where the former enter foul ground, will be difficult to avoid so long as the commerce of live farm stock is conducted as at present; for diseased animals will be sent to market, although contrary to the spirit of the law, thus not only communicating the disease to all they come in contact with, but also infecting the market-place. According to the experiments made by Mr. Simonds, acari left upon a post, hurdle, or hedge by a scabbed beast will survive for fourteen days, and at the end of that time lay hold of any animal coming in contact with it, and thus communicate the disease; consequently our weekly and fortnightly markets are little better, during the warm months of summer, than nests for propagating malignant parasites. It would be well, then, if very stringent statutes were enacted, confiscating all diseased animals offered for sale, and that microscopical examinations were enforced in all suspicious cases, so as to guarantee its healthy operation. It will thus be seen that the means for preventing contagion may be summed up in two words—improved management.

At a very early period in our history, acarus scabiei was much more common among our forefathers than among their posterity of the present day. The change which has taken place is obviously attributable to progress in dietetics, clothing, and medicine; and among our domesticated animals it is no more than reasonable to conclude that similar causes will produce similar effects. In point of fact, improved management has already greatly reduced the prevalence of scabies among both our herds and flocks, thus holding out every encouragement to persevere in obtaining further progress. There is, perhaps, no branch of husbandry where greater advances can be made than in cattle-cookery, household accommodation, including grooming, etc., and medicine; and when we contemplate how much cutaneous diseases are dependent on management in each of these three respects, it were difficult to estimate what influence further improvements in them may have on the disease in question."

LICE ON HORSES.

Lice, in a majority of cases, may be attributed to want of cleanliness; for they are rarely if ever seen on the body of an animal with clean skin. Brush and curry-comb, and clean stables, are, therefore, the preventives.

A short time ago I had occasion to visit a stable, located in Waukegan, Ill. There I saw a horse almost perfectly hairless, except the mane and tail. I asked the owner what occasioned the depilation of the hair. He informed me that the horse's body was completely infested with lice; that he had tried all sorts of remedies without success, and, at last, thought he would try kerosene oil, and, accordingly, rubbed the body all over with it, using about two quarts. He said "it killed all the lice, but came very near killing the horse. It made him perfectly crazy; his limbs became swollen to the proportions of a young elephant, and the hair all fell off."

The remedy I have found most efficacious is composed of

No. 30. Crude cod-liver oil.................. 1 pint.
Pulverized lobelia.................. 2 oz.
Mix.

This should be thoroughly rubbed all over the body by means of a stiff brush; at the expiration of four hours, carefully wash the horse all over with soft soap and warm water. It may be necessary to make a second and even a third application, on successive days, ere the parasites are all killed. The parasites can be seen with the naked eye; therefore the owner of the horse must be the judge as regards the number of applications needed.

At a meeting of a Farmer's Club, "Lice" formed the subject for discussion. Dr. J. R. SMITH explained that lice, as well as all similar insects, breathe through holes in the body. These holes are minute spirules, constantly kept open by an elastic ring, and surrounded by a fringe of extremely delicate hair, which prevents the intrusion of any solid particles. To kill the insect it is only necessary to close these breathing-holes, and this is done by smearing them with any kind of grease or oil. You may catch a caterpillar and examine him with a magnifying glass, and you will find these spirules arranged in two rows, one on each side; then, if you take a moth or butterfly, you will find the breathing-holes in the body corresponding with those in the body of the caterpillar from which it was produced—the same body, in fact, remaining after the wings are developed. If you dip a feather in oil, and smear the two spirules nearest the tail, the lower portion of the body will be paralyzed so far as these holes. Proceeding upward, you may paralyze the whole body until you come to the last two, which are situated just below the jaws. So long as these remain open, the insect will continue to breathe, but if these are now closed, he dies immediately. To exterminate lice upon any animal, it is only necessary to cover the animal completely with grease or oil. The simplest and cheapest oil is best—lard, fish oil, or any other that is at hand. When the lice have not become so numerous as to produce the disease known as pityriasis, or mange, this application, undoubtedly, is the best that can be used. In addition to being effective, it is perfectly harmless. Stock breeders should always recollect that prevention shows wisdom, and saves annoyance and expense; and, by neatness, cleanliness and comfort, the health of the animal will be insured, and the animals will not become infested with the troublesome parasites.

Poultry lice affecting horses.—Horses, when located near a hen-roost, are apt to be infested with poultry lice. The remedy is the same as in the case of the common louse.

DISEASES OF THE HEELS (GREASE, SCRATCHES, "WATER FARCY," "WEED SHOT," "MILK LEG").

In olden times grease, or scratches, and cracked heels, were not only very prevalent but very formidable affections. Before veterinary surgeons were employed in the British Army, many thousands of valuable horses were condemned as useless for active service in consequence of the prevalence of that loathsome affection known as inveterate grease; but now, in consequence of the presence of veterinary surgeons, and the consequent improved system of management in camp and stable, grease is almost unknown. Most veterinary writers contend that grease almost always arises from improper management of the horse, yet I have seen it occur in horses that were well cared for and properly treated. The truth is, some horses are predisposed to affections of the skin of the heels and the sebaceous glands of the same; and, although cleanliness and good management may stave off an attack for a long while, yet when the system abounds in morbific matter, the same is very apt to gravitate toward the heels, (they being most remote from the center of circulation,) inducing inflammation of the skin of the heels, distension of the sebaceous glands, and a stinking and unhealthy deposit on the surface, and a purulent discharge through ulcerated cracks. This constitutes grease.

Cause.—It is contended by Mr. BLAIN and others that grease has local weakness for a cause. As fluids press, not in proportion to their diameter, but to the height of their column, the venous blood must find some difficulty to its ascent. Debility is, therefore, more felt in the distended vessels remote from the influence of the heart, under which circumstance the effects productive of grease necessarily ensue. Some horses are constantly the subjects of swelled legs, and, if the heels happen to be accidentally abraded, the latter often operates as the exciting cause of a very formidable affection of the heels. Associated with swelled legs is a scurfiness of the skin beneath the fetlock. These conditions, together with the fact that the animal is of the lymphatic temperament (gray color), are sure indications that predisposition to maladies of the above character are present. Having such horses under our care, the chief object should be to ward off an attack, by dietetic and hygienic measures. Too much washing of the heels, without drying them by rubbing, is just about as bad as if the filth was allowed to remain. The practice chills the part by a process of slow evaporation, and the result is local congestion, etc.

I now propose to illustrate, for the benefit of the reader, the treatment of grease or scratches. I was requested to see a gray gelding, the property of the Transfer Company of St. Louis. On arrival, I examined the animal, and found him to be a large, flabbily-organized creature, having a large amount of loose tissue under the skin. I found that there was a bad odor arising from some ulcerations and exudations about the heels and sides of the same. Intermediate of the ulcers were dry, horny scabs; the hair about the parts pointed straight out; the heel was excessively tender, quite vascular, and blood escaped from its vessels. The moment a hand was placed upon the locality, the animal would catch up the limb and appear to suffer much pain.

Treatment.—In the first place, I had the parts well cleansed with soap and water; then, after wiping them dry, they were wetted, three times daily, with a portion of the following solution:

No. 31. Nitrate of potass (saltpeter)........... 2 oz.
         Water............................... 1 pt.
         Glycerine........................... 4 oz.

Every time the solution was applied, the parts were dusted with pulverized charcoal. This charcoal, completely covering the abraded parts, and being a good antiseptic, had an excellent effect in excluding atmospheric air, and in correcting feted odor. I administered, as an alterative, to correct the morbid habit,

No. 32. Iodide of potass..................... 20 gr.
         Sulphur............................ 4 dr.
         Powdered sassafras bark.............. 2 dr.
         Gentian............................. 1 dr.

These were mixed in food, and the same quantity was given during a period of four days, and the dressings were also continued. On the fifth day the animal had very much improved, stood fair on the foot, and seemed to suffer but very little pain. The limb was somewhat swollen, partly from want of use, and otherwise from slight effusion into the cellular tissue. I discontinued the medicine, and ordered the following mixture, to be applied twice daily:

No. 33. Pyroligneous acid................ ⎫
         Coal oil........................ ⎬ equal parts.
                                          ⎭

After each application the charcoal was reapplied.

The following case serves to illustrate that the virus of grease can be transmitted from horse to man:

Transmission of the Virus of Grease from the Horse to Man.— Drs. MAUNOURY and PICHOT have published an interesting series of experiments, tending to prove the identity of grease and cowpox. This doctrine, which was always maintained by JENNER, has received confirmation from the observations of LOY, GODINE, and others. The following is a summary of the facts, related by Drs. MAUNOURY and PICHOT:

" Francis Barthelemy B——, aged twenty-eight, of lymphatic temperament, a farrier, presented himself to Dr. Pichot on the 5th of March, 1856. He had not been vaccinated. On the backs of his hands, which were red and swollen, were several confluent opaline pustules, depressed in their center, and having all the appearance of vaccine pustules of the eight or ninth day. The inflammation with which the pustules were surrounded had ap-

peared on the second; the pustules themselves preceded the inflammation some days. This man had not been in contact with
any cow, but on the 11th of February he had shod a horse suffering from grease. There existed, at the time, numerous cracks
about the hands. The disease from which the horse was suffering
was certified by a qualified veterinarian. Various inoculations
were practiced with the liquid taken from the pustules presented
by B——, with the effect of reproducing the same disease. The
most perfect set of experiments were made by M. Maunoury, who
transmitted the virus through four sets of cases. The following
are the results of his observations:

1. That virus obtained from the hands of the farrier B——,
and inoculated on the arm of an infant, produced a full pustule,
having all the character of a vaccine pustule—form, evolution,
termination.

2. That lymph taken from this pustule, and inoculated on the
arm of three persons, had produced identical pustules, which are
truly vaccine.

3. That the transmission of the virus by successive generations
has not diminished the intensity of the force of the poison. One
of the set of cases presented large pustules, depressed in the center,
and filled with matter; each pustule served for several inoculations and the charging of several sets of glasses.

4. That from these facts it is evident that the virus taken from
the pustules of the farrier was identical with the vaccine."*

SURFEIT.

The term surfeit is used to designate an eruptive affection of
the skin, which usually makes its appearance very suddenly.
There is nothing dangerous about it, and it sometimes disappears,
in the course of a few hours, without any medical treatment. It
usually appears in warm weather, and especially among horses
that are in a state of plethora, from the use of a large quantity
of meal, or what is known as cut feed.

Symptoms.—A horse may be taken out of the stable apparently
in perfect health, except that he is fat, and, after being driven a

---

* Generales de Médecine, April, 1857 pp. 365-398, from the British and Foreign Medical Chirurgical Review.

-bort distance, he is brought back with his neck and sides covered with blotches or elevations. On pressing them with the fingers they crepitate, showing that they are gaseous elevations or distensions, originating in the cellular tissue beneath the skin, from the spontaneous generation of gas. The gas probably arises from fermentation of the food, and instead of accumulating in the intestices, finds its way to the surface of the body.

Treatment.—Dissolve two ounces of hyposulphite of soda in one pint of warm water, and drench the animal with the same. Then sponge the body with a portion of the following surfeit lotion:

> No. 34. Aqua ammonia........................ 2 oz.
>   Animal glycerine.................... 1 oz.
>   Water .... 1 quart.

If taken in time, the affection will disappear in a few hours. The patient should be kept on bran-mashes, well salted, for a few days; or a couple of weeks' run at grass, if the season permits, will be of great service.

DROPSICAL LIMBS (SWELLED LEGS, "STOCKING").

It is a very serious annoyance to the owner of a valuable horse to find that, after the animal has been standing in the stable for only a period of twelve hours, his hind limbs have wonderfully increased in size; and equally mortifying is it to see some animals with their limbs continually tumefied, and without any other apparent sign of disease.

Dropsy of the limbs generally proceeds from congestion. The serum, or more fluid part of the blood, transudes through the walls of the blood-vessels, and accumulates in the cellular tissue, under the skin. But dropsy takes place not only in consequence of venous congestion, but accompanies very many forms of disease, such as scarletina, influenza, farcy; and the cure of such diseases also accomplishes the cure of dropsy.

Causes.—The cause of that form of dropsical limbs which seems to occur without any other visible disease is now the subject for consideration. When a horse's limbs swell persistently, after a few hours' rest, and disappears under exercise, we may infer that there exists a dropsical diathesis in his system; that, by virtue of his constitution, he is predisposed; and such cases are designated as chronic local dropsy.

Treatment.—The remedies are hand-rubbing, exercise, and stimulating liniment. The best liniment for 'ocal dropsy of this character is as follows:

> No. 35. Fluid extract of wormwood............ 4 oz.
> Fluid extract of ginger... .... 3 oz.
> Spirits of camphor..................... 1 pint.

Rub the region of tumefaction with a portion of this liniment every night.

When swelling of the limbs do not assume a periodical character, and suspicion of predisposition can not be entertained, then, in addition to the application of the liniment, give the animal, morning and evening, one ounce of the fluid extract of buchu.

SCARLATINA (SCARLET FEVER).

Scarlet fever, or scarletina, as it occurs among horses in the United States, is a febrile disease of a very prostrating character; yet it is not considered by the author a contagious malady. It has only lately, however, been recognized in this country as a distinct equine disease, and, being of rather rare occurrence, we know but little about it. The cases that have come under the author's treatment were unattended by ulceration of the throat, and this may account for the non-contagion; for, in the human subject, when ulceration of the throat takes place, the case becomes malignant.

WATSON says: "The malignant sore throat may be caught from a patient who has mild scarlet fever; and mild scarlet fever may, in like manner, be contracted from one who is suffering under the malignant sore throat. The two forms graduate insensibly, in different cases, toward each other; and it would be impossible, even if it were desirable, to draw any strict line of separation between them. Many would say, and probably with truth, that the difference was this: in the one form, the poison of the disorder is seeking its vent, principally, by the throat; in the other, by the skin." It appears, therefore, that in the human subject the disease is capable of being communicated at any stage.

The cases that have come under the author's notice since he first recognized the malady have all been of a mild form; that is to say, non-malignant. The limbs, sheath, and pectoral regions were excessively dropsical; the skin was the seat of rash or minute ele-

,ations, and scarlet spots or patches were always present on the visible surfaces of the mouth and nostrils; yet the patients always drank and ate mashes freely, which they could not have done so readily had their throats been sore and ulcerated.

The author would intimate, however, that, should the disease ever become malignant in the horse, it may also become, as in the human subject, contagious and infectious. Perhaps the reason why it has not attained a malignant type in this country is, because it is of recent origin; whereas, among the members of the human family, scarlatina has been known to exist for several hundred years, and may have acquired intensity and malignancy with age. In anticipation of such an event, the author recommends that scarlet-fever patients be separated from other inmates of the stable.

The following case is from the author's note-book, and may serve to instruct the reader on the theory and practice of the malady: A short time ago, a bay gelding, aged about nine years, was admitted into the Chicago Infirmary. On examination, the following symptoms were observed:

Symptoms.—The hair about the neck and fore and hind limbs was elevated in blotches, with corresponding elevations on the skin beneath; the membranes within the cavities of the nostrils had scarlet spots, variable in size; all the limbs were more or less swollen; the breath was hot and fetid; the animal was unsteady in his gait, apparently very weak; had great thirst, yet little appetite. The elevations on the surface of the body, and the scarlet spots on the nasal membrane, constituted the diagnostic symptoms of scarlet fever.

Treatment.—During the first three days, I drenched the animal with three ounces of liquor acetate of ammonia, in an equal quantity of cold water. The elevations on the surface of the body were sponged once a day with a lotion composed of two ounces of aqua ammonia and one quart of rain-water. A sloppy bran-mash was occasionally furnished, of which the patient partook freely. This tended to keep the bowels loose, and moderate the febrile symptoms. On the fourth day all the limbs became very much swollen, from dropsical effusions, which feature of the case is very apt to present itself. I now changed the treatment— omitted the lotion bathing, and gave, night and morning, two-ounce doses of compound fluid extract of buchu and juniper berries. A slight discharge of glairy mucus now commenced running from both nostrils; the scarlet spots had become more diffused, and of a brighter tint; the appetite had returned, and a few oats and a small quantity of hay were occasionally allowed, which were eaten. The limbs were hand-rubbed often, and the body was clothed with warm blankets. This treatment was followed up for a period of five days, at the end of which time he had improved very much, yet the limbs were considerably swollen.

On the tenth day I commenced to give four ounces of fluid extract of resin-weed per day, with an occasional dose of fluid extract of golden seal, under which treatment he convalesced very rapidly, so that on the sixteenth day all medicine was discontinued. I now turned the animal into a small inclosure, where he seemed to take pleasure in rolling and exercising himself, which had the effect of reducing the dropsical swelling of the limbs, and on the fourth of December the patient was sent home—not exactly to go to work, but to receive care and attention until he had recovered from the effects of the disease, and had gained his ordinary strength.

It is very difficult to say to what extent scarlet fever among horses prevails in the United States; for very little, if any thing, has ever appeared in print of American origin, except that which has appeared from the pen of the author of this work. On the other side of the water, however, the disease is of frequent occurrence, and is well understood. The first account of it published in England was from the pen of Mr. PERCIVALL, in the year 1843, who says that "the existence of scarlet fever has received such confirmation from other quarters as to leave no doubt in my mind that, rare as the malady acknowledgedly is, and hitherto undescribed as it has remained, it will one day find a place in our established veterinary nosology."

Since the above date, Surgeon HAYCOCK and several other writers have noticed the disease, treated it, and minutely described its symptoms. They all agree that when the disease does make its appearance, it is usually the sequel of epidemic catarrh, or influenza; and this was probably the case regarding the patient the subject of this paper, for the owner informed me that the animal had previously shown symptoms of distemper.

RINGWORM.

This disease usually makes its appearance on the shoulders and sides of the horse, in the form of circular patches, attended by scurfiness and loss of hair. Small vesicular eruptions can also be detected, by means of a magnifying glass. It is usually spontaneous in its origin, but probably is generated by filth.

Treatment.—In view of treating the disease successfully, the body should be well cleansed with soap and water, and afterward rubbed dry; then apply daily, by means of a sponge, a portion of the following:

No. 36. Powdered sulphate of iron............ **4** oz.
        Fluid extract of bloodroot............. **2** oz.
        Rain-water .... **1** quart.
        Mix.

Give, as an alterative, half an ounce of the fluid extract of mandrake, in half a pint of water, night and morning.

When the disease has been neglected, the parts are apt to ulcerate; in that event, the following preparation is recommended:

No. 37. Pyroligneous acid.................... **6** oz.
        Linseed oil........................ **5** oz.
        Spirits of camphor.................. **2** oz.
        Mix.

Dress the ulcerations twice daily.

A liberal supply of cut grass, sliced potatoes, or carrots should be allowed, if they can be obtained. Particular attention should be paid to the matter of keeping the skin clean. The treatment must be continued until all vestiges of the disease have disappeared, or it may break out again.

PRURIGO (CALLED ALSO THE " ITCH ").

This is a disease of the skin, accompanied by a terribly annoying itching sensation. The torment experienced by animals suffering under this form of malady is scarcely describable. They often rub and abrade the skin until blood appears on the surface, and they are rendered perfectly miserable. There are said to be various forms of prurigo, but they do not differ in kind, only in degree. Heat aggravates the malady; therefore horses located in a warm and unventilated stable are apt to suffer most. And the same remarks apply to many other forms of disease, which shows how important it is that stables should be properly ventilated.

Symptoms.—The skin in those parts where the affection is located shows on its surface small elevations, known as papulæ; but they are difficult to discover in some horses, as they are of the same tint as the skin. After awhile the rubbing tears away the summius of the papulæ, and a secretion exudes which soon forms very minute scabs.

During the past four years the disease has prevailed very extensively among army horses, and it usually proved very obstinate, from the fact that very few of the army farriers understood its true nature; neither were they acquainted with the *modus operandi* of treatment. The authorities contend that prurigo is not a contagious affection, like itch and mange, and, therefore, it may be classed as a local affection, brought on by inattention to cleanliness. In order to prevent it, the author advises a free use of the brush and curry-comb.

Treatment.—The best mode of treatment is as follows. Sponge the affected parts thoroughly with the following lotion :

> No. 38. Powdered borax....................... 4 oz.
>   Rain-water .... 2 quarts.

After using the above quantity, at one or two dressings, then rub the affected regions dry, and lubricate the same with a portion of the following :

> No. 39. Kerosene oil..... .... 4 oz.
>   Pyroligneous acid................... 12 oz.
>   Mix.

It is usually good policy to put the patient under a course of alterative treatment; therefore the author recommends that he have, daily, thirty grains of iodide of potass, dissolved in four ounces of water, to be given in the form of drench, every day, until the disease has disappeared.

POLL-EVIL

The name of this disease originated in England, in consequence of it being more prevalent in that country than in some others; hence, was a great evil, and occurring about the nape of the neck or region of the poll, it was called " poll-evil." Much of the poll-evil occurring in former years was unnecessary, and occurred in consequence of neglect, injuries, and abuse. The disease first makes its appearance in the form of a local tumor, of an inflammatory character. Soon it runs into the suppurative stage; pus or matter is secreted, and we then have a common abscess, which may result in fistula of the poll.

Causes—The disease may arise from blows and bruises. Horses when located in low-roofed stables are apt to strike the poll against the ceiling, or a beam, or the upper part of the doorway, and frequent repetition of the act either ends in induration (hardening) or poll-abscess. Some grooms are in the habit of occasionally jerking the animal's head by means of the halter, and some horses are very restless in the stall, and are often seen to jerk their heads upward, forgetting that they have any halter around their necks, until they feel the pain which they themselves inflicted. The injury, slight as it may be, often remains undiscovered or unattended to until considerable tumefaction or even suppuration has set in. The case may then become protracted, and exhaust the patience of all concerned. Excessive friction on the nape of the neck from bridle or halter, or pressure from either one of these causes, produces poll-evil; and the halter, by being fastened too tightly, is a fertile cause in producing this affection. Dirt and filth are operative, also, in producing this malady. It is well known that the poll is a part which very seldom makes the acquaintance of the brush or curry-comb, yet it is the locality of considerable dust and scurf; and sometimes, owing to accumulations of the same, a cutaneous eruption arises, and the itching sensation consequent causes the horse to fret and rub whenever he can get a chance. The evil goes on, until what was at first superficial now becomes deep-seated. The bungling and cruel manner of sometimes forcing a small collar on a large horse often does mischief, and lays the foundation for poll-evil. Some persons are in the habit of bracing the horse's head downward with the martingale, so as to bring the lips and chest in close proximity, yet forget that the strain comes on the poll and bruises the soft parts.

Treatment.—Should the tumor be discovered in its early stage, before matter has formed, the parts should be bathed often with a portion of vinegar and water, equal parts; or else a cold-water dressing or pad may be applied, which should be reapplied as fast as it becomes dry. Half an ounce of powdered chlorate of potass should be given night and morning. This may be dissolved in half a bucket of water, which the horse will drink. In the course of a couple of days, the practitioner will be able to determine whether or not the tumor is likely to suppurate; if so, it will have increased in size, and have a soft fluctuating feel.

Just as soon as matter can be detected, the abscess must be punctured at its base, by means of a thumb lancet. It will not do to allow the matter to accumulate; for, if so, it will burrow and spread in various directions, making a very extensive and formidable abscess. It would be necessary to make a free opening into the abscess large enough to admit a man's finger, and if, in the course of a day or two, the opening should partly close or contract, it must be dilated with a knife. So soon as the abscess is laid open, all the matter must be squeezed out, which process must be accomplished by the use of sponge and hot water. We then, in view of removing every particle of matter, carefully inject the cavity with a quantity of pyroligneous acid and water, equal parts. A glass syringe is best, as the acid has a bad effect on a metal one.

It will be necessary to sponge the cavity once daily for several days, or until matter ceases to form. In the mean time the animal must be put under treatment, for it is very rare that this disease can be cured by local treatment alone. My usual custom is to give the patient thirty grains of the iodide of potass, twice daily, in half a bucket of water. The animal will not refuse to drink it. Should he be weak and emaciated, tonics and stimulants are indicated. Half an ounce of powdered golden seal and the same quantity of ginger, given with a small quantity of water, as a drench, daily, will answer the purpose. Sometimes it is advisable to put a pad of cotton cloth, or a large wad of oakum on the poll, in view of keeping the skin and sub-tissues in contact, by which means they more readily unite. The pad may be secured to the part by passing a cotton roller, five inches wide and three yards in length, around the neck. At the poll the bandage is to be further secured, by tying a lock of the hair of the forehead with a lock of the hair of the mane. This precaution prevents the bandage from slipping backward. The bandage should be removed and readjusted every day, and the parts are to be cleansed and syringed, in the manner just alluded to, as long as necessary. After the lapse of a few days, should the secretion of matter decrease, then the chances are in favor of a cure.

Cases, however, will sometimes occur which set at defiance all our skill. In such we find that either the ligamentary, tendinous, or fleshy parts, or perhaps the bones, are involved, and it often happens that fistulas or pipes run in various directions. Such cases require the services of a surgeon, who will dissect out the fistulous pipes, and remove all the diseased portions of the flesh and bone. The after-treatment is then the same as we have just indicated.

FISTULOUS WITHERS.

This disease resembles poll-evil, and, in a majority of cases, owes its origin to the same exciting causes; namely, external bruises, or injury inflicted by a bad-fitting collar or saddle. The treatment is the same as that recommended for the cure of poll-evil.

WARTS ON THE SKIN.

Warts are of two kinds. One of these makes its appearance upon, and consists in part of an increased development of, the outer scurf or scarfskin; the other kind is contained under, and, while it elevates, does not grow from the scarfskin. As it frequently becomes desirable to remove these growths, and to counteract that condition of skin on which their reproduction depends, we shall briefly describe them. The most common kind of wart, and that called the true angleberry, is the first of the two to which we have alluded. It has always a hard, horny sort of investment externally. This is formed of scales of scarfskin matted together in great quantities, and prolonged from the true skin, to which they are attached. As this outer covering increases in growth, it dries and splits up into shreds or fibers, which generally become more or less pointed at their free ends. Many of these shreds may grow from one common stock or pedicle, called the "neck." In other cases there is no well-defined neck, and a large surface will be covered with a coarse, rough, and horny crop of warts, all, as it were, united where they originated from the skin.

Cause.—Warts are always produced from what may be considered as greatly-developed knots, or papillæ, within the true skin. Blood, in greater quantities than natural, is sent to these, and the papillæ continuing to grow, forms perceptible tumors, which are somewhat sensitive, painful if exposed to friction, and form or throw off the scaly covering which eventually forms the main part of warts. We see many instances of warts on the human hands and other parts of the body, and always find that the dry, horny part grows from the skin somewhat thickened and increased in vascularity. So great, indeed, is the quantity of blood which some warts receive, that they bleed much more profusely than other portions of skin when wounded.

The other kind of wart to which we alluded does not cause any external roughness or scaly appearance. It is formed within the outer skin, and commonly appears as a circumscribed round tumor. On being examined by the hand, it is felt to be contained in a sack or pouch, within which it is easily moveable. It is almost insensible, and consists of a quantity of white fibrous and sometimes grisly substance, which does not possess blood-vessels in its interior. We often meet with this kind of encysted wart near the outer organs of generation in the horse or mare. The horny wart which we first noticed is common in horses, cattle, and dogs. The encysted wart is comparatively rare in cattle; it is more common in dogs, but is most frequently found in horses.

Treatment.—In treating warts with a view to remove them, and prevent their reproduction, it becomes necessary to destroy, in a measure, the surface from which they sprout. When an angleberry has a "neck," this should be cut off flat and close with the skin, and the raw surface thus made should be touched occasionally and freely with caustic or a heated iron. If the neck be so large as to endanger much bleeding if divided at once, a ligature of thin, strong string may be tied round it very tightly, in order to cut off its supply of blood. The ligature should be applied frequently—daily, if convenient. Where a large surface is covered by warts which do not possess any definite necks, their horny parts should be cropped or cut closely down to the true skin, with scissors or other convenient instrument, and the surface thus exposed should then be touched with a hot iron, or rubbed with caustic potass. Such application is to be continued till the horn ceases to grow. Encysted warts are removed by cutting the skin of the sack and squeezing them out. The exposed cavity should be dressed daily with digestive ointment.

PURPURA HEMORRHAGICA ("SCURVY").

A disease not unfrequently makes its appearance among horses termed purpura hemorrhagica, known in human practice as the "purples." It consists of congestion (extravasation) blood of and effusion of serum (water) into the cellular tissue. The disease probably owes its origin to a depraved condition of the blood.

Symptoms.—On making an examination of the affected animal, we find that the cellular tissue, in various parts of the body, is distended with serum and blood. Local swelling will appear in various parts of the body, more particularly about the face, lips, and limbs. The disease also affects internal parts. Blood is sometimes passed with the urine and feces; respiration is embarrassed; the heart palpitates, and abnormal cerebral symptoms set in. In the human subject the disease is considered strictly as a hemorrhage. Small round spots appear on various parts of the body and legs, of a dull crimson or purple color. Pressure upon them does not efface the color, nor render it fainter, as it does that of common inflammatory spots of the skin. There is scarcely any prominence of the purple stigmata; but they are sometimes intermixed with livid blotches, with appearances exactly resembling bruises, and they undergo, before they disappear, the same changes in color which attends the disappearance of a bruise. In fact, the anatomical condition of a bruise is exactly the same, with the diffused condition as in purpura. In each case the color is the result of ecchymosis (effusion of blood beneath the skin).

In the human subject, also, the disease is not confined to the skin. WATSON informs us that " the spots are not confined to the skin, nor to the subcutaneous tissues, but are found, occasionally, upon all the internal surfaces also, and within the substance of the several viscera (internal organs of the body). I have seen these purple spots in the mucous surface of the mouth, throat, stomach, and intestines; in the pleura and pericardium; in the chest; in the peritoneal investment of the abdominal organs; in the substance of the muscles, and even upon the membranes of the brain and in the sheaths of the large nerves, and I have known them to be accompanied with large extravasations of blood in most of the vital organs of the body."

The same appearances have been observed in equine autopsies, which accounts for the immobility and deranged condition of all the functions. Sometimes the throat is so badly affected that tracheotomy has to be resorted to. Now and then the tongue becomes livid and swollen, so that the animal can not partake of food. In such cases the tongue may be freely swabbed with warm vinegar, until its volume is reduced.

Treatment.—So soon as the disease is discovered, the animal should be put under the following treatment:

No. 40. Fluid extract of bloodroot............ 2 oz.
Fluid extract of buchu............... 4 oz.
Fluid extract of ginger............... 3 oz.
Mix.

Give one ounce of this mixture night and morning, in about a gill of water. Should there appear to be any trouble with the throat, lubricate it, twice daily, with a portion of the following:

No. 41. Olive oil........................... 4 oz.
Camphor........................... 1 oz.
Mix.

In preparing the above, it is best to pulverize the camphor, when it will dissolve more readily in the oil. A lotion must now be prepared, with which the whole surface of the body should be sponged daily.

No. 42. Spirits of ammonia.................... 3 oz.
Spirits of camphor.................... 2 oz.
Pulverized rock salt.................. 6 oz.
Rain-water.......................... 2 pints

Dissolve the salt in the water, allowing the impurities to settle; then add the clear liquor to the camphor and ammonia, previously mixed.

It occasionally happens that the skin, in various parts of the body, cracks, and portions slough off. Should there be any appearance of the kind, discontinue the above lotion and substitute the following:

No. 43. Pyroligneous acid............... } equal parts.
Olive oil...................... }

Every time the patient is watered, (say about three times per day,) half an ounce of hyposulphite of soda should be dissolved in the pail. Vegetables, also, are indicated, if they are to be had. I should give tomatoes, sliced carrots, potatoes, and green grass. Bleeding, purging, and all kinds of poisonous and sedative medicines are inadmissible.

OUT OF CONDITION (HIDE-BOUND).

Either from derangement of the digestive function, or in consequence of the animal being fed on poor provender or worked beyond his strength, debility and emaciation ensue. A horse out of condition usually loses flesh; the hair loses its glossy appearance; the skin becomes unhealthy and scurfy, and he can not perform his ordinary labor without apparent fatigue.

Treatment.—The best remedy, if the season permits, is a run at grass, taking care to give a good feed of oats every night; otherwise the grass will not improve his condition. In the stable the treatment is as follows: Give the animal one ounce of the fluid extract of camomile flowers every morning, before feeding, and at night mix one ounce of powdered poplar bark with the oats. This, together with good food and rest, will complete the cure

HERPES (TETTER).

Herpes is a disease of the skin, manifesting itself in the form of vesicles, which contain a small quantity of aqueous fluid effused beneath the true skin. Sometimes they are thinly scattered over the surface, and sometimes collected into clusters. The vesicles appear in irregular succession. They terminate, also, in various ways: by the reabsorption of the fluid, and, in bad cases, falling off of the hair, and peeling of the skin, in some places. In the human subject the malady is considered transient and non-contagious, consisting of red patches, of irregular form and variable size, upon each of which stands a crop of vesicles.

Treatment.—I have been very successful in the treatment of this malady by using the following lotion:

> No. 44. Glycerine............................ 2 oz.
>   Tannin.............................. 2 dr.
>   Rain-water.......................... 4 oz.

Apply once or twice daily, by means of a soft sponge.

In the horse the affection sometimes assumes an epizoötic form. In that event it can be communicated to man, as the following cases will show:

"At the commencement of 1849, Count Faverges invited Professor Papa to the valley of Borne, in Savoy, to see a disease affecting animals, and which even spread to men. Every facility was offered by Government to Professor Papa, and many persons having affected animals were requested to permit their inspection, and, indeed, threatened with a fine if they did not. Papa saw about three hundred horses and mules affected. The disease appeared in circular patches of furfuraceous scales, with grayish-white scabs. These patches had usually well-defined margins, about the size of a dollar or five-shilling piece. Usually they were isolated, but at other times they were confluent, or running together in groups. The head, neck, withers, shoulders, and loins were the parts chiefly affected. More rarely the upper portion of the extremities, and never on the lower part of the limbs, chest, or belly. The malady commences with a violent itching, and an eruption in small circumscribed points, about the size of a lentil, is witnessed. The scabs form, with the exudation drying and entangling cuticle and hairs. In the vicinity of the first, other eruptive spots appeared, which, widening, became confluent and run into one another, especially where the skin is folded and animals have a chance of rubbing themselves. A scab forms on the sore surface, and the surface beneath it is red and tumefied, but in a little time desquamation occurs. A very careful microscopical examination failed to indicate the existence of any acari.

The disease is contagious, and Papa says all those who come more or less in contact with herpetic horses or mules, and especially the conductors of the same, were covered on the arms, legs, chest, and face with pruriginous eruptions, limited and circumscribed, sometimes isolated, occasionally confluent, in the form of red patches covered with papulæ and vesicles, which become incrusted with brownish-yellow scabs, beneath which purulent deposits formed. In consequence of the violent pruritis attending this disease, it was believed to be scabies or itch by the people, and, though in many houses individuals were affected, they were ashamed to confess it, and it was with great difficulty that Papa collected information on the subject; but, having gained confidence on the latter, the people more freely related their cases to him. The first to be affected were those intrusted with dressing the diseased animals. The parts first attacked were the articular regions about the forearm, arm, face, and rarely the lower limbs.

Papa describes one of many cases of direct contagion. It occurred in a lad of sixteen, who had jumped on the bare back of an affected horse, to take it to a watering-place. Two days afterward.

DISEASES OF THE SKIN AND SUB-TISSUES.

on the inside of the thighs and legs, from the pubis to the calf of the leg, there was extreme pruritis. The skin became of reddish-brown color, and vesicles, full of yellowish lymph, formed, which gave way to vesicular patches or scabs, rough to the touch, first isolated, and afterward confluent.

The transmission occurred from horse to horse, horse to man, and from man to man; in fact, persons who never touched an affected horse were infected from individuals they came in contact with. A soldier, having arrived in perfect health from his regiment, slept with his brother, who suffered from the disease, and became affected. The wives of carters not only took the disease, but communicated it to their sucking infants.

Papa visited the district at a season when the fetid hellebore was growing abundantly. He caused some to be gathered, and obtained decoctions of the roots. With this he mixed equal parts of water, and distributed the lotion in different parts, to be applied to men and animals that were affected. Without any internal treatment, this application had the best effect. He tried tobacco decoction, but not with a similar amount of success. Papa regards the disease as the herpes tonsurans, which has been spoken of by many veterinarians as common in the live animals."*

---

* Edinburg Veterinary Review.

SECTION XI.

ON BREEDING, AND THE PRINCIPLES OF THE SAME

BREEDING—THE PRINCIPLES OF BREEDING—IN-AND-IN BREEDING—PERMANENT VARIETIES —ANIMALS BECOME PARENTS TOO EARLY—OBJECTIONS TO IN-AND-IN BREEDING ANSWERED—THE ADVANTAGE OF IN-AND-IN BREEDING THE IMPORTANCE OF STUDYING ANATOMY AND PHYSIOLOGY.

BREEDING.

THE choice of a sire and dam is a point of the utmost consequence in breeding horses (or indeed any other animal), as the offspring will be found, in almost every instance, to inherit the qualities of its parents. Peculiarity of form and constitution is inherent, and descends from generation to generation; hence the necessary attention to those niceties which breeders are often apt to forget. Nor is it sufficient that one of the parents be good and the other indifferent; for the perfection of the sire may be lost through the deficiencies of the mare, and *vice versâ*.

In the selection of a stallion many things should be observed. There should be general uniformity and compactness in every part. The height should depend on the occupation the foal is destined to fill. The legs should be particularly examined, and disease should pervade no part of the system. Fat, heavy horses, with thick legs, and coarse, unseemly heads, should always be avoided. Horses should be free from specks on the eye—partial or total blindness. Temper is an essential point, for vice is sometimes hereditary. Stallions that cover too many mares in a season, in the latter period produce weak offspring.

As regards the mare, it is well known that the dam contributes more to the integrity of the offspring than the sire. It is essential that she be in full possession of her natural strength and powers. The vigor of the constitution determines much in favor of the

foal. It is a great error to suppose that a mare that has once
been good, and capable of great exertion, should, when old, and
no longer fitted for work, produce offspring equally efficient as
when in her prime. The foal will certainly inherit some of the
weakness of the present nature and broken-down constitution.
Mares should never be put to the stallion until they have arrived
at maturity, which takes place about the fifth year. Mares are
bred from earlier, but it is a bad practice, for strength and beauty
are absent; and thus not only is the dam rendered inefficient
sooner than in one that is allowed to come to maturity, but the
foal can never be expected to be either healthy or strong in constitution. The period of going with foal is eleven months. After
the sixth month, great care and gentleness should be exercised toward them. Moderate exercise is essential. Hard work in harness, over bad roads, is likely to produce abortion, and mares that
have once aborted are very liable (if the same causes are in operation) to a recurrence of the same.

The proper time for copulation is when there is a discharge of
a whitish fluid from the vagina. She neighs frequently, and exhibits great desire for the horse. When she has conceived she
shows no further desire, and the discharge from the vagina stops.
Some writers recommend that the mare be put to the stallion early
after foaling. This is bad policy, for the simple reason that the
dam, if she become impregnated, has to nourish both the foal and
the embryo. This is excessively weakening, and tells on either
dam, fœtus, or foal.

Disproportionate copulations are also objectionable, as, for example, putting a large horse to a small mare. When improvement in size is the object, it should be attempted by regular
gradations. This is the best method of securing size and beauty,
as well as strength. The best mode of preserving the breed and
making improvements is, to make selections of the best on both
sides. Much judgment and circumspection is necessary, at all
times, in crossing the breed, and many errors arise in consequence
of a want of knowledge on this subject.

When the period of foaling draws nigh, the mare should be
separated from her companions. Having foaled, she should be
turned into a pasture, if possible; otherwise, into a loose box
inclosure, or empty barn. The foal may be weaned at the age
of six months, more or less, as circumstances seem to require. A

weak foal should run with its mother for a longer period than one that shows signs of vigorous health. Should the foal die at the period of parturition, humanity would seem to suggest that the mare be excused from duty for a week or two, by which time she will have recovered from the effects of parturi ior.

THE PRINCIPLES OF BREEDING.

It is a law of Nature that peculiarities of form, size, color, etc., shall be transmitted by parents to offspring, ("like begets like,") although, under certain circumstances, a modification of this law is to be expected. If, for example, we liberate an animal from domestication and its influences, which are known to operate very markedly on animal organizations and habits, the creature thus liberated loses its acquirements, and, in successive generations, gradually returns to the original type. This is a modification of the above law, and, supposing our pecuniary interests are the object of the experiment, it will be an improvement in the wrong direction.

On the other hand, take a wild animal; bring him under the influences of domestication, and he gradually loses all his distinctive characteristics of size, form, and instinct, and, in popular language, becomes a new creature, improved or not, as the case may be, under the direction of his lord and master; so that the inferior orders of creation are really the creatures of circumstances. These changes are the result of man's experience or non-experience. These are general propositions which "precede beauty and symmetry."

Beauty and Symmetry.—If we examine into the methods pursued by some of the most successful raisers of live stock, we shall see that they paid particular attention to the selection of well-formed, beautiful animals. They very naturally supposed that external conformation was transmissible; that if they happened to obtain a good calf or foal from inferior, diseased, or malformed parents, it was purely accidental, and out of the ordinary course of Nature. In selecting beautiful animals, they naturally excluded those of narrow chest, which peculiarity is indicative of predisposition to pulmonary affections and founder (the latter term signifying a worthless or ruined condition, which, in the eye of the law, renders them actually unsound), because they have that about them which may impair their future usefulness. Hence, for more reasons than one, such are unfit for breeding purposes, unless, however, the morbid phenomena be neutralized by blood, in one of the parents, of a superior order, free from taint, as it appears in the other.

In my adventures in search of the beautiful, I should pass by all animals having any peculiarity of conformation, general or local, which experience teaches me are sure to reappear either directly or indirectly, in future progeny. This appears to be the proper course to pursue; yet many, otherwise intelligent men will invest their money in the purchase of defective animals, merely because they can be bought cheap, when every body knows, or ought to know, that cheap animals, like cheap watches, cost more, in the end, for repairs, etc., than a commodity of more real value. The same rule applies to horses. Some men, if they happen to be in possession of a broken-down mare, unable, from faults, defects, old age, and disease, to perform ordinary horse duties, commence a speculation in a different enterprise. The reproductive organs must be exercised. Just at this time some cent-and-dollar calculating jockey advertises and exhibits a well-fed, prancing steed, bearing a popular name; but his whole carcass is redolent of faults and defects, which the owner is not expected to know any thing about, and if he does, it is not for his interest to point them out. The price of copulation is dirt cheap, and the foolish speculator expects to obtain a smart colt, that will eventually command a high price. But, alas for such milk-and-water calculations! the expected specimen of equine animality inherits and exhibits its parental infirmities and deformities, and ultimately becomes a living monument of the folly of its owner.

There is no beauty in the faults and defects which must necessarily occur through the sexual congress of faulty and defective animals; neither can there be grace in the motions of an overfed or ill-conditioned animal. Beauty, symmetry, grace, and soundness are defunct in the ill-conditioned creature, or otherwise defective one—a mere apology for the handiwork of Nature. The muscular system of a horse or cow may, in some regions of the body, be well-developed; but should their limbs be unnaturally curved, or any part of their system be out of proportion to the other parts of the body, then beauty has never seated itself on their thrones. Yet we can improve the appearance of such animals by artificial means and costly adornments. This artificial

20

kind of beauty is what most men seek after, and, at a high price, to their sorrow, obtain it; whereas the real beauty of form and endowments, which practical husbandmen really require, must originate unartificially in the living citadel—must be the work of Nature and not of art.

It is a fact, well known to many men, that animals of faultless form, having symmetrical proportions, are exempt from many diseases which are the heritage of faulty and inferior animals, and insurance companies take advantage of this knowledge, and insure the former at less rates than the latter. As an illustration of the above, I may be permitted to remark that the "Arabian Courser," English "blood horse," American "Black Hawk," and pure "Devon" cow—all remarkable for good points, beauty, and endurance—are exempt from many of the common and unnecessary diseases and abortions. Beauty and symmetrical proportions contribute to health, because there is a certain degree of relationship between external and internal organs and functions. For example, a fine exterior, good depth of chest, well-proportioned muscles, and graceful limbs are generally coëxistent with good "wind and bottom," and the subjects, of course, possess active respiratory organs, a finely-balanced circulation, and a digestive apparatus that can digest every article in the shape of fodder. In selecting the beautiful, we therefore obtain material for perpetuating health, long life, and valuable offspring.

One of BLAKEWELL'S great secrets was, "Breed from the beautiful." I allude to him because he was the most successful "breeder" that England ever boasted of. At the onset he paid great attention to "beauty of form." Having developed a fine exterior, he engrafted on it the useful. He was well aware that beauty and utility were not always combined, but, being in possession of the former, he could produce the latter to "order;" and he accomplished his object in the following manner. He required cows that would yield a large quantity of milk; consequently he selected those whose dams had long been celebrated for their excellent milking qualities, and from among such he chose the very best female of the family, and united her with a beautiful male. Having, in the production of offspring through the above union, accomplished his object, he then paid less attention to beauty, and more to the milking qualities, so as to render the latter permanent in the breed. Still, in all his experiments he rejected uncouth,

unthrifty, and defective animals. I am aware, however, that the
evils of domestication are operative, to a certain extent, on all
animals. Still, I contend that the " well-bred " animal can resist
the insidious encroachments of disease, and survive longer under
its depressing influence than "ill-favored" creatures. Illustrations are not wanting to prove the correctness of BLAKEWELL'S
theory; therefore, I advise the reader to follow the example of
one who has been styled the " Napoleon of breeders."

IN-AND-IN BREEDING.

We now inquire, What is breeding in-and-in? I answer, It
implies consanguinity—breeding from animals of the same blood,
or propagating in a close degree of relationship. Some persons
have an idea that this system is pernicious, and leads to degen-
eracy and premature decay; but that is a matter of argument, and,
as I shall attempt to prove, depends on the skill of the "breeder,"
and his ability to make wise selections. The human race—the
sons and daughters of Adam and Eve—afford a stupendous illustration of the practicability of in-and-in breeding. The millions
terrestrial all originated (so the Good Book informs us) from our
common parents, and, consequently, we are all "blood" relations;
and this fact goes to show that in-and-in breeding is in accordance with the laws of physiology, and does not conflict with the
intentions of "Him who doeth all things well." It is a part of
the great scheme of creation, a physiological law, the problem of
life, to solve which God has endowed us with reason, "God-like
reason," the exercise of which puts us in possession of the fruit
of the " tree of knowledge." The non-exercise of the same furnishes us with forbidden fruit.

Ill-assorted marriages—consumption mating with consumption,
and scrofula with scrofula—seeks at the altar an introduction to
an early grave. The same is true of animals. Bad selections
thwart the intentions of the Creator, mar his handiwork, and if a
friendly epizoötic sweeps them from the face of the earth, it is
more of a blessing than a curse. From the great human family
I select a branch to illustrate this principle, and I choose to refer
to the Jews. So long as they are so in faith and practice—mar-
rying their own kindred—they are a living illustration of the
above principle. Have they degenerated? No. The distinctive

characteristics of the race are just as remarkable as they ever were. The vicious and defective find an early grave, but the "race" remains pure.

Now, as regards horses and cows, the same is true; they all had a common origin. The different varieties that now exist are the work of time, circumstances, and the art of man. The famous horses mentioned by Mahomet, in the "Koran," resulted from in-and-in breeding. They have existed and multiplied for three thousand years without the slightest intermixture of foreign blood, and, from the time of Solomon up to the present day, their pedigree has been watched and chronicled with great care, so that no doubt exists as regards their consanguinity. Their fine form, splendid action, endurance, spirit, speed, and docility can only be retained by preserving the race pure, and this is an argument in favor of in-and-in breeding.

No breed can be preserved pure unless the in-and-in system be pursued. Take the Suffolk pig, for example. So long as we put Suffolk to Suffolk we get "pure" Suffolk, and, if proper selections have been made, good Suffolks are the result; but deviate from direct lineage, and the breed degenerates, for better or worse, as the case may be, and they lose their permanency of type, and cease to become pure bloods.

'From the brave descend the brave.'

The Suffolks in this country are notorious for a cutaneous disease simulating scrofula, and many suppose that this arises in consequence of the in-and-in system of breeding; but this I think is an error. It results from the evils of domestication, and our want of knowledge in making proper selections. There often is one or more animals in a litter incapacitated, by fault, defect, or debility, to perpetuate the stamina and remarkable points of the breed; these are to be rejected. If we fail to do so, the next generation, or the next to that, furnish more convincing proof of error, which I contend exists in making "bad" selections, and not in the above system.

Turn for a moment to the history of the French Merino, and we shall find that VICTOR GILBERT—a name familiar to America's most successful sheep raisers—practiced no other than the in-and-in system of breeding. A lot of Merinos were sent, in 1786, by the Queen of Spain, to the King of France. The latter, in order to benefit the agricultural community, sent half of them to Rambouillet, and the other half to Croissy. The climate happened to suit them, and they were considered superior to the existing breeds in France. In the year 1800, VICTOR GILBERT bought at Croissy a four year old ram and eight ewes. He bred from those animals during a period of ten years, occasionally, however, "renewing the stock," as he termed it, by purchasing annually from two to four sheep. In 1821 he purchased some of the Rambouillet stock, and asserts that, "from a union of the same, he obtained great advantages." Now, the reader will perceive that there was an alliance of blood. The two flocks were of the same lineal descent, belonged to the same "family," and of course the system of breeding was on the in-and-in principle.

Now, as regards French Merinos obtained from the above source, and imported into this country, we have only to refer the reader to TAINTOR, of Hartford; DEFORREST, of New York; SANFORD, of Vermont; PARKER and HOWARD, of Ohio, and many others, who are ready and willing, we presume, to testify as regards the prosperity of the above breed in this country, the result of in-and-in breeding. S. W. JEWETT, of Vermont, has purchased, up to the present time, over three hundred bucks and ewes, the offspring of the Croissy and Rambouillet breed, and this is proof positive of the correctness of our theory.

Now, as regards the advantages of in-and-in breeding, what are they? When VICTOR GILBERT purchased the first ram, he carried but 9 pounds of wool; he finally realized 24 pounds from the rams, and from the ewes 18 pounds, a tangible illustration of the advantages. A very fine French Merino, originating from the above source, was very recently exhibited at a sheep-shearing in Wisconsin; the animal weighed 325 pounds, and carried 40 pounds of wool. Now, the originals, as I have already shown, carried but 9 pounds; so that the standard, as regards a particular quality, has been raised, and that is what we understand by "improvement."

Mr. V. BARFORD, of England, has demonstrated the propriety of in-and-in breeding. He is a man, however, of superior judgment, and he culls and casts out from his flocks and herds all animals showing the least predisposition to any defect. A writer in the "Mark Lane Express" considers Mr. VALENTINE BARFORD "one of the most successful in-and-in breeders of sheep in the kingdom; for he has bred from the pure Blakewell blood for sixty-five years, without the aid of any other sort or kind, and bred from his own flock for upward of fifty years, maintaining size, weight, and constitution. I have known Mr. Barford's flock for upward of half a century, and they appear as strong in their stamina as when I first saw them. Mr. Barford does not use a ram unless he has a wide loin, a large breast and collar, and very wide between the forelegs, all of which points denote a hale constitution. Although in-and-in breeding has beat thousands to a stand, Mr. Barford still shines in it."

Therefore I contend that great advantages are derived from preserving the breed pure and in making proper selections. I have been very much pleased in perusing some remarks made by SANFORD HOWARD, and published in the "New England Farmer;" and, as they happen to have a direct bearing on our subject, I here give them:

"What is breeding in-and-in? While some apply the term only to animals distantly related, he conceived the only true idea of the matter to be that it applied to creatures of the same blood. The consequences of breeding in-and-in, he believed, might be either good or bad, and depended wholly on the skill of the breeder. As proof that breeding in-and-in was not contrary to Nature, he referred to birds, the buffalo, etc., in a wild state. It is known that they breed in-and-in constantly, and yet no deterioration takes place. He had known geese to be propagated in this way for forty years, and not the slightest depreciation in size, quality, or feather was visible in them. Still cases could be cited where breeding in-and-in had produced bad results, yet these might be attributed to imperfections in the parent stock. It is only necessary to select perfect specimens."

Having now attempted to show that in-and-in breeding is in accordance with Nature, and that the birth of inferior and defective animals, under this rule, resulted from the breeder's ignorance or neglect in making proper selections, I shall next introduce some remarks of a general character.

Permanent Varieties.—In-and-in breeding should be practiced in view of some specific object; namely, perpetuating species capable of constantly and permanently transmitting characteristic peculiarities, such as may be observed in any particular permanent variety. As an illustration of the same, we offer the French

Merino sheep, Black Hawk horse, and Devon bull. These are permanent varieties, the type of each having been established by a long course of in-and-in breeding. We are told, however, that many of the imported French Merinos (so called) have produced offspring scarcely worth the trouble of raising—far inferior to our common native stock. The principal reason is, many sub-races of Merinos, existing in France and elsewhere, are the result of cross-breeding. They are nothing less than mongrels, destitute of that permanency of type indicative of pure blood. But there are other causes for the deterioration complained of, which will be noticed hereafter.

By preserving the race pure through a certain period, we thus obtain the requisite permanency of type, which can then be transmitted indefinitely. Let us see, now, if we understand what is meant by a permanent variety, and inquire if such do really transmit their permanent peculiarities. A permanent variety (Black Hawk breed, or Devon cow, for example), are animals whose peculiarities were not coeval with the tribe, but have arisen, or been engrafted on the same, subsequent to the advent of their existence; and, therefore, what some may term distinct species are, in reality, nothing more than permanent varieties. The Shetland pony, therefore, is a permanent variety. Circumstances have made him just what he is. He, being an inhabitant of a stormy, tempestuous region, comparatively destitute of shelter and artificial food, is left, in many instances, like the cattle of Norway, (whose diminutive size is notorious,) to seek such food as the county affords; and, owing to their being secluded from other horses, frequent intercourse among themselves have rendered their small and peculiar forms permanent in the race; and, with unerring precision, "like begets like."

The permanent varieties that now exist among the cattle of the British Isles is due to local circumstances and the art of man. Most of the fine breeds now owned by English husbandmen are named after the location in which they are said to originate. These are permanent varieties, yet the most of them (in fact all, except importations) are descendants of the "Wild or White Forest breed." Hence the Highland breed, dispersed over the mountains of North Britain; the Alderney, of the Norman Isles; and the Devon, from the Bristol Channel, etc., are all permanent varieties, originally of the White Forest breed. Yorkshire is celebrated for the production of a permanent variety of horses known as "Clevelands." They are an ancient and unmixed race. Their bodies have peculiar markings, and if a foal should be born without them, the owner would give orders for its immediate destruction. I presume it is now understood what is meant by permanent variety.

Animals become Parents too early.

Victor Gilbert never allowed ewes to have lambs until they had passed their third year, and the bucks were never used until they had arrived at full maturity. He, as well as many other sagacious stock-raisers that I might name, are probably conversant with the fact that, during the period of growth and development up to maturity, the reproductive organs are dormant, while, at the same time, the nutritive function was wholly engaged in elaborating chyle and blood for the development of bone, muscle, and nerve; and that by calling into requisition the reproductive or generative organs before the animal had attained full growth, must necessarily divert the elements of matter intended for nutrition from their legitimate channel, and direct them to the reproductive organs. This is precisely what takes place. A too early use of the purely animal functions induces weakness and stunted growth.

The author remembers that, in his school-boy days, it was customary, so soon as a boy had accumulated a few pence, to invest the same in a rabbit (a favorite animal at that period); consequently we had a community of juvenile rabbit-raisers; and from the results of past experience, as observed by the older boys, a rule was adopted among us that the doe should not be put to the buck until she had attained full maturity. The reason assigned for this rule was, "Early breeding prevented the doe from growing." The facts were evident; still we knew not the why nor wherefore, but acted on the spur of experience. Physiologists have assigned the above as the true cause, and have, in their writings, pointed out the woeful results which often follow a too early use of the purely animal functions and organs. I seriously urge farmers to give this subject that attention which it evidently demands. By so doing they will prevent many of the unnecessary abortions which are almost constantly occurring. Among the higher orders of creation the same law holds good.   In fact, there can be no deviation from it without incurring the hazard of paying the penalty.

Heifers should never be put to the bull until they have attained the age of three or four years.   At this period they are in their prime.   If they happen to have acquired too much fat, their daily allowance must be reduced.

Human growth, according to the best authority, ceases between the ages of twenty and twenty-five.   In very warm regions, however, where development and decay are universally allowed to be more rapid, the inhabitants come to maturity much earlier.   A superior class of human beings can only be produced by selections and exclusions similar to those so successfully employed in rearing the inferior orders.   We may rejoice in a Fulton, Franklin, or Webster occasionally, (by mere accident,) the parents of such being absolutely ignorant of the first principles of physiology; but, in the breeder's language, such are in possession of the "pre-requisites."   In the first place, they had not entered the marriage relationship prior to the age of reason.   The parents, if we mistake not, were full-grown men and women, not boys nor girls. They possibly possessed a sound mind and healthy constitution, free from hereditary defect of mind and body, which stunted growth, aided by artificial modes of life, are almost sure to entail. Until within a few years, a wise and salutary law was in operation in the British Isles which interdicted marriage until the candidates had arrived at the age of twenty-one.   That law has been set aside, and, consequently, the mass of the population of the present day will not bear comparison with that of the past.

One safeguard, therefore, against stunted growth and ill-health is to avoid a too early use of the reproductive function.   And herein we are not safe unless proper selections have been made, and faulty animals or persons rejected.   I am persuaded that if husbandmen of these United States could all view this important matter in its right light, a very decided improvement would soon be the result.   The subject will, however, eventually command the attention of all.

Many persons urge that the offspring of blood relatives are weak, nervous, deficient in intellect, complete idiots, etc., and that further intercourse in the same direction only makes bad worse.   These are facts which can not be controverted, yet the weakness, nervousness, etc., does not occur from the relationship which exists between the parties, but is the result of their defective constitutions. Many of the cousin-marriages are entered into without the least regard to the rational theory of selection and rejection. The question is one of interest, not of utility. These blood connections are consummated merely to retain property in families, and to satisfy the caprices of doting parents, superannuated uncles, and disappointed maiden aunts; and thus many an idiotic or emaciated sprig of aristocracy unites his destinies with a fashionable cousin the very counterpart of himself, both inheriting family defects of precisely similar character. Should they be blessed (cursed, rather) with offspring, may we not expect a miniature of the originals? "Without due attention to avoiding disease, it will increase in the progeny."

Among the human species, therefore, where the objects are of such immense importance to present and future generations, the principles of selection and rejection are entirely overlooked. We pay more attention to the principles of breeding in cattle than we do to those applicable also to their lord and master. Hence the faults and defects alluded to are not the result of close affinities, but result from our indifference or ignorance of the laws of life.

OBJECTIONS TO IN-AND-IN BREEDING ANSWERED.

The two preceding paragraphs are brief answers to the objections against the sexual congress of blood-relations; yet, as other evidence than my own, and of a more general character, may be demanded by the reader, I have thought it best to introduce other testimony. The following quotations are from the pen of P. S. HUMBRICKSON, of Ohio, published as a "Prize Essay:"

"Many object to what has been termed in-and-in breeding. Indeed, with most of those who do so, their objections assume the form of a very strong prejudice. This is attributable to a certain confusion in their minds, by which reasons, wherein there is no proper connection, are made, with the aid of imagination, to assume a form of mutual dependence and coherence in support of their theory. In the human family, marriages within the Levitical degree, or, as they are also called, incestuous marriages, are forbidden. But the prohibition rests exclusively upon moral, and not at all upon natural or physical reasons; yet the direct physical calamities are imagined to pursue the infringement of that law

Now, it would seem that where the moral reasons in favor of an enactment are sufficient not only to justify but to demand it, no more should be required. And this is in accordance with the soundest rules of philosophizing. Moreover, if no such moral reason existed, the authority of the Lawgiver, in the case of the divine law, is a sufficient foundation and warrant for it ; and the human law is bound by and but follows the divine. Shall we, then, take the step beyond, and apply to the inferior animal the rule that was intended for rational, accountable man? We know that, if left to his own observation and experience, and being himself the judge, the objector must avouch the fact that, in the state of Nature, there is no restraint, and that such connections are and must be continually taking place.

The Advantages of In-and-in Breeding.

" And if there are advantages arising from the having placed in the line of the direct ancestry, near and remote, of our flocks, a great number of approved individuals, both male and female, as has been seen, it follows that there must be far greater advantages arising from the having the same one individual (if he be of marked superiority) placed in that line the greatest possible number of times. This is done by in-and-in breeding, and is the object of it. Now, it is easier to find this one unsurpassed individual than to find many ; for in the many there will, most certainly, be one to be preferred to all the rest. Then, under the operation or the principle of atavism, the chances that the resemblance of such unequaled ancestor will be obtained must be in the ratio of the number of times that he occurs in the ascending lines ; hence greater uniformity and greater excellence in all the progeny. An apt illustration of this is found in the frequent occurrence of the Godolphin Arabian, in the pedigrees of all our best blooded horses, carried back to him as their founder."

In conclusion, the author offers the following, in view not only of benefiting the horse, but his owner also.

The Importance of Studying Anatomy and Physiology.

There is an aphorism which is said to have emanated from a Divine source, and it appeals to the intelligence of this enlightened and progressive generation with peculiar force. It is, "Know thyself." To understand our moral, intellectual, and physical natures and tendencies should be the business of every one. The cultivation of, and proper direction given to, the former, bring us within the halo of health, purity, and peace. A knowledge of the physical or physiological laws of life, and in practicing fidelity to what they teach, places us on the high road to health and long life. Anatomical and physiological acquirements are needed by every one, in order that we may know ourselves, and thus be able to preserve our wondrous mechanism, "the harp of a thousand strings," in its normal condition; for without health we can not enjoy life, nor answer the purposes for which we were created.

We have the testimony of learned men, and our own reflective minds confirm the facts, that an alarming number of premature deaths, and an untold amount of physical infirmities, are the result of either our ignorance of, or indifference to, the uncompromising laws of Nature. The rational being, free from hereditary taint, of mental or physical deformities, comes into the world with all that is essential to his future life and happiness. He has within his organization a radiant volume of intellectuality, organized, compiled, and bound by the Divine hand, the first glimmer of which reveals something adapted to present and future wants and necessities. The intellectual spark, once ignited, is capable of an endless increase. We can add ray to ray, power on power, until the God-like man acquires the mental greatness of a Webster, or the mechanical skill and distinction of a Fulton. We do not expect that all can become Websters and Fultons, because we do not all practice that invariable perseverance and stern energy characteristic of giant intellect and mechanical skill. We are not all willing to toil, mentally or physically, with that perseverance and industry so necessary to success; and if we were willing, after having attained maturity or manhood, our minds and bodies being trained and molded for station and circumstances, are then not so well adapted, as in youth, for increase and power. Still, at this period of life, we are not destitute of the means of self-culture.

We are living illustrations of progressive mental capacities which often transpire in men past the meridian of life, who, before that period, never considered themselves favored in this line. How unfortunate it is, therefore, that the generality of mankind should be so indifferent about the science of life, and be so slow in making it the basis of rational action in warding off disease, and in promoting the integrity of their existence! The reason is, our education is incomplete. We are all that Nature intended in elementary ability, and only deficient in its use. The means are anatomical and physiological studies; the time is in youth, when the mind is pliant, capable of receiving permanent impressions. The place for the engrafting of the same is in our common schools and at the fireside, the mother and the teacher wielding a common scepter of instruction.

Some people do not realize that our organizations are the result of the most positive laws of Nature, and that our ailments are the result of our own ignorance or folly. The mass of mankind are not aware that our physical systems are capable of improvement analagous to the mental. No; they generally think and act on the false and ruinous proposition that our diseases, aches, and pains are so woven into the filamentary mechanism of the living citadel as to be beyond the power of either art or science to eradicate—a proposition that should never, for a moment, be entertained; for, if we live right up to the laws of life, we are then within the impregnable ramparts of physiology, where our acquired and fashionable maladies can not obtain.

Within the bulwarks of physiology certain conditions are imposed upon us, and we must observe them. For example, we require a pure atmosphere, at all times, to vitalize the blood, and thus deprive it of those defiling elements acquired by venous blood, and which would otherwise operate, as they often do, in our crowded assemblies and unventilated sleeping-rooms, as the germs to excite unnecessary disease. Next, the body should be kept erect, so that the muscles and organs may acquire and maintain natural action. We require vigorous daily exercise of all the muscles of voluntary motion; freedom from all compression by dress; apparel that shall afford needed protection; a quantity of food and drink, at intervals, that shall furnish materials for the wants of our systems, and that control of feeling which enlightened reason and virtue demand. Our time should be distributed into daily periods of labor, rest, and amusement; and, above all, we must bear in mind that infirmity wedded to infirmity is a sin against our nature—a wanton violation of the law of Nature and of our existence, to which a fearful penalty is appended, even unto the "third and fourth generations."

The faithful observance of the physiological laws of Nature result in health and strength, and promote longevity. Such is the fruit which the tree of knowledge beareth. The non-observance of these laws of life places us on a par with the poor idiot, when sporting with the implements of self-destruction, and furnishes to us the forbidden fruit. Health depends entirely on the use which we make of the various organs and functions of the body. Health never comes to us or our live stock by virtue of good luck. None are healthy by special Providence or Divine favor. Neither can health be found in the popular nostrums of the day on the proceeds of which quacks grow fat and wealthy, and patients often become poverty-stricken, and finally sink into the arms of Death, in blissful ignorance of what brought them there. Much of what is here written is equally true regarding horses and cattle. Many of their diseases spring from the same causes which are operative on the human subject; hence, knowledge of the above description is what all interested should seek after.

SECTION XII.

LAMENESS FROM VARIOUS CAUSES.

Curbs—Elbow Tumor—Capped Hock—Varix, or Bog Spavin—Atrophy, or Wasting of Muscles—Thorough-pin—Navicular Disease, or Lameness—Pleuro-dynia—Acute Rheumatism—Chronic Rheumatism—Ring-bone—Splint—Spavin—Osteo Porosis—Curvature of the Spine—Hoof-bound—Laminitis—Sandcrack—Quarter Crack—Navicular Thritis—Stifle Out—Contraction of the Hoof—Bowed Legs—Corns—Lameness.

CURBS.

CURB consists of sprain of the posterior ligament of the hock, known as the calcaneo-cuboid ligament. It is thus named in consequence of being inserted into the oscalcis and cuboid bones. The tumefaction is usually confined to the skin and subcellular tissue. At first the part is hot and painful, and the animal is lame—does not flex the hock in a natural manner; yet, when standing still, the horse keeps the lower part of the limb flexed, so as to relieve the overstrained tendon. The liability or predisposition to the disease lurks in breed. Such horses are observed to have ill-shaped hocks. Such have been denominated "cow-hocks." The parts are not symmetrical. There is an obvious disproportion between the width of the parts above and below the hock, which gives it an angular appearance; and the more angularity there is, the more liability is there to sprain, because there is too much tension on the calcaneo-cuboid ligament.

Causes.—The exciting causes of curb are sprain, occurring in the act of backing a heavy load, or in leaping; in short the performance of any feat that violently calls into play the flexor muscles and tendons of the hind extremities; hence stallions with faulty hocks, when used for procreation, are very apt to become the subject of curb.

CURBS AND ELBOW TUMOR.

EXPLANATION.

1 1. Curbs, or enlargement in the region of the back or posterior part of the hock, an affection known to veterinarians as sprain of the calcaneo-culoid ligament.

2. Elbow tumor, or enlargement at the point of the elbow. The point of he elbow corresponds to the olecranon of man.

Treatment.—So soon as the injury is discovered, the part should be fomented often with either an infusion of hops or poppy-heads.   It is best to apply the infusion cold, and by means of a sponge, the animal being kept at rest during the acute stage. Sometimes, in view of relieving the tension of the sprained liga-ment, it is customary to tack on a high-heeled shoe.   This may or may not be necessary, and depends on the severity of the case. After having modified the inflammatory action, the high-heeled shoe may be dispensed with.   Then, in order to get rid of the chronic enlargement, the part must be daily anointed, for a short time, with a portion of the following:

> No. 45. Powdered iodine...................... 1 dr.
>   Simple ointment...................... 7 dr.
>   Mix.

After a short time, the animal may be put to light work, still making an occasional application of the iodine ointment.

Should the parts about the hock degenerate into a hard callous, friction with cod-liver oil may be employed.   Should this fail to have the desired effect, then make a few applications of the acetate of cantharides, prepared as follows:

> No. 46. Acetic acid (fluid)............. .... 1 oz.
>   Water................................ 5 oz.
>   Powdered cantharides................. 1 dr.
>   Mix.

ELBOW TUMOR.

Elbow tumor (see cut on preceding page, fig. 2,) is usually re-cognized as "capped elbow."   It sometimes grows to an enor-mous size, and not only becomes unsightly, but seriously interferes with the action of the elbow and its articulation.   At first, the part is hot, inflamed, and tender.   Soon effusion takes place; then the part is soft and fluctuating; next, albumen is deposited, and this, finally, is metamorphosed into fibrine, at which stage the tumor is large and dense; the walls of the skin are also very much thickened; in fact, when dissecting out the tumor, it appears that the skin is in a state of hypertrophy, or morbid growth.   How-ever, there is usually a central orifice in the tumor which seems to indicate that the first trouble originated in a bursal sac, from external injury, either by the calking of the shoe or the toe of the hind foot; or it may arise from a bruise inflicted in the act of lying down on a hard floor.

Treatment.—In the early stage, and when there is nothing but fluid in the sac, an opening should be made into the lower part of it, by means of a thumb lancet. After the fluid has been evacuated, by pressure, inject into it an astringent, composed of pyroligneous acid and water, equal parts; and rub the exterior of the tumor, twice daily, with a small quantity of cod-liver oil. The cod-liver-oil will act as a glandular stimulant, and thus aid in the absorption of the tumor. A seton, inserted through the tumor, sometimes has a very good effect; but if the tumor be large and unsightly, the services of a surgeon will be required for its removal, by means of the scalpel.

CAPPED HOCK.

Capped hock consists of tumefaction, or enlargement at the point of the hock. Occurring, as it does, at a very prominent and exposed part of the limb, it is almost sure to attract notice; hence is not only an eye-sore, but a serious defect, and, if accompanied by lameness, the animal may be pronounced unsound, during the period of lameness. The tumor is the result of enlargement of a synovial sac which lies between the bone and the tendon gastrocnemii. This sac contains synovia, and it is furnished for the purpose of lubricating the parts for the prevention of friction. The skin, also, becomes thickened, or, rather, is in a state of hypertrophy (morbid growth), which add largely to the size of the tumor.

Causes.—Capped hock is generally produced by a blow or bruise. It is almost always occasioned by the whiffle-tree coming in contact with the parts. The animal may, however, injure the parts in the act of kicking. The parts may also be bruised in the act of lying down or rising in the stall. Therefore, not knowing any other cause for capped hock, we may safely infer that it is the result of external injury.

Treatment.—In the early or **inflammatory** stage, refrigerating lotions are indicated. Take of

No. 47. Sal ammonia......................... 1 oz.
   Nitrate of potass..................... 1 oz.
   Water..................   .... 1 pint.

Sponge the part quite often, for a day or two, until the heat and attendant pain subsides; after which the affection will run on to the chronic stage, and perhaps acquire magnitude; then the treatment must be altered. The object now is to excite absorption. In that view, I recommend the following:

> No 48. Powdered iodide of potass............ 2 dr.
> Yellow basilicon.................... 1½ oz.
> Mix.

Apply a portion of this ointment to the enlargement, morning and evening.

VARIX (KNOWN AS BOG SPAVIN).

Varix, or " bog spavin," consists of an enlargement on the inside of the hock, corresponding to a disease of the same character in the human subject known as varicose veins. It very rarely produces lameness, and most writers contend that it is nothing more than an eye-sore. Be this as it may, it is a defect, and, in the eye of the law, constitutes unsoundness, simply from the fact that soundness consists in perfection of function and structure, and, also, because sound horses do not have defects of this character. The enlargement does not generally impair the usefulness of the animal for ordinary work; yet sometimes, in consequence of over-work or sprain, the tumor acquires magnitude.

It is always dangerous to attempt to remove varix, or to puncture it; and the only proper method of treatment, in view of limiting its growth, for it can not be cured, is to bathe it often with the following astringent lotion:

> No. 49. Fluid extract of witch-hazel............ 2 oz.
> Fluid extract of bayberry bark........ 4 oz.
> Proof spirit.......................... 1 pint.
> Mix.

Bog spavin is an indication of weak and faulty hocks, unless it can be shown that the animal has been made to labor too early in life, or has performed excessive work beyond his capacity. Then the affection can be accounted for, and may not owe its origin to either fault or defect.

ATROPHY, OR WASTING OF MUSCLES IN THE REGION OF THE SHOULDER-BLADE, (KNOWN AS SWEENY.)

Atrophy signifies wasting of muscles, in which diseased parts become notably smaller than those of the natural, without other alteration in texture. The author has had several opportunities of examining horses that have died in consequence of disease or old age, who, during life, were the subjects of what is known as sweeny, and found that the muscles of the shoulder had quite a pale appearance, somewhat resembling those of a calf which had been bled to death. A condensation or wasting of muscular fiber was very apparent; the cellular membrane under the skin had almost disappeared, merely by condensation and contraction of the skin over the region of the parts, which contraction occurs in consequence of shrinkage of the muscles beneath the skin.

A FOUNDERED OR DEAD LAME HORSE. (See p. 336.)

Causes.—As regards the cause of atrophy, it is a well-known fact that inaction of a muscle produces a notable decrease in size. The muscles of laboring men decrease in size whenever, from choice or necessity, the individual leads a sedentary life. A limb kept at rest for any length of time undergoes more or less atrophy (wasting). In some cases, the foot diseases, so common, are of that character as to cause the subjects much pain in progression;

and, in order to relieve the feet, the horse spares the muscles, which, for want of free and full action, leads to diminished nutrition and wasting (atrophy). Some of the readers of this work must have noticed the fact of the shrinking of the muscles of the haunch in case of disease or severe lameness of one of the hind feet. This shrinkage often occurs, as in the case of the shoulder, from want of natural action and full play of the muscles concerned. We have ample proof that wasting is the result of want of action, in the fact that when a horse becomes the subject of a painful disease in the hind foot, he neither advances nor rotates the limb any more than he can possibly help, but holds the foot up, sometimes points it. On examination, the atrophy, or wasting, is confined to the muscles which perform these two actions; and in the case of pain or lameness in one of the fore-feet, the same faulty action is observable, and the same class of muscles (whose action is almost identical with that of the hind parts) are the ones affected.

The author deems it of great importance that husbandmen should fully understand this subject; for it will enable them to perceive that, while inflating horses' shoulders with a quill, or practicing any other absurd treatment in the vicinity of the wasted or sweenied shoulder, they are overlooking the real malady (in the foot), and at best are only treating symptoms. On careful examination, however, should it appear that the patient is free from any disorder of the foot or feet, and, on the other hand, it is clearly evident that myositis or myalgia (muscular pain) exists, then the treatment must be directed to the affected parts. The facts in either case can only be determined by a competent practitioner.

The author has often noticed that when horses have been over-driven or overworked, they become stiff and lame in those parts of the body most susceptible to the ordinary influences. For example, one horse may perform a long journey, with a weighty man on his back, and the next day show symptoms of lameness in the lumbar region. His back is arched; the limbs are brought under the center of the body, and every movement indicates that the animal suffers tormenting pains, which are located in the muscles of the back. This is myositis, or inflammation of the muscular fiber. Sometimes, however, the malady constitutes myalgia (muscular pain and spasm), unaccompanied by inflammatory action. Another horse may be put to a similar kind of work, and, in a few hours afterward, be found stiff and lame.   But the symptoms are not the same as those just recorded.   The pathology is probably identical; it is the same disease, only it has a different locality. It is myositis of the pectoral muscles, involving, also, the muscles of the shoulder-blades.   And the intelligent owner of the afflicted animal will notice that the fore extremities are unnaturally advanced; the foreparts of the body are unnaturally hot and tender; the fore-feet are feverish, and the hoofs are hotter still.   The advanced position of the fore extremities and shoulders gives a very marked hollowness to the forepart of the chest, and many men, on seeing such a case, would declare that the animal was chest-foundered, this being the name which is sometimes given to such a condition; but it is really owing to myalgia (muscular pain and stiffness).

It is well-known that all muscles are subject to inflammatory action and muscular pain from work disproportionate to their strength.   For example, let any man undertake to saw wood, practice with heavy dumb-bells, or ride on a horse—feats that he has never been accustomed to—and, whether the labor be excessive or not, the individual will shortly complain of more or less muscular pain; and if he be a weak man, the more excessive will be that pain.   Now, the muscles of the horse, being just as susceptible to pain as those of man, are just as easily operated on through the well-known exciting causes.

Symptoms.—In the case of a horse, when the work has been excessive, and of a character to bring the muscles of the shoulders, their tendons and coverings, into a state of over-exertion, it will often be found that some of the muscular fibers in the region of the shoulder are fractured, small blood-vessels are ruptured, and other pathological changes take place.   This is followed by inflammation (myositis).   Suppose, therefore, a horse shows symptoms of myositis after excessive work, and, in the course of a few days, it is noticed that the muscles of his shoulders are wasting away, and continue to waste, so that all persons who examine the animal pronounce him sweenied, we may then infer that the wasting (sweeny) is not symptomatic, but is the result of myositis.

Treatment.—The principal treatment is rest; afterward diminish the work, and, by proper diet and tonics, increase the power to do it.   The muscular parts affected should be bathed, morning and evening, with a portion of the following:

No. 50. Fluid extract of wormwood...... } each 2 oz.
Fluid extract of poppies........ }
Proof spirits........................ 1 pint.
Mix.

Should the feet prove to be very hot and feverish, frequent bathing with cold water, or a cold infusion of hops, will have a very good effect. In the mean time give the animal two drachms of fluid extract of gelseminum, once or twice per day, until the feverish symptoms subside. Then, when the case passes into the chronic stage (sweeny), the shoulders and wasted parts are to be well rubbed, often, with a portion of the following:

No. 51. Fluid extract of ginger............... 4 oz.
Gum camphor...................... 1 oz.
Olive oil........................ ½ pint.

Pulverize the gum; add it to the oil, in which it will soon dissolve; then mix with it the ginger, and the preparation is fit for use.

Should the muscles of the shoulder waste, in consequence of chronic disease or altered structure about the hoof or within it, such as ring-bone, ossification of the lateral cartilages, etc., then a strong counter-irritant must be applied, for a week or so, around the coronet. At the same time we should treat the shoulder as above. A good counter-irritant for the coronet may be thus prepared:

No. 52. Cod-liver oil.................. } equal parts.
Kerosene oil.................. }
Mix.

The great trouble in chronic cases of this character is, that it usually takes a long time to free the animal from lameness, and to restore the muscles of the shoulder to their original size; and, after all we can do, it may be necessary to let the animal have a run at grass. It will be advisable, in all chronic cases, unless the horse be very lame, to insure regular exercise, which will tend to develop the shrunken muscles.

The best way to prevent sweeny is to keep the horse in good condition by a rational use of his muscular powers, being careful not to overtax them, and by giving proper attention to his feet, both in the stable and at the blacksmith's shop.

THOROUGH-PIN.

The seat of thorough-pin is between the popliteous (fig. 2) and the point of the hock, near where the tendon is severed, as shown in the accompanying engraving. The disease is called thorough-pin, simply because the fluid contained in the bursal sac can be squeezed from one side to the other.

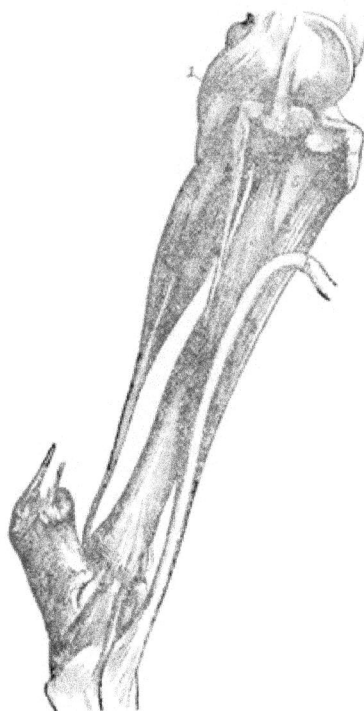

VIEW OF SOME OF THE DEEP-SEATED MUSCLES IN THE REGION OF THE HOCK AND STIFLE.

EXPLANATION.—1, Popliteous; 2, Flexor pedis accessorius; 3, Flexor metatarsi magnus; 4, The tendon, common both to the flexor metatarsi and extensor pedis; 5, The groove in which the extensor plays.

Symptoms.—The disease is similar to bog spavin and wind-galls; namely, enlargement of a synovial sac. It seldom if ever lames a horse, yet, when large, is apt to interfere with the free action of the joint.

In former years it was customary to open thorough-pins by means of a lancet, in view of evacuating the fluid; but many fatal

cases having occurred, the practice is now discontinued. Among
the many fatal cases recorded I select the following, which has
been recorded in the " Veterinarian," by W. A. WHEATLEY, V. S..

"The subject was a five-year old cart-horse, the property of the
South-eastern Railway Company, admitted on the 1st of September, 1855, with an immense thorough-pin on the off hock. The
animal being then in a very high condition, a support shoe was
placed on the foot of the diseased limb, and the frequent application of cold water to the diseased parts resorted to, the diet being
restricted to mashes.

October 6.—The horse was considered in a fit state for the operation. Accordingly, I had the animal secured by a side line, and
first punctured the enlargement with the exploring-needle, in four
different places; but, as the imprisoned fluid did not escape so
readily as I expected, I determined upon opening the part with
the lancet, which being done, the fluid escaped to the amount of a
small tea-cupful. It was of a thin consistence, and of a pale straw
color. I then applied bandages dipped in cold water, and kept
the hock constantly wet.

October 7.—The swelling had very slightly returned. I gave
some aperient medicine, and continued the bandages with cold
water, and applied a suitable truss to the hock.

I need not give a report of the case from day to day. Suffice
it to say that the cold water and the truss were continued in use,
and the animal never manifested the least inconvenience from the
operation.

November 21.—He was considered cured, not the least appearance of the enlargement being left.

November 29.—On this day the horse was attacked with abdominal obstruction, and, despite all our remedial measures, consisting
of active purgatives, as solution of aloes, calomel, oleum lini et
crotoni, with the frequent use of medicated enemata, counter-
irritants, etc., he continued to grow worse till the 1st of December,
when death took place. This was the fourth attack of this kind
that the animal had experienced while under treatment for the diseased hock."

The danger to be apprehended in cases where an opening is
made by the lancet arises from the liability to synovitis, (inflammation of the synovial membrane,) which reacts sympathetically on
the system, producing derangement of some of the organs or func-

tions of the body. Therefore I consider it injudicious to lay open a synovial capsule in any part of the body.

Treatment.—The safest method of treatment is to puncture the sac, by means of small needles. After having evacuated the fluid through the sieve-like needle-holes, the part is to be smeared with ointment of iodine—one drachm of iodine to seven drachms of simple ointment. Then apply a spring truss, so constructed as to press on both sides of the hock. By this means the internal parts of the sac are kept in contact, and they finally unite. Thus the enlargement is obliterated. If, at any time during the treatment, the parts become hot or tender, discontinue the ointment, and substitute a pad on each extremity of the truss, which are to be kept constantly wet with cold water until the inflammatory action disappears.

NAVICULAR LAMENESS

Navicular lameness is of very frequent occurrence among horses. It usually depends upon strain or injury of the tendon, known as perforans, just where it passes over the navicular bone. Sometimes the navicular bone becomes diseased, ulcerates, and finally becomes fractured; at other times the bone becomes the seat of incrustations, or roughing, and thus irritates the flexor tendon. Such affections always cause pain and lameness. Horses with narrow heels, or those laboring under contraction at the quarters and heels, suffer much pain in the posterior or navicular region, in consequence of hoof pressure on the sensitive parts of the foot.

A disease known as navicular thritis (inflammation of the parts in the navicular region) often prevails. This is an inflammatory affection, very painful, and much resembles laminitis) fever in the foot).

Symptoms.—Pointing of the affected foot; the animal, while standing, throws the affected foot forward, by which means the weight of the forepart of the body is transferred to the other limb; the animal has a short, tripping gait, and steps cautiously; the muscles of the shoulder waste; the spine of the shoulder-blade becomes quite prominent; and this condition of the muscles is often called sweeny. We usually find, also, that the heels are pinched in, and the walls are thin and weak. The liability to navicular lameness is very great in horses with narrow chests, upright pasterns, and outturned toes. Other defects in the loins exist. Hence

ıt is supposed that a predisposition to this affection may lurk in
the breed; in fact, it is known that a tendency to it exis.s in the
progeny of certain horses. It is related that the progeny of
"Young Musician," a thorough-bred horse, all showed a tendency
to navicular disease.

Treatment.—We first ascertain whether the difficulty is of an
acute or chronic character. Should it have made its appearance
very suddenly, and the part feels hot and painful, then we class
it as the acute stage. Should the trouble, however, be of long
standing, it will be in the chronic stage. In the acute stage, cold
water bandages must be applied to the region of the fetlock, which
should be kept wet with water, to which may be added a little
vinegar, merely to acidulate the former; the patient should be
kept at rest, and fed on a light diet. In the chronic stage, counter-irritants are indicated, yet organic disease may be the result;
in that event, the horse has always a sort of groggy walk, and is
never sound.

Neurotomy has been practiced in view of treating this affection,
but I can not perceive how such an operation can benefit a horse
having an organic difficulty. Neurotomy merely destroys, for a
short time, the sensibility of certain parts of the foot; but so soon
as the divided nerve reunites, sensibility is again restored. There
may, however, occur some form of lameness which might be benefited by neurotomy, but I think the same thing might be achieved
by counter-irritation, and without the use of the knife. The counter-irritant I have found most useful is as follows:

No. 53. Cod-liver oil .... 2 oz.
        Kerosene .... 3 oz.
        Cantharides .... 1 dr.
        Mix.

Apply a portion of the above to the navicular region, between
the heels and fetlock, on the back part, once or twice daily, until
the part is well blistered, which will take place at the end of a
week. The more thorough the blistering, the more chance is there
of doing good. No after-dressing need be applied, unless the
parts become very sore; in that event, I should lubricate with
olive oil.

It will be necessary, in chronic cases, to secure the services of
a good blacksmith, so that the animal may have a shoe affixed
properly adapted to the necessities of the case. If any signs of

contraction exist, the heels are to be opened, and the walls at the quarters must be rasped thin, so as to remove the lateral pressure on the sensitive parts.

It may be necessary, in the event of the patient not being benefited by the above treatment, to reapply the counter-irritant; this matter is left discretionary with those who take the case in hand. In the treatment of such cases, the owner of the afflicted animal must be prepared to exercise due patience, for the day of recovery may be far distant; often it is a mere question of time, and not one of skill.

PLEURODYNIA (PAINFUL AFFECTION OF THE PLEURA AND MUSCLES OF THE CHEST).

Pleurodynia (in human practice) signifies pain in the side, from a rheumatic affection of the pleura. In horses the malady not only affects the pleura, but also the intercostal muscles. It is a very painful malady, and some persons are in the habit of calling it "founder." Pleurodynia, however, differs in its origin from founder, inasmuch as the latter may arise from various causes, whereas pleurodynia is almost always the result of an excessive use of the muscular powers, or when the animal has been subjected to long-continued and very severe exertion.

Symptoms.—It will be noticed that the animal moves in a very stiff manner; steps short and quite slow—he appears about used up, as the saying is; the back is arched, and the skin is over the service of the body, and is not only hot but very tender; on making pressure over the regions of the sides of the chest, symptoms of pain will be elicited, showing conclusively that the intercostal muscles between the ribs and pleura, or lining membrane of the cavity of the chest, are involved; the respirations are short and jerking, and the number of pulsations range from forty to sixty per minute. Both fore-feet are usually advanced, and, on applying a hand to them, it will be found that they are unusually hot, and that they are inflamed; the appetite is not good; the urine is high-colored, and the bowels constipated. Such are the most observable and common symptoms of this painful malady. It is something like rheumatism in its pathology, and often migrates to the brain, or its investing membranes. When this happens there remains but little hope for the patient.

Treatment.—In view of illustrating the treatment, I introduce the following case: The animal had been hired of Mr. P——, of Chicago, by a person who appeared to have but very little respect or compassion for horse-flesh; for on the return of the animal to the stable, he showed unmistakable evidences of shameful abuse. He was completely jaded, and could scarcely stand on his feet. He received proper care and attention during the night, and next morning I visited him. On making a careful examination, I found most of the above symptoms present. It seemed, however, that his feet were more affected than other parts of the body, and it was impossible to get him to move. I had his feet bathed constantly with cold water, and his joints, back, and sides were rubbed with a liniment composed of equal parts of cod-liver oil and spirits of camphor. The medicine administered was one ounce of powdered niter, night and morning, for three days. The pulse had decreased to thirty-six per minute, and the respirations were more tranquil; so I discontinued the niter, and commenced an alterative plan of treatment, by giving one ounce of fluid extract of stillingia, night and morning, at the same time continuing the local applications until they were no longer needed. One ounce of fluid extract of buchu was occasionally given, in view of exciting the kidneys to action, and this is all the medicine the patient received until convalescence was established, when he was placed under tonic treatment. The tonic used was one ounce and a half of fluid extract of golden seal, daily. The treatment occupied a period of two weeks, during which time the animal stood but little on his limbs, seldom rising except to partake of food, which consisted of bran and water, a few oats, and small quantities of hay. So soon as the patient had regained some strength, he was led to pasture, where he soon got well. It appears that in the treatment of a disease of this character, light diet, a few simple medicines, and good nursing are all that are indicated.

ACUTE RHEUMATISM,
(CALLED "CHEST FOUNDER" AND "WATER FOUNDER.")

The faithful servant of man, the horse, does not enjoy immunity from this distressing disease. The same causes which operate on the system of man to develop a disease of this character are, with unerring certainty, operative on the inferior creature. Rheumatism is mostly confined to the fibrous textures, or faschia of muscles, and the inter-articular cartilages of the joints, and, finally, the articulating surfaces of the bones become involved. The disease is always characterized by grea pain, and sometimes swelling in the region of joint. It is accompanied by a quick and wiry pulse, and by other symptoms, denominated febrile. It occurs among men and horses at all seasons of the year, yet at the period of sudden transitions from heat to cold it is most prevalent. Animals when heated by exercise, and then suffered to "cool off," without ordinary care, are very apt to become the subjects of this malady, so that prevention, to a certain extent, is within our power, and, in the exercise of preventive measures, we may, n common parlance, "stave off," for a time, this disease, although it may be latent in the system as an hereditary affection. Mr. F. Dun, who is good authority on the subject, says:

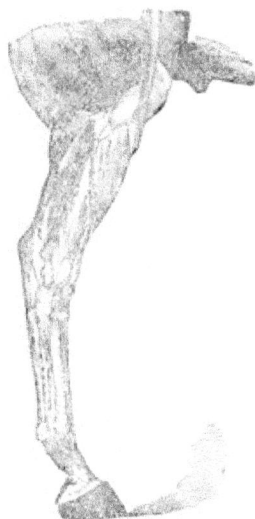

EXTERNAL MUSCLES AND TENDONS OF THE OFF-HIND LEG.

" Rheumatism is neither so common, nor are its symptoms so well marked, in horses as in cattle. When, however, it does occur in the horse, it manifests the same well-known appearances which characterize it in all animals. It affects the fibrous tissues of joints, the coverings of muscles, tendons, valves about the heart, and larger vessels, and manifests a peculiar tendency to shift from one part of the body to another, often affecting, in succession, all the larger joints—at one time chiefly in the neck, at another, in the back and loins, while, in many of its more acute attacks, it appears to involve almost every portion of fibrous and fibro-serous tissues throughout the body. In all its varied types it exhibits a full, strong, hard, and unyielding pulse, caused by the inflammation involving the serous and fibro-serous tissues of the heart and circulating vessels. During its existence, various excrementitious matters accumulate in the blood, and the fibrinous constituents of the same exceed their normal proportions, as indicated by the production of the buffy coat on the blood. In severe or badly-treated cases, the inflammation is very apt to be transformed from the joints and muscles to the heart, and its investing membranes, and it is the danger of this change in the seat of the disease that renders rheumatism so formidable, and often so fatal. It always leaves the parts affected so altered as to be extremely predisposed to subsequent attacks, and it is more than probable that this altered condition is reproduced in the progenies of rheumatic subjects, and constitutes in them the inherent tendency to the disease.

Horses sometimes suffer from rheumatic inflammation in the fibrous sheathing envelopes of the muscles of the neck, constituting what is popularly known as the chords. When thus affected, the animal is very stiff, remains as much as possible in one position, and is unwilling to bend his neck either one way or the other, or to elevate or depress his head. There is always more or less fever, with a strong, full pulse. Sometimes, as in lumbago, in the human subject, it affects the muscles of the back and loins, causing stiffness, tenderness, and pain, which are especially evinced on moving or turning the animal. These rheumatic affections are very readily produced in predisposed subjects by exposure to rain and cold, especially when accompanied by overheating or exhaustion.

Rheumatism sometimes occurs in horses as a prominent symptom of that epizoötic affection which usually receives the much-abused title of influenza. In such cases the rheumatism is of a somewhat more subacute or chronic character than common, and is accompanied by that low, debilitating fever so often the concomitant of epizoötic maladies. It usually affects all parts of the body susceptible of the rheumatic inflammation, is attended particularly by those symptoms which indicate disease of the heart and pericardium, as an intermittent pulse, etc., and often terminates fatally by effusions into the pleura or pericardium, thus causing death by arresting the motions of the heart."

Treatment.—In the treatment of rheumatism simulating an inflammatory type, our first object is to produce a sedative effect on the heart and its vessels of circulation. With this object in view, we administer one or two drachms of fluid extract of gelseminum every four hours, until the pulse becomes softer. In the mean time a few doses of nitrate of potass* should be given in the water

---

* NITRATE OF POTASSA.—In a case of synovial rheumatism this remedy was given by a Boston physician, in a single dose of one ounce, dissolved in a pint

which the animal drinks, at the rate of one ounce per day, divided
into doses in proportion to the number of times the patient is
watered. So soon as the animal evinces signs of improvement,
the above treatment is to be discontinued; then give six drachms
per day of fluid extract of prince's pine. The joints which seem
to be affected are to be bathed occasionally with camphorated
vinegar, in the proportion of one ounce of camphor to eight ounces
(fluid) of vinegar

CHRONIC RHEUMATISM

Chronic rheumatism is usually the sequel of the acute kind. It
is generally obstinate and lingering, and prone to recur. Unlike
the acute kind, pain is alleviated by counter-irritation and exercise; and when the patient gets warmed up, as the saying is, he
either forgets his pains or becomes relieved. Bathing with warm
vinegar has a good effect in the treatment of this disease, but I
have found the following liniment useful in almost all the cases
that have come under my care:

        No. 54. Oil of cedar........ .... 2 oz.
                Sulphuric ether..................... 1 oz.
                Proof spirit........................ 1 pint.

First, mix together the oil of cedar and sulphuric ether; then
add the proof spirit, and, after shaking it awhile, it is fit for use.
Give the animal one fluid drachm of the fluid extract of colchi-
cum-root, night and morning, for three days; after which give,
daily, twenty-five grains of the iodide of potass, dissolved in half
a gill of water. Should the animal show any signs of debility or
loss of flesh, tonics, stimulants, and nutritious diet are indicated.

---

and a half of barley-water. This was followed by one grain of opium. In fifteen hours the pulse was found reduced, and the pain absolutely gone; and in a
few days the tongue was clean, and the swelling entirely abated. The remedy
caused neither emesis nor catharsis, but passed off by the kidneys. In another
case of acute synovial attack, following chronic rheumatism, the same dose was
prescribed without any good effect, causing active catharsis. Again, half an
ounce of nitrate of potassa, largely diluted, was given every two to four hours,
until the patient took three ounces in eighteen hours, with two doses of opium
of one grain each, with entire relief to the pain and fever. When the potash
was reduced to drachm doses, nausea and vomiting followed, which were relieved
by vesication with aqua ammonia, the blister being sprinkled with half a grain
of morphine. The patient recovered. Another physician had used the remedy
to the extent of half an ounce in twenty-four hours, for three successive days.

The authorities teach us, and my own experience has confirmed the fact, that iodide of potass is a valuable agent in the treatment of chronic rheumatism. It is most sure to act beneficially when the periosteum (fibrous covering of the bones) is affected. There is no doubt that chronic rheumatism often cripples the joints by producing disease of the bones in the region of their articulations. This makes a poor horse dead lame, and renders him almost useless. When the bones become ulcerated or carious there remains but little hope of cure.

RING-BONE.

Ring-bone is a form of disease the pathology of which is similar to that of spavin and splent. It is located just above the coronet or crown of the hoof. The enlargement is called exostosis (a morbid enlargement of bone), and when the large pastern becomes united to the smaller one, the case is called anchylosis (permanent rigidity of the joint).

The term ring-bone is far behind the times as regards our present knowledge of nosology (the doctrine of the names of diseases); yet, to the unprofessional, it is suggestive, and, therefore, we are compelled to retain a name which all can interpret. It signifies a complete or partial eminence around pastern bones. The ring is formed by incrustations, or osseous deposits in the pastern, or involving the joint, as the case may be; yet, in order to make out a pure case of ring-bone, that comes strictly within the meaning of the term as horsemen interpret it, there must exist a complete ridge of bone around the anterior part, extending from side to side. A pure case of ring-bone generally originates on the surface of the bone, on or beneath the periosteum. Commencing at the lower margin of pastern and upper part of the coronet bone, the disease spreads, involving both capsular ligament and joint. The remote cause of ring-bone lies in hereditary idiosyncrasies, transmitted through the sexual congress. The disease itself is sometimes directly transmitted. I once bought an unweaned colt, and brought it up by hand. A short time after purchase, I observed enlargements on the hind pasterns. He ultimately turned out to be a ring-boned animal, and had stiff joints. I found out afterward that the mother had large ring-bones on both hind pasterns. The disease was, therefore, congenital.

Mr. PERCIVALL, whose opinion on such matters is unquestion

22

able, says that his attention to the hereditary origin of ring-bone was first aroused from a remark made by an extensive dealer in horses, in reply to a question put to him, How it happened that but few ringbones were met with, compared to the number that attracted notice in times past? The reply was, "Because no breeder of horses nowadays will send a mare to a horse having ring-bone." A very good example for American horsemen to follow, for the disease is very prevalent in some parts of this country. A vast number of our best as well as inferior horses are the subjects of this infirmity. The disease lurks in breed, after the fashion of scrofula and consumption in the human subject. When both parents are affected, the disease in the offspring is doubly severe.

The author just quoted remarks that "a coarse or half-breed, fleshy or bony-legged horse, with short and upright pasterns, is the ordinary subject of this disease; and there exists satisfactory reasons why we should expect him to be so. The pastern and coffin bones constitute the nethermost of the column of bones composing the limbs, and being so, they receive the entire weight and force transmitted from above. The pastern, being long and oblique in position, receives the superincumbent weight on such an indirect line that, bending toward the ground with the fetlock, nothing like jar nor concussion follows. The very reverse of this, however, happens every time the foot of a limb, having a short, upright pastern, comes to the ground. In such, instead of the weight descending obliquely upon the sessamoids, and the fetlock bending therewith, it descends directly, or nearly so, upon the pasterns, making this bone entirely dependent on the bone beneath it for counteracting concussion; and should any thing occur to diminish this, or to throw more weight on the bones beneath than they can counteract, jar of the whole apparatus ensues; and an effort of Nature to strengthen the parts, by investing them with callous and ossification, is likely to be the ultimate result; for we would view ring-bone, disease though it must assuredly be called, a recourse of Nature to strengthen weak parts, the bones being unequal to the exertions or efforts required of them."

Another quotation may possibly interest the reader:

"Ring-bone is an exostosis (a growth of bone from bone) situated around the coronet, mostly near the pastern joint, at other times just above it; and not unfrequently the joint becomes anchylosed owing to the spread of ossification; that is, the coffin bone and pastern bone become united together by bony growth. Ring-bones are the result of hereditary predisposition, structural organization, and accident. As to hereditary predisposition, there is no doubt but that some sires, the subjects of spavin, ring-bone, and other forms of exostosis, father foals which often, sooner or later, themselves become victims of the same disease. The same law applies to dams. Breeders of horses are fully aware of this fact, and are, therefore, more careful than formerly in the selection of parents for the purposes of breeding from. As to form, many living cases which now exist prove the truth of the statement that low-bred, coarse-limbed horses, with short, upright pasterns, are frequently subjects of ring-bone. Why such should be the case with animals with upright pasterns is easily explained, as follows: The pastern bones in a well-formed limb ought to form an oblique angle with the cannon-bone (metacarpus) placed above them; whereas, if the pastern is upright, the column of bones from the elbow-joint to the foot is, necessarily, placed in one and the same straight line, causing, at every step made by the animal during progression, a severe concussive shock to be produced on the whole column of bones, but more especially on the pastern bones, owing to the superincumbent weight of the bones above them. Thus one or more of the pastern-bones become injured and inflamed, and, finally, as a consequence, exostosis ensues, the coronet becomes involved, and the disease denominated ring-bone is set up."

Mr. PERCIVALL quotes from Gibson: "When a well-formed, high-bred horse happens to have a ring-bone, we may conclude it to proceed from some accident rather than from any natural fault. The disease may also be produced by a blow to the part, and from the efforts made to increase speed, when concussion causes injury, inflammation, and consequent exostosis to the bones of the pastern. As to treatment, we should recommend the proprietors of young horses, subject to ring-bones, to place their horses in a state of nature, by removing the shoes from their feet, and by turning out to grass. If this, after a fair trial, fail, then the treatment for spavin and other exostosis must be resorted to, such as blistering the affected part and keeping the animal on a low diet. No disease is more troublesome to treat, or more demands the strict attention of a skillful veterinarian."

Causes.—The exciting causes of ring-bone are ligamentary

sprains, brought about by overwork, extraordinary feats of strength in drawing heavy loads, etc. In short, either sprain, injury, blow, or bruise may act as the pathological spark to bring about the disease.

Regarding the treatment of ring-bone, it would be interesting, and perhaps amusing, to call attention to some of the heroic methods of combating this malady; but the animal has nerves of sensibility, and is as keenly alive to pain as one of us; therefore it is not really pleasant to dwell on the atrocities which are not unfrequently practiced on this noble animal. A very common error exists in the minds of some people that ring-bone is fed by a bladder at the posterior part of the pastern, which has about as much to do in the production of the disease as the author of this work. This error would not amount to much, however, only it has led to the infliction of a cruel operation, without the least advantage; namely, the cutting out of the bladder. Now, this so-called bladder is, in reality, a bursal sac, the use of which is to secrete synovia, to be used for the purpose of guarding against friction; therefore, the extraction of it must be disadvantageous to the parts.

It seems that ring-bone is not enough torment for the poor brute to suffer, but he must, in addition, submit to a species of cruelty unheard of in the annals of human medicine. Such treatment is wrong, and every man who loves a horse should set his face against it, and use every effort for the purpose of putting a step to the practice of all barbarity in the treatment of domestic animals.

Treatment.—The question to decide, when we undertake to treat a case of this character is, whether the disease is in the acute or chronic stage. Experience teaches us that nearly all diseases may be divided into two classes: First, the acute, or those which come on suddenly, attended by pain and fever, or, rather, febrile symptoms; for there are no primary fevers in the horse. Equine fevers are always called into existence, or accompany a pathological condition which frequently existed, and the fever is in exact ratio to the intensity of the disease. Cure the disease and the fever ceases. Acute diseases also tend to termination within a short period of time, or else run into the chronic stage. Secondly, the chronic stage. Chronic diseases are those of slow progress, and exceed in duration the ordinary existence of the acute class. They may arise in consequence of acute disorders, or be developed originally in the form which they long afterward preserve. For the most part, chronic disease and lameness are not marked by any violent symptoms, nor are they accompanied with much pain, if the horse be kept at rest.

Suppose the disease to have first made its appearance, the parts will be hot and painful; but the idea of curing ring-bone—that is, restoring the parts to their normal condition—is really absurd; for, when once the bones are anchylosed, the joint never can perform its natural function again. The new growths (exostosis), and the changes of structure that occur within and about the region of disease, are a part of Nature's own handiwork, in view of strengthening a weak bone or joint. Our object in the treatment, therefore, is merely to aid Nature. Excuse the animal from work, and apply sedative lotions (cold-water dressings) so long as the acute stage lasts; then, when the case passes into the chronic stage, apply counter-irritants for a week or two, and, finally, turn the patient out to grass. The following will be found an effective sedative lotion:

> No. 55. Spirits of camphor.................... 2 oz.
> Fluid extract of wormwood.......... 1 oz.
> Vinegar............................ 5 oz.

To be applied twice daily.

The following is an excellent counter-irritant for ring-bone in its chronic stage:

> No. 56. Cod-liver oil .... } equal parts.
> Kerosene ....

Apply once or twice daily, until a crop of pustules appear; then omit the application, and reapply it at the expiration of two or three intermitting periods.

The remedies used by veterinary surgeons generally, are oil of cantharides, acetate of cantharides, blistering ointment, ointment of biniodide of mercury, and, lastly, the actual cautery, or heated iron. I have tried all except the latter, which savors too much of barbarity, and I never found any thing to equal the preparation here recommended—namely, cod-liver oil and kerosene.

NAVICULARTHRITIS (NAVICULAR DISEASE).

Navicularthritis is an acute disease, supposed to be confined to the navicular region, or joint, which is in the posterior part of the foot. The navicular bone (sometimes called the shuttle-bone) is so placed as to strengthen the union between the lower pastern and the coffin bone, and to enable the flexor tendon, which passes over it, in order to be inserted into the bottom of the coffin bone, to act with more advantage. There is a good deal of weight thrown on the navicular bone, and from the navicular bone to the tendon, and there is a great deal of motion or play between them in the bending and extension of the pasterns. It is very easy, therefore, to conceive that from sudden concussion, or from rapid and unrestrained motion, the delicate membrane which covers the bone, or the cartilage of the bone, may become diseased. The bone sometimes becomes ulcerated and fractured, so that it renders the horse almost perpetually lame.

Symptoms.—The principal symptom of this affection is pointing of the foot. If the horse be watched, when standing still in the stable or on the street, it will be noticed that the affected limb is advanced or pointed in such a manner as to relieve the navicular bone of any pressure from the flexor tendon, and this occurs in the chronic stage as well as in the acute. It will be observed, also, that the animal, when traveling, takes short steps, goes on the toe, and tries to favor the heel, or posterior parts, as much as possible. The favoring of the heel tends to destroy the function of expansibility of the hoof, and contraction of the same is the result.

Treatment.—Should the disease be discovered early, that is, in the acute stage, the foot should be placed in a boot lined with a sponge. The latter is to be kept constantly wet with a portion of vinegar and water, equal parts. This may be continued for a period of twenty-four hours, at the end of which time bathe the coronet and heels with tincture of arnica. Light diet and perfect quiet, as in every other variety of inflammatory disease, are always necessary.

In old chronic cases we resort to counter-irritation, through repeated applications of the acetate of cantharides. Should there be any appearance of contraction, the heels of the hoof must be well opened, and the walls of the hoof, near the heels, must be rasped thin, so as not only to favor the expansibility of the foot, but also with a view to relieve the navicular region of any undue pressure.

STIFLE OUT (DISLOCATION OR LUXATION OF THE STIFLE BONE).

The patella of the horse, or stifle bone, as it is familiarly known, corresponds to the knee-pan of man, and answers the same purpose; that is, to facilitate motion, and protect the two extremities of the bones which compose the stifle-joint. It is retained *in situ* at the anterior part of the inferior extremity of the thigh bone and superior extremity of the tibia.

An accident of this character is generally accompanied with considerable pain and uneasiness; but the most notable symptom is, the bone is thrown off its pulley-like articulatory surface, and thus the recti and vasti muscles (the latter being direct extensors of the leg or thigh, and the action of the former being similar) are, for the time being, paralyzed; that is to say, their action is suspended. If they act at all, it is under great disadvantage, and at the expense of their muscular integrity. The liability to luxation of the stifle bone (patella) is not so great as some persons would have us believe. The bone is secured in its position by a number of strong ligaments—has a complete articulatory joint of its own, whose capsular ligaments have strong attachments to surround them. The muscles, tensor vaginæ, rectus, vasti faschia, panniculus carnosus, subcellular substance and integument, all operate favorably against the chances of luxation. The accident, however, does sometimes occur—not, perhaps, from any disarrangement of the mechanical order of the parts by violence, such as might be operative in dislocation of the thigh or arm bones, but often from spasm, cramp, etc., of certain muscles in the immediate vicinity of the parts.

"I think his stifle is out," is a very common complaint made by horse-owners when a horse becomes lame, and the "neighbors" can not discover the exact location of the same. They hastily conclude, if the lameness is in the hind extremity, that the "stifle is out" (luxated). Here is a case in point. I received a message some time ago, to visit a horse said to have "his stifle out." On making examination of the foot on the lame side, I found it much hotter than its fellow on the opposite side. The owner, at the moment, remarked that there was "no trouble in the foot." He, however, was mistaken; for I found that a nail had entered the posterior part of the cleft of the frog, which I extracted, and this was the sole cause of lameness. I might relate a number of mistakes of this kind, all going to show how very apt people are to be deceived regarding the seat of lameness in horses. In short, judging from my own experience, and the testimony offered by other professional men, it may be confidently asserted that luxation of the patella is quite a rare accident. These remarks are intended for the instruction of the non-professional reader. Persons conversant with the anatomy of the horse are not expected to make such a mistake as the one just alluded to; hence they do not stand in need of the author's advice.

Method of reducing dislocation of the patella.—The means of reduction are very simple. The patella will be found on the lower and outer condyle of the thigh bone. With one hand let the operator grasp the leg, just below the hock, and raise it from the ground in an upward and forward direction; the other hand embraces the stifle bone, and by a simultaneous action of pressing the latter inward, and the bones that compose the joint backward, the reduction is soon effected. The bone, in returning to its natural situation, emits a sound such as usually follows that of a dislocated bone just at the moment when slipping into the socket. A recent dislocation, when reduced, needs no after-treatment, except the occasional application of an astringent and a couple of days' rest; but, should the bone have remained unreduced for many hours, we may infer that the muscles and ligaments have been stretched or distended too long in one direction, and that, on reduction, the fibers of the muscles concerned may not immediately contract; hence the bone is apt to slip off its articulatory surface again. To guard against this mishap, a man must be stationed beside the horse. He must keep his hand on the bone, and by pressure, when the horse moves, retain it in its proper place. At the same time, the part is to be bathed with some astringent. An infusion of white-oak bark will answer; or, failing to obtain this article, apply alum-water, in the proportion of two ounces of alum to one quart of water.

CONTRACTION OF THE HOOF.

Contraction of the hoof is so easily recognized by persons accustomed to handle horses, that very little needs to be written on that subject. A contracted foot "speaks for itself," as the saying is. The heels are approximated too closely, and this mars the beauty and form of the foot. Contraction is almost always, in its first stage, accompanied by lameness. Faulty shoeing is said to have considerable to do in the production of contracted feet; yet there are plenty of horses, having strong, thick, walled hoofs, very wide at the heels, often shod in the worst possible manner, and their heels never contract. It is certain, therefore, that a predisposition to contraction lurks in some breeds of horses, which is shown by their having weak quarters and a thin, shelly hoof. A contracted hoof, which is very thin at the heels, may not occasion lameness, because it is more elastic than the thick, walled hoof, and gives to the lateral pressure of the foot when it comes to the ground.

PERCIVALL contends that "shoeing must be regarded as the main cause, indirect though it be, of contraction; and to the modification or correction of it must we look for the prophylactic. Shoeing, as it respects horses, has been said to be 'a necessary evil.' Without shoes, upon our artificial roads, we can not make use of horses; and no shoes have been found to answer save such as are hard and inflexible or metallic, and as are fixed to the hoofs with nails. Here, therefore, we find ourselves in an awkward dilemma. We can not do without horseshoes; and from the moment we nail them to the hoofs, the feet begin to undergo more or less alteration in form, and, in too many instances, to experience harm from their application. The art of shoeing has given rise to a wonderful deal of difference of opinion and controversy, some thinking one shape of shoe answered best, some another; while some prefer one mode of nailing it upon the hoof, others a different one. As far as my own experience has served as a guide to me through this labyrinth of opinion, I have ever found that method of shoeing the preferable one which approached the nearest to Nature, or, in other words, which interfered the least with the economy of the foot. If we could do without them, horseshoes would, undoubtedly, be best abolished altogether; but, since this is impracticable, let us adopt such shoes and modes of attaching them to the hoofs as are found to work the least mischief to the feet. On this principle it is that a half shoe is to be preferred to a whole shoe, and for the same reason it is that tips, of all the horseshoes that were ever invented, are the best; that is, the least objectionable. If those in the profession would come forward and inform us of their experience (if they have had any) of tip-shoeing, I believe it would uniformly be found that, whatever objection might be urged against the use of tips, no one would deny their tendency the least **to** interfere with the operations of the foot. **If** there be any horse-shoe calculated to prevent contraction, and navieularthritis as well, I feel no hesitation myself in pronouncing that horseshoe to be the tip. In saying so much, I am fully aware that tip-shoeing can not be introduced into general practice for reason of the roads horses have to travel and work upon, and of the numbers of horses having hoofs of too weak and brittle a fiber to stand work without chipping and breaking and wearing too rapidly away. On horses, however, whose hoofs are strong and hard enough, and whose work is light enough to admit of their wearing tips for any length of time, or in situations where the roads or parts of the country they have to do their work upon enable them to wear tips con-stantly, no wholly-shod horses' feet will ever bear a comparison with theirs.

Pressure to the frog.—Coleman's favorite prophylactic against contraction (considering shoeing to be an indispensable evil) must certainly be regarded as next in importance, as a preventive, to getting quit of the shoe itself, or of part of it. The frog being a body which in action operates in the expansion of the hoof, the removal of it, or even the impairment of it, must, necessarily, give facility to contraction. It therefore behooves us, in ordinary shoe-ing, to look well to the preservation of the integrity of this im-portant part of the foot.

The cutting away of the bars in shoeing, through robbing the hoof of a couple of stays operating against the closure of its heels, conduces to its contraction. Nature gave the bars as a sort of buttress against either heel of the hoof, to oppose its drawing in-ward, while the frog, placed between the heels, is operating in forcing them asunder; consequently, if the bars be removed, the expansive or counteractive powers of the hoof lose an agent they can, in many cases, ill afford to be deprived of.

The contracting effects of heat and drought on the hoof may be guarded against by keeping the horse's stall free from ferment-able litter, while the atmosphere of the stable is maintained cool and unpolluted. The practice, also, of stopping horses' feet (or, what I believe to be better, of wearing swabs in the stable) will likewise tend to guard against the contracting effects of these agents. We now come to the

Treatment of contracted feet.—The first thing to determine, whenever a case of contraction is submitted to us for treatment, will be whether it be one of the pure or mixed description. If the former, the horse not therefore lame, and his feet be submitted to our inspection simply from the apprehension of his becoming lame, and the contraction be on this account desired to be removed, the simple and best means of doing so will be to substitute tips for the horse's ordinary shoes, and to order that he stand with his contracted feet in cold water (or, what is better, in a bed of clay) for a couple of hours, once or twice a day, he being allowed to lie down, as usual, at night. By such a simple plan of treatment as this will his hoofs, giving sufficiency of time for Nature to carry out her operations, become restored to their pristine condition."

The author's method of treating contraction is, to soften the hoof by soaking it often in warm water; then open the heels by means of the drawing-knife; rasp the quarters of the hoof; then pare the foot, but do not touch the bars nor frog; then put on a light shoe, and use the animal, daily, at light work. It will be proper, also, to anoint the hoof, frog, and sole occasionally with a portion of fish oil and spirits of turpentine, equal parts. This will soon soften the hoof, and render it elastic.

HOOF-BOUND.

This term is usually applied to hoofs that appear to be dry, brittle, unyielding, and having the appearance of more or less contraction at the heels. The hoof, under such circumstances, "pinches the foot," as the saying is, or, rather, embraces the soft tissues too closely. The consequence is, pain, fever, lameness, and perhaps, finally, organic disease of the foot. The condition of the foot termed hoof-bound usually occurs in horses laboring under some predisposition to the same; that is to say, they have faulty feet, which fault may be either functional or structural, and thus the foot may be affected accordingly whenever the ordinary causes of disease are in operation.

Causes.—The ordinary or exciting causes of diseases of the feet are said to be faulty shoeing; allowing horses to stand on plank floors, or in any situation which is likely to pervert the function of the foot; sometimes the feet are very much neglected, and scarcely, if ever, make the acquaintance of sponge and water, and it is very certain that uncleanliness has a bad effect upon the feet.

as it is known to have on other parts of the body. "No prince ever died of the plague," shows that the ablutions daily practiced by persons of refinement, occupying an exalted position in society, operate as a preventive against even a mortal malady. Yet, not-withstanding all this, it is proper for us to remark that some horses are occasionally the subjects of foot maladies, notwithstanding the very best system of shoeing is practiced on them; and others, that never stood on a plank floor or inclined plane, are in the same predicament; so, also, are some that have their feet washed often This goes to show that if animals under the very best care and management get hoof-bound, or contracted feet, it is due to a predisposition which they have either inherited or acquired, and, as the old saying is, they would not enjoy bodily health unless they were hoof-bound.

It is very evident that many diseases of the feet are traceable to bad shoeing. Some persons seem to suppose that the foot of a horse is as insensible as stone; and, when we see the blacksmith wrenching off shoes by main force, often fracturing the hoof or breaking off a portion of the bars, frog, and sole, with as little care as a man would rough-trim a log, and afterward applying a red-hot shoe to the thinned sole, we might be led to infer that the foot is really insensible. The Hon. MARSHALL P. WILDER thus denounces such bad practice:

"I know of no greater evidence of inhumanity to that noble animal the horse, than the almost universal custom of paring down the foot and burning on the shoe. It has ever, to my mind, been among the barbarisms of the age, which I trust will be no longer countenanced by civilized society; and I sympathize with the efforts of every one engaged in correcting such dangerous practices."

The practice, also, of using a pair of pinchers on the feet, in view of discovering some real or supposed lameness, is equally barbarous, and often results in bruises of the sole, and should, therefore, be discountenanced.

Treatment.—The blacksmith should so prepare the foot and fit the shoe that the frog may be brought in contact with the ground over which the horse travels. This not only prevents undue concussion, which would otherwise occur, but tends, also, to expand the heels, and preserve the natural tread and function of the foot. Having prepared the foot in a proper manner, we then, if the season of the year permits, apply cold water to the hoofs, often and freely. At night the feet are to be wiped perfectly dry, and, by means of a sponge, a portion of the following mixture must be rubbed on : Kerosene oil and honey, equal parts ; or neats-foot oil and turpentine, equal parts. Notwithstanding the very best treatment, the restoration of the foot to comparative or natural soundness will be a work of time, and sometimes, in bad cases, or one of long standing, an impossibility ; yet the above treatment is the best we are acquainted with. Some persons prefer to use moistened clay, cow-dung, etc., but they are of no value, only in consequence of the moisture they contain, and clean water is far preferable.

The Mechanism of the Hoof.—A brief exposition of the mechanism of the hoof may possibly interest the reader. The external covering of the foot may be divided into four parts ; namely, the wall, bars, sole, and frog. The external portion, or wall, serves to defend the sensitive tissues within. The wall, or hoof, is composed of small filaments, or hollow tubes, consolidated in such a manner as to preserve their canals distinct. These canals constitute the excrementitious outlets of the hoof, from which morbid materials find an outlet, and in these canals are found the vessels by which horn is secreted or produced. The small vessels arising from the vascular and nervous tissues of the foot proper enter, also, into these canals. The bars are a continuation of the external portion just described. They form an angle at the heels, which terminates toward the toe. They thus serve to give strength and durability to the hoof, and also to prevent contraction at the heels. When these bars are cut away or demolished by means of the butteris, then the foot often goes to "rack and ruin." The sole is much more elastic than the crust, and is a medium of the sensitive faculty, through which, together with its powers of electricity, the percussion of the foot against the ground is regulated. The frog is much more elastic than either of the parts just described. Any unnecessary paring of the same is a monstrous evil.

On the internal portion of the above-named parts we find a beautiful set of leaves (laminæ), resembling those found in the under surface of a mushroom. Their number is said to be five hundred. These articulate with a like number given off from the sensitive tissues of the foot proper, each laminæ having two sides and an edge, from a series of articulations, numbering three thousand. The whole surface of these articulations measures four square feet; hence, the horse having four feet, his body rests on sixteen square feet of surface, all contained within his hoofs.

The reader will probably realize that a knowledge of the anatomy and physiology of the foot is absolutely necessary, in order to practice, with any degree of success, the art of shoeing. How is this knowledge to be obtained? Can it be furnished at the forge, in the blacksmith shop? Can such a complicated piece of mechanism be thoroughly understood by the craftsmen of the forge, who have never made the subject one of scientific inquiry?

Let it not be understood that I am in a fault-finding mood. The fact is, the husbandmen of this country have failed to put their shoulders to the wheel for the purpose of establishing and endowing veterinary schools in this country, and if blacksmiths and "vets" do not understand the anatomy of horses feet, the former, who have vast interests at stake, must come in for a share of the blame. On the other hand, we are disposed to be charitable toward blacksmiths; and, taking into consideration the unyielding nature of our paved streets, and the rough ground over which some poor animals are forced to travel, the wonder is, that their feet are not oftener diseased.

LAMINITIS ("FEVER IN THE FEET," OR "FEET FOUNDER").

Laminitis, or fever in the feet, is an acute affection, confined to the sensitive tissues within the hoof. It is one of the most painful maladies to which the horse is subject.

Symptoms.—It manifests itself, very frequently, after a long drive or race. The horse may be put into the stable at night all right. In the morning he is observed to be all in a heap, as the saying is—his flesh quivering, pulse very active, respirations quick and jerking; the flanks are tucked up, the back is roached, and the fore-feet are sent forward, so that the animal treads on his heels; and, when compelled to move, he endeavors to throw the weight of his body on the hind legs. He is thirsty and feverish, but has no appetite. All appearances about him indicate that he suffers great pain. When a hand is placed upon the fore-feet, they feel unnaturally hot. Such are the most prominent symptoms of laminitis, or fever in the feet.

Treatment.—The principal objects to be accomplished are, to keep the feet cool by frequent sponging with cold water; next, the bowels must be kept loose my means of bran-mashes, or, if necessary, a dose of Glauber salts—dose, twelve ounces, dissolved in a pint of warm water, to which add half a gill of syrup—and an occasional enema of soap-suds, and also a few doses of fluid extract of gelseminum, say two drachms night and morning. This is the kind of treatment that the author has found most successful during a long period of practice.

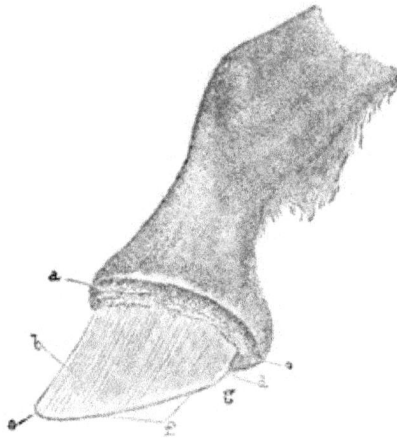

EXTREMITY OF ONE OF THE FORE LIMBS.

EXPLANATION.—*a*, The region of the coronet; *b*, The sensitive laminæ; *c*, The point of the toe; *d*, The quarters; *e*, The heel; *f*, The sole; *g*, The solar border.

Many who treat such diseases are apt to do too much. YOUATT, and several other writers, recommend repeated bleedings, blisters, and purges, and even tell us to bleed in the chronic stage. This is all wrong, and such outrageous treatment is almost sure to end in suppuration, founder, or ruin. (See article on Inflammation.) The patient must be kept at rest, and, if he should lie down, must not be disturbed. I never remove the shoes, because the patient is in so much pain that he can not stand on the frog or sole, and the shoes are a protection to the frog.

SAND-CRACK.

" The name of sand-crack seems of questionable application.　It .s, evidently, a compound of the word sand and crack, as though it denoted a crack with sand in it, or a crack occurring in a sandy country, or in a dry, sandy season, which several derivations have been ascribed to the term.　May not the word sand admit of resolution into its primitive signification, and mean in this, as in other instances, a sundered crack?　A sand-crack may be defined to be a longitudinal division in the fibers of the wall of the hoof, amounting to a flaw simply, or else to a cleft or fissure through the substance of the horn.

The direction of the crack is slanting, from above downward, and from behind forward, following the course of the fibers of the hoof.　A sand-crack in the side of the wall slants more than one in front, owing to the greater obliquity of the course of the horny fibers, as we proceed from the toe to the heel of the foot.

There are two kinds of sand-crack, quarter sand-crack and toe sand-crack, the former occurring in the fore, the latter in the hind foot.　At least this is generally the case.　It is rare to find the reverse, though there are occasions on which we meet with sand-crack in the toe of the fore-foot and the quarter of the hind foot. It is possible for cracks to occur in other parts of the hoof; but in these two situations it is that veritable sand-crack occurs, and there are here, as we shall find hereafter, special causes for their production.　Let us first consider

QUARTER CRACK.

The situation of this crack is the slanting line of the wall of the hoof, directly opposed to the extremity of the wing of the coffin bone; and it is oftener found in the inner than in the outer quarter, added to which the hoof in which sand-crack occurs is always a contracted one, quarter sand-crack, no more than toe sand-crack, never happening in a hoof disposed to obliquity and flatness.　The same description of foot which is predisposed to contraction is, for the same reasons, predisposed to sand-crack. There is an obvious connection between contraction and quarter sand-crack.　The light, near-the-ground stepping horse, with strong, narrow, upright hoofs. will be equally likely, under certain conditions, to have the heels of his fore hoofs becoming contracted, and exhibiting quarter sand-cracks. Hot stables conduce to this, but, more still, hot climates. Hurtrel· d'Arboval informs us that at the time the French army was in Egypt, their horses were continually having sand-cracks; and he adds that long voyages on board of ship are, on some occasions, attended with like results.

The proximate cause of quarter sand-crack is, then, contraction. The horn, from dryness, or other cause, rendered cracky and fragile, breaks at the quarter of the wall, from being at that part bent across the edge of the ala of the coffin bone, during the curving in of the heel, the result of contraction; and this oftenest happens to the inner quarter, from its being the thinner and weaker one, as well as from being·the one which is the first and most disposed to contract—not that sand-crack is the necessary consequence of contraction, but that contraction becomes a necessary precursor to sand-crack. If this were not the explanation of the case, sand-crack would be as likely to occur in any part of the wall as at the quarter, and on the outer as often as on the inner side of the wall. This likewise accounts for the comparative unfrequency of quarter sand-crack at the present day among our nag and cavalry horses, since that which has tended to diminish the frequency of contraction has had the same effect in regard to the occurrence of sand-crack. Greater attention to shoeing, and increased care about the condition of the hoof itself, has, no doubt, had a very beneficial effect in the prevention both of contraction and sand-crack.

The origin of sand-crack is usually sudden, both in the fore and hind feet, though in the former case, from its situation in the inner quarter, a part not exposed to transient view, it is possible for a crack to exist for some time, unless lameness happens to arise from it without being discovered. The crack first takes place through the superior or coronary border of the hoof, that being composed of newly-formed horn, and, consequently, thin and fragile. Quickly it extends downward, through the thickest part of the wall, stopping, as it does in some instances, at least for a time, half way down, and afterward reaching nearly or quite to the bottom.

Sand-crack is either penetrant or non-penetrant. It is usually penetrant, by which we mean the crack extends completely through

23

the t.i. kness of the hoof, and produces lesion of the sensitive parts underneath the horn, giving rise to some little hemorrhage 'n the first instance, and subsequently to more or less inflammation—occasionally, even to suppuration, depending, of course, on the extent and nature and duration of the lesion, as well as on any treatment, or aggravation from non-treatment, it may have received since being occasioned. When the crack does not extend through the substance of the wall—which it sometimes does not at first, though it may do so afterward—it may be said to be non-penetrant; and this, when it happens, seems to be referable to a sort of natural fusion there certainly exists between the solid horn, as it descends from its secreting gland, (the coronet,) and the horny laminæ (new matter), which become glued to it in its course downward. It is in this uniting horny medium that seedy toe commences and progresses.

Lameness is the usual accompaniment of penetrant sand-crack, but not the invariable one. When penetrant it is said to arise from the sensitive laminæ getting pinched between the sides of the crack. However this may be, inflammation following the lesion has certainly its share in causing pain during action. In general, the lameness does not amount to much, nor is it of long duration, provided the sand-crack receive proper and timely attention; but if the crack is not heeded, action of the foot will much aggravate the malady, and end in lameness increased so much as to forbid further use being made of the animal.

TOE SAND-CRACK.

Occurring in the hind foot, as this crack almost invariably does, occupying a different situation in the hoof, and arising altogether from a different cause, toe sand-crack may be regarded as almost a distinct affection.

The subject of it is not the light horse, but the heavy one--the cart and wagon-horse, the dray-horse, the latter especially; and in particular, I believe, in paved cities, they are the sufferers from this disease. I say 'sufferers,' because it is only those veterinary surgeons whose practice lies among horses of this description that have any conception of the amount of pain and irritation with which sand-crack—simple as it may seem to those who are in the habit of meeting quarter crack only—is found to be accompanied.

The cause of toe sand-crack is violence. Shoeing, also, may have something to do in its production. The horses who are the subjects of it are those employed in laborious and straining draft. The toe of the hind foot is the grand fulcrum through which the hind limbs, the propellers of the body, exert their power; and it is in some violent and forcible effort that the hind hoof, strained as it is to its uttermost, and in particular at the toe, splits, commonly first at the coronet, the same as in the fore-foot, where the horn, but newly-formed, is then unresisting, the crack subse· quently extending gradually down the wall, even as far as the point of the toe. Digging the tip of the toe into the ground, or stamping it hard down upon the pavement, and especially when this stress upon the forepart of the wall is at all times promoted by high caulkings to the shoe, must certainly, one would think, be the main producer of toe sand-crack—an opinion still further favored by the observation which has been made of shaft-horses in drays being more subject to the accident than trace-horses. Still, however, for all this, it behooves me to say that, with the best judges of such matters, the point is one not yet set free from doubt and difference of thinking. Short and upright pasterns, with clubby prominent hoofs, indicate a predisposition to toe sand-crack, the disease being in no instances seen in flat, shelvy, oblique hoofs. It is said sand-crack may originate in tread. Undoubtedly any lesion of the coronary body, sufficient to injure or destroy its secretory apparatus, may occasion imperfect or morbid formation of horn, or loss of horn altogether; but I do not believe this to be a very common cause of sand-crack.

The consequences of sand-crack in the hind hoof are, as I have before hinted, apt to be of a much more serious nature than any usually arising from a quarter sand-crack. Whether the crack extend to the bottom of the wall or not, being uniformly of the penetrant description, lameness, to greater or less degree, is the invariable result. And when the fissure does reach down to the toe, the wall opens and exposes the laminæ, probably the whole way from the coronet downward, the consequence of which is inflammation and suppuration of those parts, and sometimes even mortification and sloughing of them; and not of them alone, but of the bone to which they are attached as well, which not unfrequently runs into a state of caries, ending in defalcation of substance, to be filled up by the effusion of callus, which usually terminates in exostosis, coated with some tissue very imperfectly representing the original laminated structure.

Mr. Braby, the intelligent veterinary surgeon to Messrs. Barclay and Perkins' establishment, to whom I am indebted for much of the information I possess on this part of my subject, has had many cases of this description, one of which, of extraordinary character, I shall relate here. One of his dray horses had suffered long and severely from toe sand-crack in one hind foot, but, at length, had recovered, and returned to work. Some time afterward, however, during the season of influenza, he was attacked with a violent laryngitis, which increased to a degree to call for the operation of tracheotomy, to save him from suffocation. Notwithstanding this temporary salvation, however, the patient, in the end, succumbed to the disease. His post-mortem examination became doubly attractive, owing to the circumstance of the long-standing and obstinate sand-crack he had suffered from heretofore, and the result in this latter respect proved extremely interesting. The coffin bone, along its front, occupying the line of the surface between the coronal process and the toe, exhibited a channel or loss of substance half an inch in breadth, and fully the same in depth, thereby robbing it of a quarter of an inch of its solid diameter. This, of course, left the bone considerably weakened, the result of which subsequently was, transverse fracture in two places, the fractures commencing upon the articulatory surface, whence they extended directly crosswise through the middle of its body, so as to become apparent upon its concave surface underneath. In addition to this, growing from the laminated interior of the wall of the hoof, opposite to the middle or deepest part of the channel in the coffin bone, was a projection of hard, horny, callous substance, having a covering of imperfectly-formed horny laminæ. At the time this horse was suffering in the greatest degree from this extraordinary product of sand-crack, constitutional irritation ran so high as even to create alarm for the animal's life. The treatment of sand-crack, whether it be in the quarter or in the toe, will have to be conducted upon principles applicable to both forms of the disease, though one must be regarded as of much more consequence than the other. The treatment of quarter sand-crack, generally speaking, is but comparatively a simple affair; indeed, so lightly is it looked upon by horsemen in general, that we should run some risk of their displeasure, and our own reputation as

well, were we to talk about laying a horse up for so trifling an
accident. Nay, some horses, with non-penetrant sand-cracks, or
with sand-cracks that have been penetrant, but have become horned
over, showing little or no lameness, continue to work on without
evincing any pain or inconvenience from them. Whether a horse
be lame or not, however, should he have a sand-crack, and we be
consulted about it, it becomes our duty to arrest the extension of
the crack so long as it be but partial; and, besides that, to take
measures for the cure or permanent removal of the crack. The
owner of the horse should be given to understand that no flaw or
crack in the hoof can, by possibility, unite the same as a wound
in a vital part does, but must, as the saying is, grow down; that
is, must be replaced by new horn, and be itself, by degrees, renewed, as it continues to come under the operation of the drawing-
knife every time the horse is fresh shod; so that, in fact, the cure,
or obliteration of the crack, is necessarily a work of some months,
though the lameness may be cured in as many hours or days.

Paring out the crack, the shoe being taken off the foot, is the
first thing to be done. The cutting cautiously away of its rugged
edges, and the scooping out, with a light hand, (probably with
the back of the drawing-knife,) of its cavity, will enable us to
examine into the condition of it. Should there be no lesion or
exposure of the laminæ detectable, nothing further will be needed
from the drawing-knife than the cleaning out of the crack.

Firing the crack is the next operation. Cross-firing will be
advisable below or above, or in both situations, according to circumstances. So long as the crack has not reached to the bottom
of the wall, it will be requisite, with an ordinary firing-iron, at a
red heat, to burn a deep but short fissure, or 'mark,' across its
lower extremity; and whether a similar operation be required
across the superior termination of the crack, must depend upon
its extension or not through the horn at the coronet. If there be
any interval of sound horn between the hair and the crack, of
sufficient breadth and substance to bear firing, a very slight burn
may do good. In all cases it is the practice to finish the firing
with running the sharp edge of the iron down the crack; and this
certainly proves beneficial in destroying any tendency there may
be (supposing the laminæ to have become denuded) to abnormal
action, as well to stimulate any vascular parts exposed to issue
horny matter to cover in the bottom of the crack.

Binding up the crack is a good practice after firing. With a wax-end of sufficient length (such as shoemakers use) bind round the wall of the hoof, so that any tar or pitch-plaster it may be deemed advisable to place in or upon the crack may be maintained there; at the same time the hoof itself is, by the tight binding restricted in any tendency it may have to expand, and thereby open wider the crack.

A bar-shoe is the preferable one for a sand-cracked foot. By it, the bearing being taken off that part of the wall which is opposite to the crack, the pressure and jar, so continually splitting afresh the new-formed horn over the crack at the coronet, is put a stop to, the formation of an undivided coronary horny band being the commencement of the radical cure of the sand-crack. As I said before, horn being an inorganic substance, no union whatever can take place in the crack itself. Permanent cure can be effected only through obliteration by the growing out or down of the crack. This, I repeat, is the reason why a sand-crack occupies so long a time in its removal; though, by way of compensation, a horse is not kept out of work while cure is being effected; for, after the crack has been bound up, and the hoof shod with a bar-shoe, it is quite surprising to find how soundly and firmly the animal sometimes steps upon the foot of which he had but now been so lame.

The treatment adopted by the late Mr. Read, V. S., of Crediton, carries the same object into execution, through a different method of procedure. This, as detailed in the volume of the 'Veterinarian' for 1848, consists in simply isolating the fissure within the segment of a circle, by means of an ordinary firing-iron. The best plan is to operate with the heel of the iron, beginning at the coronet with either extremity of the segment, and bringing the iron to a finish at the center. The iron should be at a strong red heat, and be carried through the horny crust until it touches lightly the sensible laminæ, and so throughout the entire semicircle. As you recede from the coronet, so, in proportion, you will require to deepen the fissure in the crust. The iron ought to be applied every week or ten days. The first effect desirable to be produced is a bulging of the crust around the coronet within the segment, and when once this is fairly established, the cure may be said to be effected, it being seldom necessary to apply the cautery afterward. The old method of making a line with the iron across the fissure can not prevent the opening and closure of the fissure during the action of the foot; whereas, isolating the fissure (or part of it) within the segment of a circle completely effects the object. No tar-cord or strapping, or alteration of the shoe, is required to limit the motion of the crust, all motion being suspended within the segment, and especially after bulging has commenced.

The objects of treatment, after what has been stated, will clearly appear to be, first, to place the hoof in such a condition as shall not render it liable to crack again; second, to remove that state of it which, in the first instance, disposed it to crack, if it were not of itself the immediate cause of cracking. The way in which the first object is effected has been already shown, and when this has become accomplished, past all risk of return, we may set about to bring into effect the second. A bar-shoe, from its taking the bearing off the quarters and placing it upon the frog, will, in a measure, give facility to what we now are desirous of preventing; namely, the expansion of the heels of the foot; but a tip (providing it can be worn, which it frequently may, with great advantage, after a bar-shoe) will bring about greater reform still—will, in fact, by persistence in its use, bring about, in time, that improved form of hoof which will be no more liable to quarter sand-crack.

The treatment of toe sand-crack is, in some respects, a different affair from that of quarter sand-crack. This disease is not only different in its relative situation as regards other parts of the foot, but it arises, as we have seen, from a totally different causation. When once it has occurred, it becomes, compared with the other, a serious affair. The horse is lame, too lame to continue his work, probably, and we have a penetrant crack to deal with, extending all, or nearly all, the way from the coronet to the toe, discharging blood, or, perhaps, matter of some ichorous, offensive description, plainly calling upon us for, not binding up, etc., as in the former case, but for opening, cleansing, and dressing the same. The shoe being taken off, let the crack be pared out, and freed from all horny rags and asperities, and laid completely open to view, so that the bottom can be inspected and dressed with whatever may be deemed requisite. In fact, when once the fissure is dilated into a clean and open channel by the drawing-knife, warm baths or poultices, or dressings of any kind, as may become necessary, are now applicable, the case being in this stage no more than one of dilated sinus in the foot, similar to what might, in another situation, be called quittor. As with the quarter sand-crack, the cure will, of course, be tedious in its duration, since we know that all complete repair can only come from the coronet. The sensitive laminæ having the power of secreting horny laminæ, may, as in the quarter crack, issue a sort of horny covering-in of the bottom of the fissure; but fissure or cleft will ever remain so long as an integral formation of horn does not grow down from the source of secretion.

Whenever the horse is in a condition for work, a bar-shoe, so made that at the toe the sides of the shoe remain unjoined together, an interval being left of sufficient width to receive the crack in front of the hoof, is, perhaps, the best. Binding the hoof up with circles of wax-end, as in the case of quarter sand-crack, with some plaster or dressing underneath it, will also now become advisable. It will restrict the spreading inclination of the hoof, as well as keep dirt and wet out of the crack. Repetition of this, and the continuance of the bar-shoe, will be required so long as there appears any risk of an extension or renewal of the crack.'"*

The diseases of the hoof here alluded to are very apt to occur in horses whose walls of the hoof are thin and brittle. They are, to a certain extent, non-elastic, and, therefore, do not ward off the necessary amount of concussion when the animal is traveling over hard roads. At the same time there is always a liability to incipient and permanent contraction in such hoof. More especially is this the case when the foot is mutilated by paring or cutting away the bars (the main stages or supports which prevent the heels approximating). These are the kind of feet on which the butteris-knife and rasp should be used with great caution; for, being naturally weak, and perhaps congenitally defective in the function of secreting horn, and the necessary amount of moisture to keep the part pliant, any unnecessary tinkering with such tools is inadmissible. Such hoofs lack moisture; therefore they should be well soaked night and morning, in warm water; afterward wiped dry, and then smeared with a portion of the following:

No. 5   Neats-foot oil................. 8 oz
         Kerosene..................... 4 oz.
         Lamp-black................... 1 table-spoonful.

Mix well, and apply twice daily, by means of a common varnish-brush.

Very many diseases of the feet, which are of daily occurrence, might be prevented by proper attention to the feet. " No foot, no horse ;" hence the feet need to be kept as free from filth as other parts of the body. I think that a great majority of English grooms pay as much, if not more, attention to horses feet than to other parts of the body, and thus many forms of foot lameness are avoided.

CURVATURE AND DISEASE OF THE SPINE ("SWAY BACK," "HOLLOW BACK").

The annexed engraving, represents a case of curvature of the spine in a downward direction, known in stable language as hollow back. It usually, when it makes its appearance, occurs in aged horses, and probably arises from putting too much weight on the animal's back.

SECTION OF THE SPINE.

It will be observed that two of the superior spinous processes of the vertebræ are united (anchylosed), and ossific deposits are thrown out on the surfaces of the spine. On the lower part, beneath the diseased spine, a cleft is seen, showing that there is an incomplete dislocation, and the capsular ligament was probably dislocated or torn.

It may be interesting to the reader to know, if he does not already, that the bones, like other parts of the body, are composed of arteries, veins, absorbent vessels, nerves, and a cellular texture. They are endowed with vitality, are nourished, grow, waste, and are repaired, and undergo various mutations, according to the age of the animal ; and they are subject to disease analagous to the soft parts. To the phosphate of lime, which is, more or less, distributed in their texture, they owe their solidity; and perhaps it is to the same earthy substance that the difference in their vital properties, and in their diseases from those of the rest of the body, is to be referred. In fact, this particular organization and inferior vitality of the bones are generally supposed to account for the small number, peculiar character, and general slow progress of their diseases.

Treatment.—So soon as a horse shows any symptom of undue curvature of the spine, he must not be used under the saddle; but, provided he can work, he may be used lightly in a buggy. If unfit for work, the owner knows what is best to do with him. I should give an animal afflicted with this disease a wine-glassful of cod-liver oil, two or three times per week, and one table-spoonful of flour of bone, every night, in the food. If this article can not be procured, substitute for it phosphate of lime.

OSTEO POROSIS (KNOWN AS BIG HEAD AND BIG JAW).

The terms big head and big jaw are used by husbandmen and others in Ohio, and, indeed, throughout the Great West and South, to designate a disease of the bony structure of the horse—a disease about which very little of a reliable character has ever been said or written; yet the subject is one of great importance to the people of this country, for the simple reason that the evil is one of alarming proportions; and should the disease multiply in a ratio equal to that of the past few years, it will be a terrible blow to the laudable and hitherto profitable enterprise of raising colts in sections where this disease prevails.

We do not wish to create any unnecessary alarm, but urge the husbandmen of the South and South-west to give this subject their earnest consideration; for the disease probably has an hereditary origin—is "inbred"—and that, as the saying is, "constitutes the root of the evil." The disease having been pronounced incurable, our only hopes of success in putting a stop to its propagation and development is, to adopt preventive measures, by rejecting, as breeders, all animals (sire and dam) that show the least predisposition to or for this affection.

Let the reader understand that the disease itself is incurable in so far as the enlargement, dilatation, and softening of the jaw-bones are concerned; yet, like spavin, ring-bone, and various other diseases, it is accompanied, in certain stages, by pain and lameness, and our services as physicians are only secured in view of removing this pain and lameness, so that the animal may be enabled to perform the ordinary equine duties. This is what some persons please to term a cure, whereas it almost always happens that some alteration in the structure of the parts remains, which actually renders the horse unsound, because he has that about him which may, from overwork or other exciting causes, impair his usefulness.

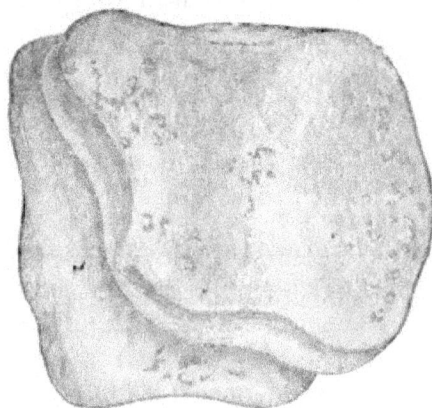

THE PATELLA, OR STIFLE BONE.

This cut shows the appearance of disease on articulatory surfaces when the animal is the subject of Osteo Porosis, or Big Head.

As regards the "alarming" features of big head or big jaw, we would inform the reader that almost every lame horse examined by us, during a period of six months, in Ohio, had either one or the other jaw enlarged. In some cases one angle of the lower jaw was the seat of thickening and enlargement, but among the majority both angles were affected. These remarks do not apply to other obvious or accidental lamenesses which are constantly occurring.

A short time ago we visited, in company with a practicing veterinarian whose attention had never been called to this affection, three horses, the subjects of lameness, which was said to be occasioned by eating Hungarian grass. To the astonishment of the parties concerned, we demonstrated that they were all the subjects of enlargements of the lower jaws. From the history of the cases,

and the character of the lameness, it was very evident that it was
of an inter-articular character, and this is a peculiar feature of this
affection. Then, again, the reader must bear in mind that the jaws
of animals do not dilate nor enlarge suddenly, nor make their appearance all at once, like a meteor in the regions of space. The
affection is insidious, progresses in an almost imperceptible manner,
and usually escapes the notice of those who daily have charge of and
handle the animal, as the following case, which is selected from
among others, will serve to illustrate:

A few years ago a gentleman purchased what then appeared as
a fine family horse—a sorrel gelding, aged five years. About the
beginning of March, one year after the purchase, in attempting to
walk the animal from one stable to another, but a short distance
off, he became suddenly lame in one of the hind limbs, low down,
toward the foot, and it was found almost impossible to urge him
forward. Finally, assistance was procured, and the animal had,
literally, to be carried into a stable. Our services having been
secured, we visited the sufferer, and found him standing on three
legs, the near hind one being placed resting on the toe. The parts
around the coronet were hot, and the patient was very unwilling
to have the parts handled. He seemed to suffer much pain, was
breathing hurriedly, and the pulsations were quick and wiry. The
near approach of any person seemed to terrify the animal, and he
actually trembled, from fear and pain, when urged to move. On
making an examination of the jaws, we found that both angles
of the lower jaw were enlarged to about two inches in thickness.
This explained the nature of the sudden and otherwise mysterious
lameness. The articulating surfaces of various bones, and the
bones themselves, had become diseased, as is often the case in
big jaw. The owner assured us that the horse had never before
been lame, and he was very much surprised to find the lower jaw
enlarged.

Here the reader will perceive that the animal had been in the
owner's possession one year; had performed ordinary duty, yet
was the subject of a gradual enlargement of the jaw and a constitutional disease, which probably existed, yet in a slight degree, at
the time of purchase. Hence we contend that the disease, in its
early stages, is of a hidden or insidious character, and requires some
tact and knowledge of the subject in order to detect it.

Nature of the disease.—Big head and big jaw are ut one form of disease, only located in different parts. When the upper jaw becomes the seat of affection, an enlargement of the facial bones will be observed, and the examiner's attention will be attracted by the unnatural appearance of the face. As the disease progresses,

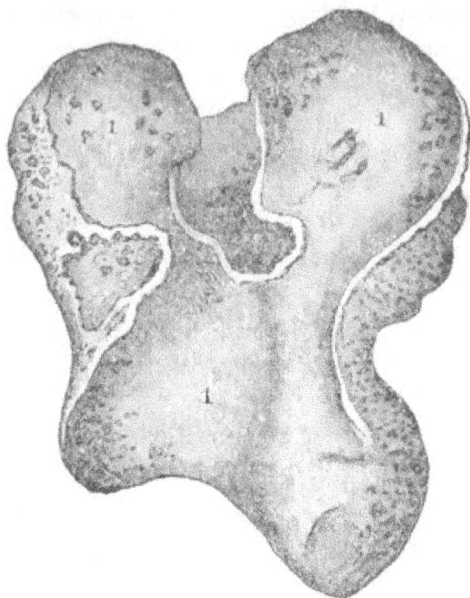

ARTICULATING SURFACE OF THE THIGH BONE.

EXPLANATION—1 1 1. Articulating surface of the lower or inferior part of the thigh bone occurring in the disease known as Osteo Porosis, or Big Head.

the face looks more like that of an elephant than a horse; and now, the disease being incurable, the subject should mercifully be put out of his misery. The enlargement of the lower jaw is discovered by manipulation. On removing the skin from the sides of the face of such a subject, and dissecting the muscles and periosteum (covering of bone), the bones appear to have undergone a softening process, and particles can be separated by means of the finger-nails. At the same time, if a transverse section of any of the shaft-bones be made, they will be found to have undergone structural changes. For example, they will not only appear softened, but honey-combed, and many of the laminæ and cartilaginous braces will have been decomposed or dissolved, so that the bone looks like fibrous net-work. In most cases, the vacuities and canals are filled with material resembling soft cheese, which is probably fatty matter. This occasionally degenerates into pus.

Dr. GORDON forwarded me, some time ago, the head of a horse who had long been the subject of this malady. The transverse diameter of the bones of both upper and lower jaw was very much enlarged; the periosteum was very highly organized with blood-vessels; the bones were softened and elastic, and could be easily cut with a knife. On submitting them to the acid test, earthy matter was found in excess, and, by burning them, it was found that they contained very little animal matter; hence we were led to infer that the disease was one of mal-nutrition. The articular surfaces of various bones, and sometimes all the bones of the body, are often ulcerated and studded with incrustations. This explains the why and wherefore of the sudden and tormenting pains which animals suffer, and goes to show how little of benefit can be expected from the ordinary treatment, which is principally directed to the enlarged jaws.

On removing these diseased cartilages from the articulating surfaces of the bones, the latter are found to be affected—in fact, ulcerated. In some cases the ligaments and tendons are separated by decomposition, or necrosis of bone, and the animal " breaks down," as the saying is.

The subjects of this disease sometimes show lameness in the vertebral region, and in that of the head of the thigh bone. In such cases, we may safely infer that the lameness is of an articular character, and the interticular cartilages, as well as the surfaces of the bones, are incrustated and ulcerated. In this condition, the animal is liable to become useless, or " break down " at any moment. The following case will illustrate what is meant by breaking down:

We were requested, a short time ago, by Prof. MUZZY, to examine an aged gray gelding, the subject of lameness in the hind feet. Examination revealed considerable heat, tenderness, and tumefaction around both the hind coronets, and both angles of the lower jaw were much enlarged in their transverse diameters. We advised the owner to let the animal have a run at grass. Accordingly he was sent into the country about thirty miles, which was too long a journey for him. A few days afterward, Prof. MUZZY received a letter that the horse's strings (tendons) became loose, and he broke down and traveled on his fetlocks, and was then dea l. We infer that necrosis took place, and the extensor tendons lost their at achment, which accounted for his breaking down.

Prof. VARNELL lately consulted Dr. HARLEY, of London, in reference to this formidable malady, which is said to have been almost unknown in England until November, 1859, when several cases occurred among some horses the property of Mr. Champion Calcot, near Reading. Three of this gentleman's animals had died, a fourth was not expected to live many days, and two others were laboring under the same disease, in a less acute form. The most singular feature of the disease, as it appeared on the premises of Mr. Calcot, was, that it did not appear to owe its origin to any perceivable cause, and it was not known to be prevalent in any other part of England. Still, our opinion is that it does exist in that country, only, not being suspected, is not sought for. It prevails very extensively among horses brought to Ohio, and hundreds of horses have been purchased for the American Government having more or less enlargement of the lower jaw, yet we do not believe that either seller or purchaser suspected the same. This arises from a lack of knowledge on the subject, and the same remarks, as far as our ignorance of the subject is concerned, may apply to England.

In answer to questions put to Mr. Calcot on the subject of management, etc., he contends that the affected animals had plenty of exercise. There were no chemical factories or works in the vicinity. One horse affected was purchased at a distance; the rest had been bred on his farm. The sires and dams all appeared free from the disease, and they were not all got by the same sire. The food consisted of good grass, hay, pollard, oatmeal, and roots. Other horses were on the farm, fed in the same manner as the affected ones, yet they had no symptoms of the disease. (Perhaps they will show symptoms when the jaws shall be examined. They may have the disease, yet not be lame nor apparently ailing.) Mr. Calcot never saw a case of the kind before.

From the above testimony, we may infer that the causes of this affection are not yet discovered in England; but our impression is that overfeeding has a good deal to do in the production of the malady. It appears to me that this disease, as I have already stated, is one of malnutrition, or defective nourishment; hence may be considered as a scorbutic affection, like that affecting the human subject, which is known to be the result of faulty nutrition, and which often results in division of bony parts which were once immovable (the epiphysis of the pelvis, for example); also in the separation of cartilages from the ribs, and the shaft bones softened and ruined by caries (ulceration or death of bone). Big head and big jaw are terms sufficiently explicit in ordinary conversation; but, in view of scientific inquiry, it is necessary to employ terms indicative of the pathology of the disease. We can not expect, however, to select any one name that shall apply to all the pathological conditions during the rise, progress, and termination of this peculiar malady. The condition of the bones is a state of enlargement, (porousness and softening;) hence the name osteo porosis.

Causes.—We have already informed the reader that the disease may have an hereditary origin; otherwise, we can not account for its universal prevalence. In certain localities, under the ordinary modes of feeding and general management, we grant that at first the disease might have had an accidental or spontaneous origin, and finally become permanent and transmissible. For example, glanders and farcy afford illustrations of a spontaneous disease becoming contagious and transmissible. There must have been a me when neither of the two latter diseases existed; hence, when the first subject became glandered, he could not have taken it by infection or by contagion, but it must have had a spontaneous origin, and, finally, became permanent. We do not pretend to urge that the active disease itself is transmissible; yet, in certain cases, a predisposition is transferred to the progeny. This may be called the predisposing cause; the ordinary exciting causes are those which disturb and derange the digestive function. The digestive or nutritive function is deranged, both by excessive and defective functional labors, or by the animal existing on food that does not contain the necessary amount of nitrogenous or muscle-making matter. This is the case when Indian corn is used as food for a great length of time. It is hard to digest, is deficient in nitrogen, and almost always over-distends the stomach; for, when submitted to the action of heat and the gastric fluids, it increases in bulk to about five or six times its original capacity. I have noticed that where much whole corn is fed, as in Ohio and Indiana, the disease is most prevalent. Associated with the predisposing causes are others—for example, hard usage, severe abuse, and bad stable management.

This disease is rarely ever heard of in England, probably from the fact that the food of horses is rich in phosphates and nitrogen, while corn contains more of starchy matter, which, instead of furnishing material for the preparation of muscular or animal matter, merely furnishes that which is consumed in process of respiration.

Treatment.—The ordinary treatment, as practiced by some persons, is to bore into the jaw-bone, and inject the same with some corrosive poison. Others expose the jaw-bone, and saw out a section of the same. Some persons blister, or apply preparations of iodine. Such treatment, I think, only tends to create unnecessary irritation and pain, and can not possibly be of any benefit, for I contend that the disease is not local, but constitutional, and the reader will probably, after perusing this dissertation, come to the same conclusion. The disease has extensive ramifications in various parts of the bony fabric, and, therefore, the local treatment must fail in curing the malady. Cases may arise which require surgical operations, and if so, I have no objections to offer. I have often been told that horses, after having all sorts of barbarities practiced on them, have recovered. This, so far as the treatment is concerned, is in accordance with the spirit of the old error, "He got well after taking my medicine; therefore, in consequence of taking it." This is assuming a falsehood for a fact, and then giving fanciful reasons for it.

In view both of prevention and cure of this organic disease of the bones and their articulations, more is to be accomplished by regimen than by medicine. The animal should always be provided with wholesome diet, and whenever green vegetables can be obtained, they should be fed liberally, or, what is much better, provided the season permits, allow the animal to roam in a pasture. Whenever I have a case of this kind under treatment, I furnish apples, beets, carrots, cabbage, or any other kind of vegetable that I can procure. I find that such articles are usually devoured with good relish. The object in feeding green food is to combat the scorbutic diathesis which usually exists. The patient should also be allowed from five to seven quarts of oats per day.

When a horse with an enlargement of the lower or upper jaw is suddenly attacked with acute lameness, he should be placed in a wide stall, and the parts where the lameness appears to be located, as well as the jaws, should be diligently rubbed, twice daily, with a portion of the following:

24

No. 58. Spirits of camphor.................... 6 oz.
         Cod-liver oil..................... 4 oz.
         Oil of cedar...................... 2 oz.
         Diluted acetic acid............... 1 pint.
         Mix.

Then procure the following:

No. 59. Chlorate of potass................. 2 oz.
         Powdered ginger.................. 4 oz.
         Gentian.......................... 3 oz.
         Podophyllum...................... 2 oz.
         Poplar bark...................... 6 oz.
         Mix.

Dose, one ounce night and morning, to be incorporated in the food.

This treatment usually palliates the lameness. Should it not do so, the owner must be patient, and give Nature time to restore the animal to comparative usefulness. Flour of bone or phosphate of lime may be given, at the rate of a couple of ounces per day. Such agents are always indicated in the treatment of the above malady.

SPAVIN.

Spavin consists of an exostosis of the hock, commonly located on the inner side, whereby bones before moveable become cemented and fixed together, and which is often productive of ulceration of the articular surfaces of one or more of the joints of the hock. Sometimes, however, spavin makes its appearance on the anterior or front part of the hock, and sometimes there is no enlargement to be discovered. Such cases, unaccompanied by enlargement, are called inter-articular spavin.

Symptoms.—"The symptoms of spavin," says Mr. PERCIVALL, "are, in general, plain, simple, and unequivocal. The horse manifests lameness in one of his hind limbs, and, on examination, a circumscribed spheroid tumor, of the magnitude of half a walnut or more, is evident enough both to the sight and feel of the man practiced in such matters. Lameness, however, the effect of spavin, may be present without any detectable tumor.

My own observation has led me to note two kinds of lameness in spavin—one, the effect either of the ordinary subacute or chronic inflammation, or of partial anchylosis of the parts, is but comparatively slight, and consists, for the most part, in stiffness on motion,

or in defective flexible power of the hock-joint; the other, the effect
of acute inflammation of the spavined parts, or else of ulceration
of the joints, consists in a sort of spasmodic catching up of the
spavined limb the moment the heel of the foot comes down upon
the ground, something after the manner of string-halt. Stiffness
may not unfrequently be observed even in the horse's side move-

SPAVINED IN OFF HIND LEG—RINGBONE IN ALL THE PASTERNS.

ment in his stall. With such characteristic lameness as this, and
with spavin present as well, evidently hot and tender to pressure,
there can exist no doubt about the case. Where, however, the
lameness is but slight, although a spavin is present, yet, from the
absence of any heat or tenderness in the swelling, as well as from
its duration, many doubts arise as to the cause of the lameness.
In such a case as this, we should take advantage of the well-
known fact of the fluctuating or evanescent character of spavin
lameness, to which end it is advisable to have the horse under
examination hard ridden or driven, or otherwise exercised, until
he be in a profuse sweat, and afterward kept standing tied up in
a stall, until he be cold and stiff in his joints, and then trotted
out again. A knowing vender of a spavined horse would take
care to 'warm' him by a good ride or drive before he took him
to show to the presumed purchaser; and then, while under ex-

amination, by dint of whip and spur, and management in the bridle-hand, he might pass his merchandise off to an unwary buyer as sound. Indeed, so much is sweating work, or exercise approaching thereto, apt to prove a foil to showing lameness, that one is almost inclined to say no horse ought to be examined under such circumstances; certainly no horse suspected of spavin. The time, of all others, that a spavined horse will be apt to manifest his lameness will be the day following after a hard day's work; and when he makes his first egress from the stable in the morning is the critical period for examination. Horses that go limpingly lame from spavin, lame at all times, and lamer still when they work, often experience pain in the seat of disease to a degree which, in the language of Solleysell, causes them 'to pine away, especially about the flanks.' They have probably been blistered and fired, perhaps setoned; have had their hocks frightfully scarred, and yet are lame to that degree that they are unable to do more than gingerly put the toe of the foot of the spavined limb to the ground, and so painfully hobble along; and, although they may still maintain their appetite, yet they are low in condition, tucked up in their flanks—evidently, in short, 'pining away.'

Such pitiable subjects, it is true, may be kept at work. The little, however, they can do, when put to any thing requiring strength of action or pull, together with the wretched condition they are generally in, is a fact so well known to coach and omnibus proprietors, and horse-keepers in general, that at the horse auctions such animals fetch little or nothing. Even for agricultural work such laborers as these prove of but little worth. Now and then, however, it happens that the spavined horse, although treatment has failed to render him sound, continues, in respect to his disease, in that state in which he appears to suffer no local pain at all while at rest, and but little while at work, and so is able to do a considerable   : of some kinds of labor, lasting in it perhaps for year:. .still, such a horse is more likely than another to receive injuries, to experience aggravation or relapse of disease in his already diseased hock; and, under such return or augmentation of ailment, unless great care be taken, and frequently with all the care we can take, may and will fail altogether.

Spavins exist which occasion no lameness. How this comes to pass will appear when the time arrives to consider the reasons why spavins in general cause lameness, and, on occasions, very great

pain as well, which can not be done before we come to treat of the
pathology of spavin.   It is sufficient for our purpose here that we
note and establish the fact that lameness is not a necessary consequence of spavin.   Nothing is more common than to meet with
horses, colts even, who have what the dealers call 'knots' in their
spavin places; and the time was when such 'knots' (which have
always been regarded as spavins) were certificated as constituting
unsoundness.

Lameness arising from spavin is sometimes present without the
outward appearance of spavin.   This is a form of disease better
known to veterinary surgeons in general, I believe, under the de-
nomination of occult hock lameness.   My own attention to the
subject was first drawn so long ago as in the year 1815, though
then I was quite in the dark as to the nature of the case.   On
my return from Belgium, after the battle of Waterloo, I had in
my possession a bay blood mare, who was lame in one of her hind
legs (I forget which), but whose lameness was of that nature that
no external sign whatever was apparent to account for it.   The
limb had been searched over and over again, by myself and some
other veterinary surgeons, and the mare had been trotted and
walked, circled and paced, and put to all other known trials and
tests, without the examinations ending in any thing like concurrent opinions respecting either the seat or the nature of her lameness.   The mare returned home, marching with the troops, led
by a man on horseback—for, notwithstanding her lameness, she
walked very well—and, as soon as she arrived at head-quarters
(Woolwich), I showed her to my father, at the time senior veterinary surgeon of the Ordnance Department.   He examined her,
and without hesitation pronounced her 'lame in the hock,' and
she was treated accordingly; and the result was, at no great distance of time, her complete restoration to soundness.

It is true, so far as the case above related goes, that the only
proof that the mare's lameness was in the hock, was her restoration to soundness after the application of remedies to that joint.
There is, however, to be said, in addition, to induce us to believe
that it was so, that, of all the joints of the hind limb, no one is so
frequently or so likely to be deranged as the hock; and, consequently, from this fact alone, is a *prima facie* case made out.
Moreover, we have, to assist us in our diagnosis, the stiff or imperfect flexion of the hock-joint in action, and the wearing away

of the toe of the shoe, showing that the heel is rarely or but very gingerly put to the ground; also the circumstance, often observable by the groom, of the animal resting the lame (hind) limb in the stable; on occasions, perhaps, knackling over upon it, and so bearing the weight upon the toe alone. And it has happened before now that, while doubt was impending as to the locality and nature of the lameness, a spavin has made its appearance, and dissipated all further conjecture; and with this development of the spavin, the lameness, so far from being augmented, is not unlikely to become better. This is an observation made so long ago as the time of Solleysell. This admirable observer, in one part of his chapter on spavins, says: 'At their first piercing, they, generally speaking, make a horse halt, and afterward the swelling growing bigger, the horse halteth no more with it.'"

I have noticed that spavined horses are usually the subjects of ill-shaped hocks; hence this peculiarity of conformation would seem to indicate that such animals have a certain amount of predisposition lurking in their systems. The ordinary exciting causes are sprain or strain, and overwork.

Now, what will cure spavin? This question was propounded to the author, some time ago, and the following answer was elicited. The answer is as good to-day as when first written. "What will cure spavin?" I answer: Time, rest, and the application of remedies adapted to the various stages of the disease. The usual remedies, in the early stage, are rest, frequent use of the shower-bath, and refrigerating lotions. In the chronic stage, and when the disease has commenced on the articulating surfaces of the hock-joint, or an ossific bony enlargement has made its appearance, then we resort to counter-irritation, by using some one of the preparations of cantharides or iodine. "What will cure spavin?" is a tough question for a medical man to solve; for, if he has been properly educated, he knows well enough that bone spavin, like hip-joint disease in the human subject, is, to all intents and purposes, incurable—I mean in so far as the physiology and pathology of the parts are concerned. Soft tissues, once having been changed into bone, never can be remetamorphosed into the original condition so as to possess their inherent qualities of elasticity (contraction and expansion). At the same time we are all horsemen enough to understand what the term "cure" means when applied to the disease known as spavin. It signifies freedom from sensibility and lameness, and a partial disappearance of the enlargement, or "eye-sore." As the reader may like to refer to the opinion of some well-known author, as well as what I here advance, I select a quotation on the treatment of bone spavin from the pen of Mr. BLAINE:

"This disease does not differ from splint, except that it is much oftener a cause of serious lameness. Occurring, however, as it commonly does, in older horses, it also proves more obstinate, and the treatment required, therefore, should be more active. Among the old farriers, who, like some of the moderns, thought nothing too strong for a horse, violent mechanical operations were resorted to, as the mallet and chisel to chip it off, boring the exostosis with a gimlet, punching it with a hot iron, or applying caustics. The first removing it mechanically, and the three latter methods destroying its vitality, promoted its exfoliation. As might be expected, for one case which succeeded (and in some it certainly did succeed) in many it increased the lameness, or ended in anchylosis and sometimes death. It is not improbable that instruments may yet be devised which will operate on these bony enlargements without risk, though the chances are fewer in the hock, from its connection with capsular and bursal ligaments, than in any other parts.

Treatment.—The treatment pursued by veterinarians of the present day varies somewhat. Those bordering on the old school still rub with some violence, and then stimulate them with oil origanum, oil terebinth, etc. Those of a later day blister and fire. At the Veterinary College setons are used, by ripping up the integument and pushing a needle around, with tape through it, so that the seton within the skin exactly opposes itself to the spavin. If the skin is tender or tumefied, it is more proper to make an opening above and below the exostosis, and to push a blunt seton-needle or eyed probe from one opening to the other, armed with a tape smeared with mild blistering ointment, or common turpentine. In this way, report says, the College practice has proved very successful. Blistering we have, however, found, when repeated over and over, commonly prove equal to all the benefit these obstinate cases can receive; for, when the bone deposit is fully formed, it is in vain to expect its entire absorption. Even its partial absorption is frustrated. The removal of acute lameness is generally the only benefit which can be anticipated. Our principal hope lies in preventing the increase of the bony deposit, and likewise in removing that inflammation which is the existing cause of much of the pain, stiffness, and tumefaction in the surrounding ligaments. Our hopes of success must also greatly depend on the time the evil has existed. When it is early attended to—that is, before the bony deposit has gained its full solidity—stimulants act more favorably, and lessen it more materially. It remains to add that, when repeated blistering fails, the firing-iron generally concludes the business, but commonly answers no further purpose than being a publication of the animal's infirmity, and a warning to practitioners as regards the previous treatment."

The remedies and means used by the author of this work for the treatment of spavin are the same as recommended for ringbone. (See article on Ring-bone.)

FRACTURE OF THE PELVIS ("HIP SHOT," OR "PINHIPPED").

My attention having lately been called to two cases of fractured pelvis, I am led to make the following remarks for the benefit of those who feel an interest in matters pertaining to veterinary science. The accident of fracture of the pelvis is of very common occurrence, and I think that many such cases might be prevented by the exercise of a little judgment on the part of horse-owners. For example, the great majority of cases with which I am acquainted have occurred in the stable, after the animals have been put up for the night "all well," as the owners have informed me, and to them the matter seemed very mysterious and unaccountable. But I think that the mystery can easily be explained.

If a horse is put up for the night all well and free from lameness, and in the morning is lame, and it is observed that the hip is knuckled down, as the saying is, (fracture of the pelvis,) or, in other words, he is " hip-shot," the evidence then seems conclusive that the accident occurred within the stable. Then the question arises, How did it occur? We all ought to know that the stalls in our city stables are too narrow; many of them do not exceed four feet in breadth. Now, let a large horse throw himself violently down, as they often do when tired or in pain, and he is very apt to strike the point of the hip against one side of the stable. The point of the hip being but imperfectly protected from external or lateral injury, except by the common integument, the jar or concussion thus received often occasions fracture. I know that this is the case; for, on making examination of some of these fractures, I found the skin over the region of the point of the hip bruised and abraded. Then, again, horses frequently get cast at nigl t, and are the subjects of colic, or other aches or pains. They struggle violently, get imprisoned, as it were, in their narrow apartments, and, when released by assistance, fracture of the pelvis is often discovered.

This is not all. Sometimes sufficient bedding is not furnished to protect the animal from the effects of hurriedly reclining, or slipping down, on a hard plank floor, and the result is often the same as that just alluded to. Then, in view of prevention, these facts suggest the propriety of having wider stalls, and a liberal supply of shavings or sawdust for bedding; that is, when straw can not be obtained; yet I think that when sawdust can be obtained, it is the most wholesome and economical bedding for horses. Taking a pathological view of the matter, I have to inform the reader that a great proportion of these fractures occur among horses well advanced in years; and it may be that their bones, in some cases, are brittle and very easily fractured, as is the case with some aged members of the human family, whose thigh bones have been easily fractured by a slight fall, or jumping out of bed in a hurry, many such cases being on record.

The horse is occasionally subject to a disease of the bones known as fatty degeneration, which is said to be partly occasioned by an exclusive Indian corn diet. In such cases the vacuities in some of the bones are filled with oily matter, are light, and easily fractured. He is also subject to a disease known as rickets, which may depend either on hereditary predisposition or imperfect nutrition. The bones are defective in early constituents, and, consequently, give way under the weight which they ought otherwise to sustain. The remedy is, a nutritive diet, partly composed of oats, and occasional doses of phosphate of lime.

Horses very frequently fracture the pelvis by falling in harness, but for this I know of no remedy except careful driving; yet sometimes, in spite of due care, they must fall when traveling on slippery pavements, and when improperly shod.

Treatment.—As regards the treatment of fracture of the pelvis I have but little to offer. It is all contained in the words rest and counter-irritation (seton or blister), yet requires time for

Nature to unite the bones after her well-known fashion. If the
horse is a valuable one, and the fracture slight, he should be
placed in slings. The hair may then be shaved off the region of
fracture to the extent of one foot in diameter; then smear on,
while warm, a strengthening plaster, composed of pitch, resin,
and beeswax. But should the fracture be of a very grave character—the bones broken in several places, the animal down and
in great agony—then the sooner he is put out of his misery the
better.

SPLENT, OR SPLINT.

Splent, or splint, makes its appearance on the inner and lower
part of the knee-joint. It consists of a bony tumor (exostosis),
and, finally, the small bone known as the metacarpus purvus
becomes united to the cannon or shank bone. This constitutes
anchylosis, so that the disease is of the same nature as ring-bone.
Splent sometimes appears on the outside of the leg ; but the disease is oftener found on the inside, from the fact that it is nearer
the center of the weight of the body than the outer, and receives
the greatest amount of concussion The inner bone is said to
receive nearly the whole weight transmitted to the small bone of
the knee. A splent on the inside is often the result of a blow
inflicted with the opposite foot, which bruises the soft parts and
periosteum, and, finally, the bones become affected. Faulty shoeing has also a direct agency in the production of splent. For
example: if one caulk of the shoe be higher than the other, or
should the foot be pared more on one side than the other, the
effect will be to throw the limb into a false position, which may
result in ligamentary lameness and splent. Splent, however, appears sometimes without any assignable cause. On dissecting the
limbs of aged horses, we frequently find the splent bones firmly
united to the shank. The following is the opinion of Prof. Coleman in regard to splent:

"'A splent once is a splent always.' The two bones being once
united by ossific matter, this connection, whatever we may do or
apply, will never be altered. It used to be the custom to remove
the splent with a hammer and chisel, but the production of a
large blemish, from the destruction of skin, was the consequence
of this barbarous method. Another practice, as ancient and barbarous as the former, was the application of corrosive sublimate,

A PERFECT FORE-LEG.

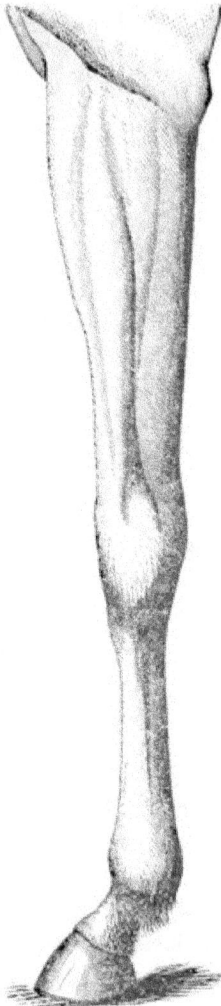

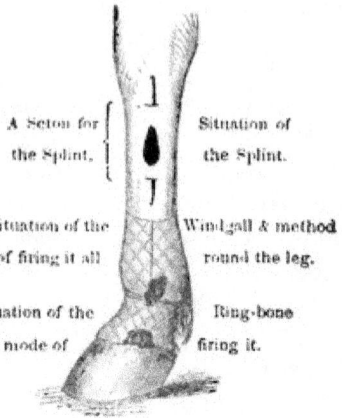

A Seton for {   Situation of
the Splint, {   the Splint.

Situation of the   Windgall & method
of firing it all   round the leg.

Situation of the   Ring-bone
& mode of   firing it.

Situation of the   Thorough-pin
& method of firing.

Situation of Blood Spavin.

Situation of the Bone Spavin   & method of
firing the same.

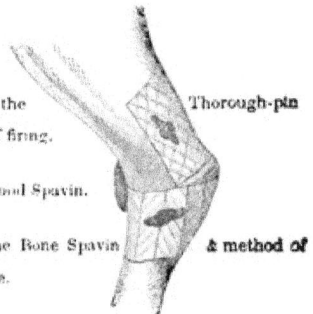

BOTTOM OF THE HOOF.

a a the crust.   c c the bars.
b b the sole.   d d the frog.

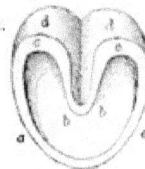

or other strong caustic, so as to occasion a slough. In this case, too, the remedy was as bad as the disease; the horse was disfigured for life. Better and more recent treatment than this was either firing or blistering. The objection to the former, however, was the marks it produced; and to the latter, in common with the former, that though the lameness was removed, it very frequently returned with the resumption of labor.

The more modern, successful, and scientific treatment of splents consists in the operation of periosteotomy, or division of the periosteum, which covers the bone.

Professor Sewell has introduced, within the few past years, an excellent mode of performing the operation. The horse is cast, the leg properly straightened and secured, and then a small opening is made just below the splent sufficient to introduce a long, narrow, convex, probe-pointed knife, the edge of which is on the convex side The knife is then passed up under the skin, and by drawing it backward and forward on the splent, pressing firmly at the same time, the periosteum is completely divided. A small opening is then made through the skin above the splent, and a narrow seton passed from one orifice to the other, after which a bandage is placed on the leg, and the horse released. The seton should be moved and dressed daily with digestive ointment, and at the expiration of a week removed, and the wound permitted to heal. I have found the operation succeed whenever I have adopted it. In the very numerous cases that occur of splents being unattended with lameness, it is 'better to let well alone;' but when the lameness is slight, and the horse can not be spared from work more than a few days, it is well to apply a mild blistering application, such as the tincture of cantharides, about a tea-spoonful of which will be sufficient for one application, which may be repeated according to the action it produces and the benefit it occasions."

Treatment.—Our practice in this country is to treat splent on the same general principles that obtain in spavin and ring-bone— namely, in the acute stage, when the accident of striking has happened, we apply sedatives and refrigerents, and in the chronic stage, counter-irritants and absorbents. For the treatment of the acute stage, a selection from the following articles may be made: Arnica, infusion of hops or poppies, cold water, or equal parts of vinegar and water. In the chronic stage, and in view of lessening the tumefaction. I recommend the following:

No. 60. Powdered iodide of potass.............. 2 dr.
       Simple Cerate...................... 1 oz.
       Cod-liver oil...................... ½ oz.

Rub these ingredients together in a mortar, and when thoroughly mixed it is fit for use. A daily application of this ointment for a short time will suffice. There are many cases of splent that do not occasion the least lameness · therefore they had better be let alone.

BOWED LEGS OR SPRUNG KNEES.

The following case will illustrate the method of cure for sprung knees. The subject of this operation had met with an accident, which resulted in "bowed limb," and had so progressed within the period of a few months as to render the limb the most deformed one I ever saw; for when the toe touched the ground, the heel was elevated about four inches. At the same time, the inferior extremity of the cannon overlapped the large pastern something like three-quarters of an inch. Such a deviation from the perpendicular line would, as a matter of course, bring a tremendous strain on the extensor tendons in front and the suspensory ligament, and this appeared to be the case; for the tendons and ligament were rigid as a bowstring.

The horse was cast on the off side, and, after securing the limbs, the near fore one was released from the hobble and held by assistants. A puncture in the outer lateral region of the flexor tendons was made through the integument. A convex, probe-pointed bistoury was then introduced, and, by a gentle sawing movement, the flexors perforatus and perforans were divided. On straightening the limb, which was done without the least difficulty, the edges of the tendons had separated almost two inches. These would, of course, after awhile, become reunited by granulation and interstitial deposits. Thus tendons would, finally, gain in length something less than two inches, which would render the leg almost straight again. The puncture through the skin being small, a single suture sufficed to close the orifice. This being accomplished, a cold-water bandage was neatly applied, and then the patient was permitted to rise. He walked to the stable tolerably well, yet still had a slight "knuckling over" at the fetlock. This was probably owing to the distended or lax state of the capsular ligament and extensor tendon. A shoe was now applied to the foot, having a straight piece of iron, eight inches in length, welded on to the ground surface of the toe of the shoe. The object in applying this contrivance was to prevent flexion, to keep the divided ends of the tendon apart, and to guard against direct union of the same. The operation finally proved successful.

The following case occurred in the practice of Professor BRACHET, as given in the " Veterinary Record : "

" This most extraordinary case was observed in September, 1856, three months after the inundation of the Rhone, on a farm near this river. The subject of the disease was a filly, three months old. She was born perfectly straight on her legs, but about fifteen days after birth the fore-limbs became bent, and the deformity increased so rapidly that when Brachet saw the animal it moved on its knees. It was with great difficulty that, by efforts to extend the limb, the acute angle formed at the knee could at all be widened, and such an experiment gave great pain to the animal. The dam being taken out of the stable, Brachet was astonished to see the filly walking after her on her knees, almost as fast as if progressing naturally. On the front of the knee the skin was hard and callous, the tendons behind it were rigid and immovable, and Brachet decided on performing tenotomy. The animal was cast and the near fore-leg operated upon. An incision was made, about an inch in length, at the back of the arm, in a line with the upper part of the carpus ; the tendon of the flexor metacarpi externus was first cut, then one of the medius, and, lastly, that of the internal flexor of the metacarpus, when the limb straightened. A splint was placed in front of the leg. Eight days afterward the off limb was operated upon ; fifteen days later the splints were removed, and, in the course of a month, the filly stood well, but in moving carried her fore-legs forward in a piece. Flexion at the knee could not be effected, but the young animal improved rapidly ; the limbs became strong, and though slightly bowed, their action was perfect. Brachet believes the deformity arose from rheumatism, induced by the damp on the occasion of the overflowing of the river near the farm."

The following is Mr. PERCIVALL'S method of performing tenotomy for sprung knees:

" The operation of tenotomy, though a formidable one for the patient, is not a difficult one for the operator. The object is section of the flexor tendons, the effect of which, as we have seen, is to let down the heel of the foot, not the fetlock, to the ground. The flexor tendons support the pastern and foot-joints principally, the fetlock-joint having the additional strong support of the suspensory ligament, which it still retains after the tendons have been cut through. This accounts for the heel of the foot, without the fetlock, being let down by the operation of tenotomy.

Having cast the horse, and so secured the limb to be operated on that there is not much chance of any interruption being occasioned through its motion, I recommend that a longitudinal incision, three or four inches in length, be made through the skin, along the back of the leg, down upon the middle portion of the flexor perforatus tendon. This incision freely and boldly made at once, the operator will be able to stretch the mouth of the wound he has made round to the inner side of the leg, in which stretched position the skin is to be held by an assistant, while the operator introduces the fore-finger of his left hand, to push back the blood-vessels and nerve (which run along the inner borders of the tendons) against the suspensory ligament, so that they be safe out of the way, while with the right hand he insinuates his bistoury between them and the flexor tendons. Opposing now the cutting edge of the bistoury to the tendons themselves, he commences incising them by a steady but firm and strong sawing movement, until both be completely divided. I say he is to hold the bistoury (which ought to be a stout one) firmly, and to use it with some force, since such is the dense and tough texture of these tendons that they are not cut completely through without some determination. Complete division being made of them, the heel of the foot may not (in all probability will not) come down without some extension of the leg; and this must be made, not by such violence as Mr. Goodenough found himself compelled to use, still with such force as will, if possible, make the limb straight, by stretching or even tearing through adhesions of moderate standing, so far as this can be effected without the risk of rupturing ligament or fracturing bone. The giving way of adhesions, in such cases, is frequently attended with a sort of snap or jerk, denotive of their being overcome, and by this the end is known to be answered, such adhesions and impediments to extension being commonly situate about or in the vicinity of the fetlock-joint. The usual and most effective mode of accomplishing the extension is, to place the knee against the front of the fetlock, and, grasping the back of the foot with one hand, and the upper end of the leg with the other, to use such steady and moderate force as will accomplish the object without doing harm thereby. This done, the divided ends of the tendons recede from each other, leaving a gap between them of one or two inches, or even more, dependent upon circumstances."

CORNS.

What is known as corns, occurring in horses, is very different from that affection in the human subject, which bears the same name. In the horse the affection is caused by bruise of the sole, and consists of extravasation of blood. On making an examination in the angle between the bars and quarters, the horn of the sole will appear red. This may not be seen, however, until the blacksmith has pared away a flake or two of the sole.

Treatment.—Neglected corns are very apt to end in suppuration (formation of matter within the hoof). This matter sometimes burrows in an upward direction, and breaks out at the top of the hoof. It then forms what is known as a quittor, the remedy for which is to make a small opening into the sole, and thus allow the matter to escape in a downward direction. After making the orifice in the sole, it should be dressed with a paste composed of equal parts of sugar and brown soap. As a dressing for common corn, without suppuration, I recommend equal parts of tar and olive oil. This I conceive to be one of the best dressings ever used for corns.

Many corns are produced by suffering the shoe to remain on too long. It thus becomes imbedded in the heel of the foot; the external crust then grows down on the outside of the shoe, and the bearing is thrown in the angular portion of the sole. In unusually troublesome cases of corns, recourse should be had to the bar-shoe.

YOUATT says: "Mr. Spooner, of Southampton, very properly states that the corns occasionally fester, and the purulent matter which is secreted, having no dependent orifice, ascends, torturing the animal to a dreadful extent, and breaks out at the coronet. These cases are very troublesome. Sinuses are formed, and the evil may end in quittor. A large and free dependent orifice must then be made, and a poultice applied, to which should succeed a solution of sulphate of zinc, with the application of the compound tar ointment.

The cause of corn is a most important subject of inquiry, and which a careful examination of the foot and the shoe will easily discover. The cause being ascertained, the effect may, to a great extent, be afterward removed. Turning out to grass, after the horn is a little grown, first with the bar-shoe and afterward with the shoe fettered on one side, or with tips, will often be serviceable. A horse that has once had corns to any considerable extent should, at every shoeing, have the seat of corn well pared out, and the butter of antimony applied. The seated shoe should be used, with a web sufficiently thick to cover the place of corn, and extending as far back as it can be made to do without injury to the frog.

Low, weak heels should be rarely touched with the knife, or any thing more be done to them than lightly to rasp them, in order to give them a level surface. The inner heel should be particularly spared. Corns are seldom found in the hind feet, because the heels are stronger, and the feet are not exposed to so much concussion; and when they are found there, they are rarely or never productive of lameness. There is nothing perhaps in which the improvement in the veterinary art has relieved the horse from so much suffering as shoeing. Where corns now exist of any consequence, they are a disgrace to the smith, the groom, and even to the owner."

LAMENESS.

We have already considered the nature and treatment of various forms of lameness, occurring in consequence of strain or sprain of different parts of the body; also that attending rheumatism and diseases of the feet. It only remains to offer a few remarks on some special forms of lameness. Lameness occurs in one of two forms. We either find it in the acute stage, when, from injury or other causes, it comes on suddenly; or else in the chronic stage, that form which has existed for some time. Therefore, there being only two forms of lameness, there are only two indications to fulfill, namely: in the acute stage, we endeavor to lessen the activity in the circulation, heat, and pain of the parts by rest and cold-water bathing; and when pain exists, we mitigate it by bathing the affected parts with cold infusion of hops or poppy-heads. When the affection assumes a chronic type, we apply stimulants and counter-irritants. The following is the best remedy in use:

No. 61. Aqua ammonia...................... 3½ oz.
        Spirits of camphor............... 3 oz.
        Salt (bay or rock salt)............ 8 oz.
        Water........................ 2 pints.

Dissolve the salt in the water, with a few drops of aqua ammonia; allow the impurities to settle, and add the clear liquor to the camphor and ammonia, the latter having been previously mixed. A portion of this should be rubbed on the region of lameness morning and evening. Sometimes the actual seat of lameness may be somewhat obscure, yet we can generally find out which limb it is in; then, to make sure that we shall hit the mark, the whole limb may be treated. Should counter-irritation be necessary, strong liniments and blisters, such as are in general use, are recommended.

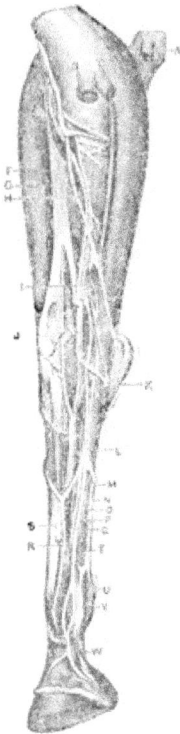

This plate represents the fore extremity, cut off a little above the condyles of the humerus, with the flexor metacarpi internus medius and the ulnaris accessorius muscles removed

A. The origin of the ulnaris accessorius muscle.

B. The flexor metacarpi internus.

C. The flexor metacarpi medius.

F. The radial nerve.

G. The radial artery.

H. The radial vein.

I. The superior suspensory ligament, or band.

J. The tendon of the flexor metacarpi internus muscle.

K. The cut edge of the posterior annular ligament turned back.

L. The inferior suspensory band.

M. Oblique branch of nerve crossing the perforatus tendon.

N. The perforatus tendon.

O. The perforans tendon.

P. The inner metacarpal nerve.

Q. Metacarpal artery.

R. Suspensory ligament.

S. The metacarpal vein.

T. One of the lumbrici muscles.

U. Cut edge of the thin, dense membrane which invests the perforans from a little above the front of the fetlock downward.

V. The perforatus tendon.

W. A portion of the tendon lined internally with synovial membrane, removed, to expose the perforans as it passes through the bifurcation of the perforatus.

SECTION XIII.

HORSE MANAGEMENT, ETC.

PROLAPSUS ANI—HEMORRHOIDS, OR PILES—PRICKING AND NICKING HORSES TAILS
—DOCKING HORSES—ANÆMIA—THE USE AND ABUSE OF THE CURRY-COMB—
HORSES AND CATTLE SHOULD HAVE A SUPPLY OF PURE WATER—GALLED BACK
—THE HEART—THE QUANTITY OF BLOOD CONTAINED IN A HORSE'S BODY—
RAPIDITY OF THE CIRCULATION—THE TRANSFUSION OF BLOOD—MANAGEMENT
OF THE HORSES PUT IN THE STABLE—STRANGLES.

PROLAPSUS ANI (FALLING OF THE FUNDAMENT OR RECTUM).

PROLAPSUS ANI, or protrusion of the rectum, is a calamity much to be dreaded. Fortunately it very rarely occurs, yet when it does, a valuable horse is almost sure to be the subject; hence it is a matter of importance that persons who own valuable horses should know something about the nature and treatment of this mishap. The best authority on this subject is Mr. PERCIVALL, from whose works the following is quoted:

"It consists in inversion and prolapse of the mucous coat of the rectum, either confined to the mucous membrane itself, or otherwise involving more or less of the entire substance of the intestine along with it. The aspect of the protrusion will mainly depend on the nature and volume of the parts ejected, and upon the time such evolution has been in existence. When recent, so readily does tumefaction (swelling) of the parts follow their protrusion, that this speedily opposes all attempts at return, should the animal make any, though, in truth, his efforts in general have the effect of straining more gut out, and through his straining, in some cases, tearing the membrane, and thus augmenting instead of diminishing the evolution as well. In this manner arises a large (sometimes enormous) rotund, red, wrinkled redness of skin, consisting of cylinders of mucous membrane, having an aperture closed through constriction in its centre, though, when open, issuing a mucous and sanious matter, especially at such time as the animal is straining afresh. The circumflex action of the sphincter ani around the neck of the swelling, together with its distension with gas, and the action of the air upon its surface, all add to its increase of volume and change of color; while the straining, at the same time, increases the difficulty of any effort the practitioner may make toward effecting a return of the protruded gut. With its augmentation of volume comes a deepening redness, turning, through the constriction of the sphincter, to a darker, and even purple hue, the membrane becoming, at the same time, inflated and swollen, and gradually changing its color from red to yellow and to brown, though still having a humid shining aspect, now becoming glairy from albuminous secretion, which is not only mucous, but at times purulent.

Causes.—These may be summed up in irritation, either direct or sympathetic, of the mucous membrane lining the anus and rectum. It is possible that prolonged constipation, from giving rise to violent straining efforts at expulsion of dry and hardened feces, might bring it on, or even the strains of parturition. The action produced by excessive purgation may cause it, but more especially irritation of a mechanical kind, arising from direct injury to the membrane or rectum during the operation of raking or manual exploration of the gut, or of clystering. It may prove an accompaniment of an enteritic, colicky, or diarrheal condition of bowel. It is possible it may ensue on nicking, though I never knew it to follow that operation. Violent struggles of any kind, as in such a case as is mentioned in the 'Veterinarian,' by Mr. J. Brown, V. S., London, wherein the horse 'had forced out nearly a foot of the rectum in struggling violently to release himself,' while being cast.

Treatment.—The formidable, and indeed awful, aspect of this disease is apt to operate in the mind of the owner of the animal favorably for the veterinary surgeon, inasmuch as it gives rise to his being called to the case sooner than he otherwise would have been. Should the summons be an early one, at the time that the protrusion is recent, and its volume nothing so very alarming, judicious exercise of the taxis, if employed at the moment, may succeed in the return of the gut. Distributing the fingers of both hands over the wrinkled and swollen body of the protrusion.

steady, firm, and forcible pressure ought to be maintained against it for such length of time as appears to afford any chance of success, augmenting the force used whenever there be any remission of the straining. Should the first efforts of this description fail, the protruded mass may have its bulk lessened and constricted as much as possible by local application of some sort, than which none offers a better chance of succeeding than such as follows from sudden and intense cold, though to a part so sensitive and vascular the practice is not devoid of danger. Ice may be powdered and inclosed in a linen bag, so as to form a sort of bolster, that may be held or braced with firmness upon the tumefaction. The contrary of such treatment as this, however, though the object be the same, is generally preferred—namely, fomentation, medicated or not, with scarification of the exposed membrane—though in other cases astringent applications are employed, with a view of causing contraction and diminution of the mass, such as lime-water, decoction of oak bark, solution of alum, etc.; and while such proceedings are going on, it is advisable to have the hind parts raised. Supposing we succeed in accomplishing the reduction, the chances are in favor of the relapse of the prolapsus, and these chances seem great in proportion to the facility with which the return of the inverted gut has been effected. Indeed, so constant and troublesome does the return in some of these cases become, that it is necessary to contrive some sort of truss to oppose the descent, the same as is done for rupture in man. Mr. Dycer's truss consisted of 'a new wet chamois leather, a breast-plate, and a hip (human) truss.' Should constipation be thought to operate against reduction, in addition to clysters, we may exhibit cathartics and aloes in solution, and should there be plethora and great irritability in the membrane of the rectum and anus, a blood-letting will be advisable, while an opiate or belladonna clyster will be advantageously administered, in order to allay all local irritation as much as possible.

As a dernier and effective mode of procedure in prolapsus ani, when the case proves either irreducible, or, after return, continually protrudes on any slight effort, such as coughing afresh, we have recourse, with safety and certainty, to an operation consisting in excision of more or less of the inverted membrane enveloping the protruding portion of the gut. This is an operation of ancient date, though to Dupuytren is ascribed the credit of introducing it into human practice, and to the French veterinarians that of transplanting the same into veterinary practice. Formerly the actual cautery was the instrument employed for the removal of the parts obstructing reduction; but in our own day this has been thrown aside for the scalpel, an instrument quite as effective, while the simple act of cutting with a knife gives so much less pain than cutting or dividing with the actual cautery. In general, it is considered necessary to cast the horse for the operation; but Mr. Gregory, V. S., did not cast his patient, but merely put on a side-line, and had one leg held up and the tail kept on one side; and, most assuredly, the standing position of the patient is one which, in such an operation, offers no mean advantages to the operator, provided he can avail himself of it without any personal danger. A great preservation against refractorness in an operation of the kind, at the moment when any pain is felt, is a twitch, well put on, and well and timely turned. The horse secured, with his tail turned out of the way, an incision, forming a circle, is to be carried around the protruding mass, at sufficient distance from the constricted part of the neck of the protrusion to leave behind it such parts as are distended from tumefaction or infiltration, whose removal will render the retraction of the gut an easy and natural effort. In making incision, it may be advisable to take up and tie with a silk ligature any vessel met with of size enough to issue a current of blood, though sometimes no such precaution is required. In the case mentioned by Mr. Gregory, not more than a quart of blood was lost. The incision ought to be made of sufficient depth to penetrate completely through the substance of the mucous membrane, however morbidly thickened that may be, without, however, running a risk of wounding the muscular coat beneath it, the object being to dissect the former away and strip it off the latter, so as to lessen the bulk of the mass to that degree that return becomes a voluntary and facile action of the animal himself. So soon, however, as this denudation has been carried near to the inverted anus, care must be taken not to dissect or otherwise injure the sphincter of that part, lest we leave the horse with an imperfection in closing an outlet of so important a function. A soft or mash diet should, for a few weeks, succeed the operation, with abstinence from hay, which, from its fibrous, prickly nature, must be particularly offensive to the denuded gut. Occasional emollient clysters are also

recommendable, especially when there appears any pain or difficulty in giving exit to the feces."

HEMORRHOIDS, OR PILES IN THE HORSE.

The following case of piles, occurring in the practice of Mr
P. WALS, V. S., is here introduced from the "Veternarian:"

"On the evening of March 13, 1851, I was summoned hastily
to attend a brown carriage-mare, belonging to an employer about
a mile distant from Norwich, which was reported to have something 'bloody' hanging out of her rectum. The coachman who
came said it had made its appearance two or three times during
the day after dunging, but had always returned again, until just
before he left home, when a portion of it was left out, and the
mare became very uneasy. On my arrival, I found, as reported,
the mare in pain, indicated by a peculiar switching or lashing of
the tail, and stamping of the hind feet; but she had not lain
down, although she made several attempts to do so. It first struck
me as a case of prolapsus ani, but, on examination, it proved to be
a true case of piles. The tumor protruding was about the size of
a duck's egg, presenting precisely the appearance of a pilous grape
in the human subject, only, of course, much larger. I had some
difficulty in returning it, in consequence of the powerful action
of the sphincter ani, which, no doubt, accounted for the pain; but,
having accomplished it, the mare became easy, and continued so
for some little time. I thought it right, however, to have her
removed to my infirmary, as I anticipated a return of the tumor
when she again dunged, in which I was not deceived, since, on her
arrival at my stable, I found it out again. On making a more
minute examination, I discovered smaller tumors round and about
the larger one, the latter, however, being the only one that protruded. The mare being naturally of an irritable temperament,
and consequently dunging more frequently than one less so would
have done, induced me to have a careful assistant remain up with
her during the night, for the purpose of returning the tumor
whenever it made its reappearance, which it invariably was said
to do after each evacuation, pain being constantly present at such
time.

Reflecting on the case next morning, and observing considerable mischief was being done by the necessary manipulation, I

foresaw I should have a very troublesome case, unless I could hit upon some plan of dilating the sphincter so as to allow of a return of the tumor without so much handling of it. This I accomplished in the following manner: I had the mare placed in a narrow stall, the back part of which was well bedded with litter, so that, in standing, her hind quarters were elevated six or eight inches higher than the fore ones. I then procured a pair of glove-sticks, such as glovers use for opening the fingers of gloves, and, having dipped their ends in oil, gently introduced them about an inch or an inch and a half into the rectum, underneath the tumor, letting it, in fact, rest upon them. In this way, by pressing the handles, dilatation took place, and the tumor receded with little or no difficulty. This treatment I persevered in for four days, in conjunction with a constant application of cold water to the parts, at the end of which time such great improvement had taken place that the use of the sticks became no longer necessary. In four days more the parts appeared to have become quite reinstated, which, of course, rendered any further treatment unnecessary. I gave no medicine, except a very mild aperient in the first instance, but kept the bowels lax with bran, linseed, carrots, etc. The mare being sold shortly afterward, I lost sight of her, and am unable to say, in consequence, whether any return of the difficulty has taken place, but I should think such an occurrence not improbable."

Pricking or Nicking Horses' Tails.

The English method of nicking horses is, instead of inserting a knife and making a subcutaneous section of the muscles of the tail, known as the depressors, to make three straight incisions at equal distances right across the tail. By this method the muscles of the tail are effectually or completely divided. The first incision is usually made about two inches from the root of the tail, and the other two at equal distances from the first, in a direction toward the end of the tail. The tail is then kept perpendicular, in the usual manner, by pulleys. It is said that by this method a more graceful curvature of the tail is secured, which I believe is a fact, yet the operation is any thing but pleasant to the horse. It is one of the fashionable barbarisms of civilization.

The English method of nicking, although the most unsightly and barbarous, is still the safest, as there is less liability to lock-

; w. It is well-known that lock-jaw more frequently ensues from a punctured wound than a clean cut or incision; yet when nicking or pricking, after the American fashion, (if care be taken to make the orifice through the skin sufficiently large, so that pus, afterward secreted, may escape,) I think there is no danger. The danger arises from the imprisonment of pus by closure of the orifice in the skin. In healthy subjects the orifice made through the skin will often unite in the course of a few hours; then, should morbid matter accumulate within the substance of the tail, it burrows, forms an abscess, or else becomes absorbed, producing lock-jaw. Therefore, I would advise persons having pricked or nicked horses under their care, to pay strict attention to such, and see that the orifices are kept open, which may easily be done by occasionally inserting the point of a penknife. The skin should be the last to heal. The healing process must be perfected in the interior first, and, lastly, the skin. Should fungus or " proud flesh " appear, apply powdered bloodroot, burnt alum, or red precipitate. When an abscess forms at the base of the tail after pricking, apply a poultice of flaxseed; and when the tail is much inflamed and swollen, it should be bathed frequently with the following:

No. 62. Acetic acid...................... 1 part.
　　　　Water.......................... 7 parts.
　　　　Mix.

When horses are nicked in warm weather, the tail should be moistened with tincture of aloes; this will keep the flies off.

Persons who are not experts in the art of pricking are apt to wound, and sometimes completely sever the coccygeal arteries, thereby causing an unnecessary and dangerous hemorrhage. In such cases, I should advise the operator to crowd into the orifice a small piece of dry sponge. This will arrest the hemorrhage, and, at the same time, keep the divided ends of the muscles apart, so that they can not reunite; hence it is not necessary to put the tail into pulleys for some hours, or until bleeding has ceased. When this is the case, the sponge may be removed by means of a small pair of forceps.

The success in securing what is denominated a fashionable tail depends altogether on the skill of the operator, who must make a complete division of the muscles beneath the tail only, as their action is to compress and depress that useful appendage; for.

snould the incision be made too high on the side of the tail, the curvators may be partly or wholly severed. Then the tail will uitimately diverge laterally, or, in stable language, it will be said that "the horse does not carry his tail straight." To remedy this, the curvator on the opposite side must be divided; then the tail has to be forcibly drawn and kept in an opposite direction from the curved side, so as to prevent the muscle from reuniting. In the course of a short time granulations are thrown out from the surface of the divided ends of the muscle. It then acquires length, and thus the tail becomes straight.

ANATOMY OF THE TAIL.

For the benefit of those who wish to practice the art of "nicking or pricking" understandingly, I now propose to give a brief exposition of the anatomy of the tail.

Bones of the Tail.—The bones of the tail, numbering fifteen, sometimes sixteen, are situated at the back of the rump-bone, or sacrum. They are not a continuation of the vertebral bones, but are an appendage to the haunch bone. The bones entire are termed *coccyx* (from the Greek, which signifies cuckoo, the bill of which bird the appendage to the sacrum, or haunch-bone of man, is said to represent); and, in order to prevent confusion in anatomical terms, veterinary surgeons have applied the same name to the tail bones of the horse. Hence, when the terms coccygis or coccygeal are used, they have some relation to the tail. The bones are roughened on their surfaces, and have depressions and eminences or the insertion of ligaments, tendons, and muscles.

Muscles of the Tail.—These muscles are divisible into four pairs. They are most distinctly seen at the root of the tail, but, in their course toward the end of the same, they become blended one with another.

The first pair of coccygeal muscles are situated on the upper part of the tail. They are attached to the sacrum and to the bones of the tail. Their action is to raise or erect the tail.

The second pair are termed *depressors coccyx*, and are situated at the under part of the tail. They are attached to the underneath part of the sacrum, and to the same locality on the bones of the tail. They are antagonistic to the first pair, and their direct action is to depress the tail. These are the muscles which should be divided in the process of pricking or nicking.

The third pair are termed *curvators coccyx*. These muscles are situated on the sides of the tail, between the first and second pairs. They are attached to the fourth and fifth lumbar vertebra, and to the sides or transverse processes of all the bones of the tail. Their action is to curve or flex the tail toward the quarter and sides, and the division of one of these muscles causes permanent flexure of the tail in an opposite direction. Considering the tail as a means of defense in switching off flies, etc., these muscles are the principal ones called into action.

The fourth pair are termed *compressors coccyx*. They are located at the sides and root of the tail, and are inserted into the back part of the hip bones (ischium), and into the five first bones of the tail. The action of these muscles is to maintain the tail forcibly against the anus. In the mare, these muscles protect her from ravishment.

The arteries of the Tail are termed *coccygeal arteries*. Those located at the sides are called *lateral coccygeal*, and the one found in the lower or inferior region is called *inferior coccygeal*.

Nerves of the Tail.—As with the muscles, so with the nerves. There are four pairs of coccygeal nerves. They are not a continuation of the true spinal cord, but are given off from it. They are distributed to the muscular filaments, to the skin, and at the extremity of the tail.

DOCKING HORSES.

All persons familiar with the old tactics of "jockeyism" and the tricks of horse trade, will understand what is meant by the above caption. It is a most cruel and barbarous outrage on one of Nature's menials, and I feel inclined to indorse the sentiments of the profound poet Tupper, when he says, "In that great day, when crime is sentenced, cruel man shall not go uncondemned because no complaining brute stands there accusing, for our Father careth when a sparrow dies."

Horses should never be deprived of any part of that valuable weapon of defense known as the tail. It is a part and parcel of the organization of the horse, and to deprive him of what Nature gave him, after the fashion of "doing all things well," is an outrage which should be considered beneath the dignity of a gentleman to perpetrate. Besides the cruelty which is acknowledged to b identified with the operation, there are other reasons, on the grounds of physiology, why the tail should be left entire, especially in the case of a mare; but, as the reader is probably conversant with the facts in the case, I shall not enter into a discussion of the same, but prefer to introduce the following paragraph, written by a person who aims at reform:

"I am glad that the abominable practice of docking and nicking horses is going out of fashion. It prevails in no country in the world besides England and the United States. We got it from the mother country, and the sooner we leave it off the better. It is wonderful how any body but an ignorant, narrow-minded blockhead of a jockey should ever have thought of it, being as offensive to good taste as it is a violation of every humane feeling. Has Nature done her work in such a bungling manner in forming that paragon of animals, the horse, that a large piece of bone should be chopped off with an ax to reduce him to symmetry, or that beauty and grace can be obtained only by cutting a pair of his large muscles? 'The docking and nicking of horses,' says an intelligent writer on farriery, 'is a shockingly cruel practice, and ought to be abandoned by the whole race of mankind. Every human being, possessed of feeling and a magnanimous mind, must confess that it is cruel and entirely uncalled for; but that creature called man attempts thus to mend the works of his almighty, wise Creator, in doing which he often spoils and disfigures them. What is more beautiful than a fine horse, with an elegant, long tail and a flowing mane waving in the sports of the wind, and exhibiting itself in a perfect state of nature? Besides, our Creator has given them for defense as well as beauty.' The same author relates an instance of a fine hunting horse, owned by an Englishman, which would carry its rider over the highest five-barred gate with ease; but he thought that the horse did not carry as good a tail as he wished; he therefore had him nicked, and when the horse got well, he could scarcely carry over two bars. 'Thus,' said he, 'I have spoiled a fine horse; and no wonder, for it weakened his loins.' Any man of common sense would give ten per cent. more for a fine horse whose tail had never been mutilated, than for one which had been under the hands of a jockey."

Anæmia.

This signifies an impoverished state of the blood, usually occasioned by want of nutritious food and exposure to inclement weather. I was called, a short time ago, to see a yearling, said to be ailing and unthrifty, and had been so for some time.

Symptoms.—The principal symptoms were as follows: The head, face, and eyes were in a dropsical condition and much swollen; underneath the heart, and along the lower part of the abdomen, dropsical swellings were also observed; the membranes of the mouth and nose appeared pale, almost bloodless; the pulse was quite slow, and the action of the heart feeble; the respirations were somewhat hurried, yet the lungs were not expanded to their full capacity. I was informed that the urine was thick, and of a cream-like color, and the feces hard and scanty. The muscular system was spare and wiry; the ribs could be easily counted, and the animal was unsteady on his legs. From such symptoms as these the reader will perceive that the case was one of debility, and from the fact that the animal had been confined to an innutritious diet, or, rather, like a hen, had to scratch for a living. It will also be inferred that the debility was the result of partial starvation.

Treatment.—I gave the animal one drachm of sulphate of iron, and three ounces of the fluid extract of resin weed per day, during a period of ten days. At the same time, he was liberally fed on oats and good timothy hay, under which treatment he rapidly gained in health and strength, and soon recovered.

Anæmia (Thinning of the Blood).

I do not advocate that the curry-comb shall be entirely dispensed with; for, in the first place, it is needed for the purpose of removing scurf and dirt which has been removed from the external surface of the body by means of the brush, and, probably, this was the intention of the man who first invented a curry-comb, and, in the next place, the curry-comb may be needed to scratch off or remove mud that may have become dry and stiff on the limbs of horses; and for these reasons, if for no others, I suppose we can not dispense with the curry-comb. There are some parts where the curry-comb should not be used—for example, on the inside of the thighs, flanks, chest, and face, where the skin is not so thick, and more sensitive than some other regions of the body. Some horses, especially those of a strongly-marked, nervous temperament, have very sensitive skins, and can not even bear a fly to light on their bodies without evincing signs of pain and irritation. On such the curry-comb should be used gently and in a spirit of mercy. There are other horses, however, especially those of the lymphatic temperament, and whose skins are less sensitive, who can bear the action of the curry-comb without much flinching, and it is the duty of a merciful man to consider this matter when cleaning a horse. Thoroughbred horses, which include a vast number of the sanguine temperament, have a very active circulation over the whole surface of the body, and this vast circulation, through the superficial vessels, argues a delicate sense of touch. To them the curry-comb, unless used very gently, is severe torture.

HORSES AND CATTLE SHOULD HAVE A SUPPLY OF PURE WATER.

Although few persons think it worth while to pay much attention to this department of husbandry, yet a little reflection will convince any reasonable person of the value and importance of furnishing animals with a constant supply of pure water. Pure, cool water is said to be a God-send to a thirsty throat; and as animals are apt to have thirsty throats, they should be permitted to enjoy a luxury which costs but a trifle, and operates very favorably in promoting their health. All classes of domestic animals have as great an aversion to impure, filthy water as ourselves, and the former will often turn away with disgust from the filthy stuff called water which is often found in water-troughs on the roadside, and within the precincts of the barn and in some pastures. The common, stagnated pond-water, which many poor creatures are compelled to imbibe, is often the exciting cause of disease, especially in the Western States, where decayed vegetable matter abounds.

Pure water will never injure an animal. I do not believe the stories which are told about horses becoming foundered in consequence of drinking pure water. In a majority of cases, Mr. Fastman is blameable. He has probably either overdriven or over-

worked the creature, or else has suffered it, when heated, to cool
off without the necessary care and attention which should always
be observed when animals are fatigued or perspiring freely.   Hard
usage, willful neglect, and wanton cruelty are more likely to produce disease than the universal beverage, so acceptable to the palate
of a weary or thirsty horse.   How often do we see a " let " horse
.ome into the stable all exhausted and used up, scarcely able to
2 lvance one limb before another !   Examine into the facts, and we
.l.all find that the powers of the subject have perhaps been overtaxed.   He has been driven too far, or at too rapid a rate, for the
present state of his constitution to endure, and perhaps he has not
had sufficient nourishment to repair the waste incidental to the
livii.g mechanism under the states of rapid and protracted labor.
Is not this enough to account for the used-up condition?   Is it
not more rational to suppose that abuse of the respiratory organs
and those of locomotion operates far more unfavorably on the
horse than water?   It is.   But Mr. Fastman must, if there be
any blame rightly belonging to him, try to shift the same from his
sh oulders, and, therefore, he avails himself of a popular error—
" He drank too much water ;" yet the individual had no means of
ascertaining the precise quantity needed.

We might say, as regards some horses, whose labors are very
fatiguing, that they come from their work, and, as soon as unhar-
nessed, go to the trough, and imbibe from one to three buckets
without any bad effect.   Some animals need more water than
ethers.   The kind of work, the temperature of the atmosphere,
and the nature of the food, whether it be wet or dry, all tend to
diversify an animal's wants.   The domesticated horse requires a
bountiful supply of good water.   His body is composed of seventy-
five per cent. of the same, and he can no more exist without it
than he can without food.   A cow or ox is probably the best judge
as regards its own wants as to the quantity of water needed.   It
is not the quantity which a rational animal imbibes which does
harm, but it is the quality that demands our attention.

Thirsty people drink all the cold water they need.   Then why
deprive a cow or horse of what they actually require?   Consider
the condition of the inhabitants of populous cities during the summer season.   Thirst amounts almost to a disease, and, in view of
quenching it, the thirsty are continually imbibing water, rendered
cold, hot, sour, sweet, or alkaline, just as fancy dictates or as fashion

prevails. Cold icings and other fixings are called into requisition to smother the fire of thirst which rages within. Every body partakes freely—the young and the aged, the exhausted and vigorous, the laborer wearied by a hard day's work, and the rich man of no work—each and all are doing their best to see the bottom of the pitcher, and to pitch their bodies into the watery element. Yet, after all, how few persons complain of any bad effect from it.

Inquire into the history of some of the acute maladies that are supposed to arise from water-drinking, and it will be found that many of the sufferers have a peculiarity of constitution which renders them amenable to the laws of primogenital disease, which, although latent under-ordinary circumstances, can, by disturbing the life forces, through neglect, cruelty, and overwork, be developed at almost any time of life.

Animals should never be allowed to drink pond-water. They should either have access to a running stream or a clean water-trough.

GALLED BACK.

Galled back, or sore back, is usually occasioned by bad-fitting saddle or harness. If that be the case, the shape of the same must be altered so as to correspond to the shape of that portion of the back which comes in contact with it. This can be done by padding the concavities with lamb's wool, and in chambering the convexities so that there shall be no unequal pressure nor causes for irritation. Yet, after all, it may be more economical to procure a new article and a better fit; and this may also be the most humane policy.

Treatment.—So soon as an abrasion is discovered on the back, the animal should be excused from duty for a few days. The abraded part should be lubricated two or three times daily with common tar and olive oil, equal parts, and this will soon restore the parts to soundness. Should there be no abrasion, yet tumefaction, heat, and tenderness, a cold-water pledget, renewed as occasion seems to require, will improve the condition of the parts, when recourse must be had to the tar and oil lotion. Sometimes, in consequence of neglect, the integument becomes callous or indurated. This condition was termed by ancient farriers "sit-fast." The treatment for this consists in smearing the callous with the following :

No. 63. Iodide of potass...................... 1 dr.
         Simple ointment.................... ... 6 dr.
         Glycerine........................... 2 dr.

Mix these ingredients well together, and they are then fit for use. A few applications of this ointment will have the effect of removing the callous, when a healthy surface will appear.

Some animals, owing to a peculiarity of constitution or of form, will chafe in those parts which come in contact with the collar and saddle, and no human foresight nor mechanical contrivance can always succeed in preventing the same. For example, some horses are humory, as the saying is—perhaps have a taint of scrofula about them—and if they perform a hard day's labor in the warm season of the year, notwithstanding they be harnessed in the best possible manner, they will come home at night with either galled shoulders or sore back. For such cases as these the harness-maker, with all his skill and ingenuity, has no other remedy than that of a palliative character.

THE HEART.

The current of blood is said to be put in motion by a muscular organ known as the heart, which is located within the thorax, or chest. The blood is distributed to all the various ramifications of the body by means of vessels called arteries, and it is returned to the heart by the veins. The red color of the blood is due to the presence of red corpuscles, which are always to be found in the blood of the vertebrata. In some of the invertebrata the blood is white, or resembles the color of chyle.

Mode of Circulation.—The blood, having been propelled from the left or posterior ventricle of the heart to all the various ramifications of the body, passes through the extreme vessels (capillary) into the veins. These veins terminate in a vessel known as vena cava, anterior and posterior, both of which sections terminate in the right or anterior auricle. This auricle contracts, and the venous blood thus finds its way into the right ventricle, where the pulmonary arteries have their origin. By aid of the muscular power of this ventricle, the blood is sent through the pulmonary arteries to the lungs (aeration and oxygenation.) The color of this blood is dark, almost black. It is highly carbonized, abounds in morbific matter and the worn-out materials of the body, and in this state is a non-supporter of vitality. In this condition,

26

then, venous blood enters the pulmonary tissues. The arterialization of the blood which follows is a process more essential to vital integrity than the assimilation of food; for animals can live a long time without food, but they can only exist a short time unless the blood be oxygenized.

The pulmonary arteries branch off into minute ramifications on the surface of the air-cells of the lungs, and where these minute divisions terminate others commence, termed venous radicles, which ultimately become large veins. By the pulmonary veins the oxygenized blood is returned to the left auricle and ventricle of the heart. This route of the blood from heart to lungs, and back again from lungs to heart, is termed the lesser circulation. The left ventricle delivers the blood into the great aorta, and thence to every part of the body. This is called the greater circulation.

Difference between Veins and Arteries.—The veins differ from arteries in being less dense, having no muscular nor ligamentary tunics like the arteries, and, therefore, offer but little resistance to tight collars and tight-fitting harness and circingles, which are apt to interfere with the return of blood to the heart. The incapacity of veins returning blood to the heart in the same ratio in which it is carried by the arteries, is proved from the fact that the former are twice as voluminous as the latter. The veins also differ in their internal arrangement from the arteries, the former being furnished, at proper distances, with **valves**, which **guard** against a retrograde venous action.

THE QUANTITY OF BLOOD CONTAINED IN THE BODY OF A HORSE.

Mr. PERCIVALL, who is our chief authority on this subject, contends that, for many reasons, the quantity of blood contained in an animal body may be made matter of speculation, but can not, for many reasons, be ascertained with any degree of precision; for, says he, " if we attempt to draw all the blood out of the body, the animal sinks and dies long before its vessels are evacuated; and as we possess no means of measuring what remains behind, any calculation we may make from the quantity that has flowed must necessarily turn out vague, if not altogether incorrect." He contends, however, that the following experiment offers a datum, or guide :

"The weight of an animal being ascertained to be seventy-nine pounds, a puncture was made with a lancet into the jugular, from which the blood flowed in a very free stream, and was collected. The vein having ceased to bleed, the caroted artery of the same side was divided, but no blood came from it. In a few seconds afterward the animal was dead. The weight of the carcass was found to be seventy-three and one half pounds; consequently, the animal had sustained a loss of 5.12 pounds, precisely the measure of the blood drawn.

It appears, therefore, from this experiment, that an animal will lose one-fifteenth of its weight of blood ere it dies, although a less quantity may so far debilitate the vital powers as to be, though less suddenly, equally fatal. In the human subject, the quantity of blood has been computed at about one-eighth part of the weight of the body; and as such an opinion has been broached from the results of experiments on quadrupeds, we may fairly take that to be about the proportion of it in the horse; so that if we estimate the weight of a common-sized horse at about 12 cwt., the whole quantity of blood will amount to 84 quarts, or 168 pounds, of which about 45 quarts, or 90 pounds, will flow from the jugular vein prior to death, though the loss of a much smaller quantity will sometimes deprive the animal of life."

The Rapidity of the Circulation.

In 1827, HERING,[*] a German physiologist, performed the experiment of injecting into the jugular vein of a dog a harmless substance, which could be easily recognized by its chemical reactions, and noted the time which elapsed before it could be detected in the vein of the opposite side. This gave the first correct idea of the rapidity of the circulation; for though the older physiologists had studied the subject, their estimates were founded on calculations which had no accurate basis, and gave very different results. The experiment of HERING is often roughly performed as a physiological demonstration, and we have thus had frequent occasion, in a general way, to verify its accuracy. If, for example, we expose both jugulars of a dog, inject into one a solution of ferro-cyanide of potassium in water, and draw a specimen of

---

[*] Milne-Edwards.

blood from the other, with as little loss of time as possible, it will be found that, in twenty or thirty seconds after the injection, the salt has had time to pass from the jugular to the right side of the heart, thence to the lungs and left side of the heart, and from this, through the capillaries of the head and face, back to the jugular on the opposite side. Its presence can be determined by the distinct blue color produced on the addition of the perchlo-ride of iron to the serum, if the specimen be allowed to stand, or a clear extract of the blood be made by boiling with a little sulphate of soda and filtering, treating the colorless liquid thus obtained with the salt of iron. The experiments of HERING were evidently conducted with great care and accuracy. He drew blood at intervals of five seconds after the commencement of the injection, and thus, by repeated observations, ascertained pretty nearly the rapidity of the circuit of blood in the animals on which he experimented. Others have taken up these investigations, and introduced some modifications in the manipulations. VIERORDT collected the blood as it flowed, in little vessels fixed on a disk revolving at a known rate, which gave a little more exactness to the observations; but the method is essentially the same as that employed by HERING, and the results obtained by these two observers nearly correspond.

HERING made observations on horses by increasing the frequency of the pulse, on the one hand, physiologically, by exercise, and, on the other hand, pathologically, by inducing inflammation. He found, in the first instance, that in a horse, with the heart beating at the rate of thirty-six per minute, with eight respiratory acts, ferro-cyanide of potassium injected into the jugular appeared on the vessels on the opposite side, after an interval of from twenty to twenty-five seconds. By exercise, the number of pulsations was raised to one hundred per minute, and the rapidity of the circulation was from fifteen to twenty seconds. The observations were made with an interval of twenty-four hours. The same results were obtained in other experiments.

THE TRANSFUSION OF BLOOD.

J. FARRELL, V. S., has lately been experimenting in an interesting department of veterinary science; namely, the transfusion of equine blood in diseases attended with low, vital action. Transfusion signifies the transfer of blood from one living animal to another, by means of a hollow tube inserted into the jugular veins of each horse. His method was as follows:

"I commenced a series of experiments, so as to discover the simplest, safest, and most effectual method of conveying blood from one animal to another. I first tried the transfusion syringe, which has been used by medical practitioners for this purpose; but, whether from want of skill in its use, or from some defect in the instrument, (which had been recommended to me as one of the best,) or from some other cause, I can not tell, but certain it is I was in no case so successful with it as I was with a more simple apparatus. I fancied that the blood lost much of its vitality by being exposed to atmospheric action, and also by its being forced and compressed within the cylinder of the instrument. These impressions as to the cause of failure induced me to undertake several experiments, with a view to the construction of an instrument which would fulfill the requisite indications; namely, to allow the blood to pass freely from the vein of the healthy into that of the diseased subject, without coming in contact with the atmosphere, and without alteration of its temperature. I at length adopted an exceedingly simple apparatus, which may be described in a few words. It consists of an India-rubber tube, some two and a half feet long, and three-eighths of an inch in diameter; that is, about the caliber of the vein in the adult horse. To either end of this is fitted a silver tube, curved somewhat like a syphon, so that one end, which is slightly rounded at the point, might be easily passed into the vein, both tubes being exactly alike. A narrow zinc or tin trough is required, to contain hot water, in which two-thirds of the tube should be immersed during the time that the blood is flowing. This completes the apparatus; and, being prepared, and the horses ready, held by assistants, the jugular vein of the healthy horse from which the supply of is to be taken may be opened, and into it one of the silver tubes carefully passed, point upward, so as to receive the current of blood as it flows back from the head—the operator holding the other end, and, having previously opened the corresponding vein in the patient, he should wait till the current is passing freely down the tube from the healthy horse, and then bring it in contact with that which is now flowing slowly from the patient. He should pass the end of the tube carefully into the vein, point downward, by which means the possibility of any air getting into the tube is avoided. The quantity to be transfused is readily ascertained by watching the expression of the eyes and noting the pulse carefully. So long as there is no dilatation of the pupils, and so long as the heart's action is not very much affected, the blood may be allowed to flow on uninterruptedly; but as soon as the pupils become dilated, it is necessary to lessen the supply gradually, by compressing the tube with the finger and thumb. If the dilatation disappear after a minute · i two, the blood may be again allowed to flow; but if it increases, further transfusion must be stopped, or, otherwise, injurious consequences will result.

In the cases of two animals which I purchased for the purpose of experiment, I purposely allowed the blood to flow after dilatation of the pupil had manifested itself, in order that I might see the result, and I found it to increase; and in one of the cases, after winking both eyes three or four times, in rapid succession, the horse reared up and fell back. In the other case, I forced the blood in from an India-rubber enema bottle. The pupils first became still more dilated; the breathing very quick and difficult; the eyes assumed a wild, agonized look, and the animal, with a sudden bound, fell forward dead. I had my finger on the artery during the entire time, and observed a great unsteadiness and fluttering of the pulse, which increased in frequency until the instant before he fell. In the first of these cases, I did not force the blood into the vein, but allowed it to flow until I perceived the injurious effects upon the horse. He tottered and fell, but in a short time recovered, and was walked back to his stable. He died in the course of the night, and, on examining his head the next day, I found considerable congestion of the brain. The tube used for the experiments transferred about three quarts of blood in eight minutes; at least it might be so inferred, from the fact that when used as a syphon, it passed three and a half quarts of water (and a little less of oil) from one vessel to another in that time.

I think, however, that I am correct in saying that this disease is by no means so likely to occur in horses as in human beings, and is certainly not so formidable; but, nevertheless, I am quite sure that it may happen as a consequence of the operation, if the instrument be not kept scrupulously clean, and also if great care be not taken in its introduction in the vein. Having, from these and other experiments, not necessary to detail, determined on the mode of operation most likely to be successful, I shall now proceed to state the result of four cases in which I have operated, and which, I think, I may look upon as having been followed with entire success. In three of the cases the patients had been bled a short time previous to my having seen them, and were so much weakened that they could scarcely walk. In the fourth case, a drastic purgative had been administered, causing superpurgation and great prostration of strength. In each of the four cases, the condition of the patient was so similar that the description I have already given may answer for all. Having selected a healthy young horse from which to obtain the blood to be transferred, I opened the jugular vein in the patient and in the healthy subject; and having inserted the tube, as before described, into the vein of the healthy horse. I placed the India-rubber tube in the tin trough containing hot water, to maintain the right temperature, and the other curved tube into the descending portion of the vein in the patient. As soon as the current from the healthy horse had completely expelled all atmospheric air, the instrument being thus arranged, the blood flowed freely from the vein of one horse into that of the other in an unbroken current. The average quantity of blood transferred in each of these cases was about three quarts. I observed no particular symptoms to follow from the transfusion, until two quarts or more had passed from the healthy to the diseased subject; but as soon as about that quantity had flowed, there appeared to be produced an amount of stimulation, indicated by an increased action of the heart; at the same time the pupils began to dilate, and the countenance evinced an anxious expression. My former experiments led me to watch with great care the progressive dilatation of the pupil, and I deemed it expedient, in each case, when this symptom was well developed, to compress the tube so as to diminish the current, and allow the transfusion to proceed more gradually and slowly. Occasionally I almost completely interrupted the current until the subsidence of this symptom, and I found that when about three quarts had been transfused, any additional quantity was followed by unpleasant symptoms, which indicated the necessity of stopping the operation. On removing the tube and closing the vein, all symptoms of irritation gradually subsided, and the pulse, from being rapid and irritable, became slower, stronger, and fuller, gradually approaching the healthy standard.

In each of these four cases the reaction was steady and progressive. The natural warmth of the extremities was gradually restored, and, in the course of ten or twelve hours, the patient presented other equally unmistakable symptoms of amendment, such as returning appetite, more quiet and steady respiration, cheerfulness of countenance, and a willingness to move about. From this point there was a gradual improvement, and, in a short time, they were pronounced cured.

I have been induced to submit these few remarks, much less with a view to record any little success I may have had in performing the operation myself, than with a hope that others, far more capable, will take up the matter, and test it thoroughly; for, whatever obstacles and objections there may be to its performance in the human subject, there are none to prevent its becoming a most valuable agent in veterinary science."

Soon after the discovery of the circulation of the blood by HARVEY, transfusion was attempted on animals, and afterward on man. In some cases the experiment was successful, while others failed to afford relief; in fact, a majority of the cases terminated unfavorably; and of late years transfusion has been abandoned, and regarded as "useless and even dangerous." We can conceive, however, of cases which might occur—when, for example, a horse has lost a large quantity of blood accidentally—to warrant us in resorting to transfusion; and if any of our veterinary friends are disposed to make experiments of this kind, we urge them to do so, notwithstanding the opinion of the Parliament of France against it, and that of some scientific men who have pronounced it "useless and dangerous."

One among the many reasons which may be urged against transfusion is the fact that venous blood is a non-supporter of vitality. It is saturated with the excrementitious material absorbed from the intestinal canal, fibrous membranes, and cellular tissues, and a vast amount of morbific matter is mixed with the venous blood in the course of its circulation, so that it can not be of any benefit to the system of a sick or dying horse until it shall become oxygenized or vitalized, or changed from venous to arterial blood, which change can not possibly take place until this fluid has reached the air-cells of the lungs. It would appear, therefore, that the transfusion of vitalized arterial blood, if the process can be conducted with the relative amount of safety attending that of the transfer of venous blood, would be a procedure more consistent with the known laws of physiology than is the case when venous blood is transfused. It is possible that some animals might be benefited by the transfusion of venous blood, provided the heart and lungs of the patient are in a healthy state, so as to insure sufficient oxygenation; otherwise, the rapid consumption of the little oxygen absorbed would be likely to end in asphyxia. If the function of the lungs be impaired by disease, then the process of aeration of the blood is checked, and the blood can not circulate, notwithstanding the healthy state of the heart. In verification of this fact, as regards the very important function of respiration and aeration, we may mention that Dr. AUSTIN FLINT, JUN., has lately demonstrated that the condition of oxygenation of the blood is necessary to the performance of the vital functions, and that a want of capillary power (occasioned by imperfect oxygenation) throws all onus on the heart, and that the heart is insufficient for the labor. In one of his experiments, after capillary circulation had entirely ceased, the chest was opened and the heart found beating regularly.

MANAGEMENT OF THE HORSE'S FOOT IN THE STABLE.

It is my firm belief, based on a knowledge of the physiology of the foot, that soft bedding, containing the usual amount of filth, is, in many cases, the exciting cause, not only of tenderness and contraction, but also of many muscular affections. I rejoice to find horse-owners exercising their reason in this department of stable management, because it is one step in the right direction, and will put a stop to those glaring absurdities which are constantly practiced just because they have the sanction of antiquity, or because Neighbor So-and-so recommends them. Reason teaches that the health of the whole animal fabric is best promoted by cleanliness and ventilation (by which a pure atmosphere is insured), and by food suitable in quantity and quality to the special wants of the animal. Hence, what reason teaches, man should practice.

It is unreasonable, therefore, to suppose that a horse can be benefited by standing on a soft bed, composed in part of the defiling excrements of his own body, which are constantly decomposing and forming gaseous vapors unfit for respiration. If dirt and filth are innoxious, in what does the value of our sanitary regulations consist? Why do our city authorities spend so much money to purify the pestiferous cesspool and sewer, and to rid our streets of accumulating rubbish and filth? Let human beings wallow knee-deep in muck, and revel shoulder-deep in an atmosphere saturated with ammoniacal and carbonic acid gases (as some horses are compelled to), and death would run riot—our cities would be converted into immense charnel-houses, fit receptacles for a race of beings that would not adopt the means which reason and experience suggest for averting the calamity.

Horses that have no better care than that alluded to, are in close proximity with disease. That they are often found dead in their stalls from the effects of carbonized blood I can testify, and many more would die, only they are permitted to take a little of the breath of life during the day, which, to some extent, dilutes the poisonous gases with which their system has been saturated during the night, and thus their life, which, under the best circumstances, is a weary toil, is prolonged.

Without attempting to prove the general effects of impure air and filth on the system of a horse located in a stall from one to two and a half feet deep of soft bedding, let us consider, in a brief manner, the local phenomena. Our readers are all aware that the combined action of heat and moisture tends to relax—enervates the tissues of the body, and, if carried beyond a certain point, ends in decomposition. Take, for example, a common poultice, apply it to a horse's foot, and renew it as soon as it becomes dry. In the course of two or three days the hoof will separate from its matrix, the frog and heels soften, the tissues be in a state of relaxation, and, if the poultice is continued, the hoof will separate from the sensible parts; if the foot is already diseased, the separation is accelerated. Warm water has the same effect. Applied externally for any length of time, it relaxes and prostrates; applied internally, it relaxes and vomits. Hence the soft, (which implies moist,) hot bedding, tends to create morbid action in the feet, and whatever disease the horse may be predisposed to in those parts will generally manifest itself. Some animals, however, escape the evils alluded to, owing to their insusceptibility; for disease of the foot can not occur without a susceptibility to it and the application of a cause. Soft bedding, cow dung, and other unmentionable filth, are often resorted to as remedies for contraction, but they are classed among the irrational barbarities of the past, with which they ought to have sunk into oblivion. Veterinary science has discovered that if such articles possess any virtue, it is owing to the moisture they impart to the foot. Therefore, on the score of decency, and in view of relaxing the contracted foot according to the principles of reason and modern science, we should avoid every kind of filth, and resort to pure water or wholesome poultice; or take off the shoes, wash the feet daily, and let the horse run to grass, so that he may bring the sole and frog in contact with the ground, and thus promote expansion of the heels.

As regards the manufacture of tender feet, in both the young and aged, it makes but little difference whether they have the range of a filthy barn-yard or are confined in a bedded stall; the effect is tenderness of foot, and subsequent lameness; and the same, or a disposition to it, may be transmitted through the sexual congress to future offspring. In view, therefore, of preventing diseased feet, strict attention must be paid to cleanliness. The stall floor should be composed of brick or plank, having just sufficient declivity to conduct fluids into a gutter, running along the entrance to the stall, which should terminate exterior to the stable, so that the ammonia, in which the urine is abundant, is carried beyond the stable atmosphere. The bedding, which, according to long custom, is stowed under the crib, there acting as a sort of noxious smelling-bottle to the horse's nostrils, should be spread out in the open air, sorted, the refuse and excrement removed to a dung-heap, located as far from the stable as possible; for the common manure receptacle, under the stable floor, is one of the worst features of stable economy. The stable floor should be washed clean as often as circumstances permit.

STRANGLES ("DISTEMPER," OR "HORSE AIL.").

This disease is most prevalent in young horses. After the age of eight the disease is more rare I have, however, seen some well-marked cases occurring in horses whose ages varied from five to eight years. Strangles is, without doubt, one of the evils of domestication, and often results from errors in diet and management, which, if we make some exceptions, will generally bear improvement.

Strangles is supposed to be a disease to which all horses are subject once in their lives, yet Mr. PERCIVALL contends that many horses escape the disease. Hence, if many escape, it is very natural for those who know the value of pure air, natural food, and exercise to conclude that the colt, while enjoying these great luxuries in the open air, by the side of its mother, guided by her superior instinct, is not liable to be attacked with a disease which, as already stated, we believe to result from depriving animals of those blessings which Nature has in store for them in their unrestrained state. But it often happens that young colts, after running a season with their mother, partaking of the invigorating country air, grow up to be strong and robust, and then the period arrives for weaning them. How changed the scene! Instead of being permitted to gambol in their native element, they are confined to a small space, not large enough to swing a cat round, and perhaps as dark as the grave; and the animal, after fretting for a season, and making unsuccessful efforts to escape from its prison-house, tamely submits to the discipline, not, however, until he has cut and bruised and otherwise injured himself. I was called, a short time ago, to visit a young colt that had lacerated his head, breast, and fore-legs in a most shocking manner, in making an attempt to escape through a window from the horrors of confinement. His companions were about a dozen cows, more calculated to alarm and render his position a perilous one than otherwise; and the impure atmosphere, rendered so by the emanations from the excrements and from the lungs of his companions, was a source of great mischief. Then, who can blame such an one for attempting to escape and regain liberty? If strangles should appear in such a subject, it would not be surprising.

Then, again, take a colt from its mother, whose milk contains all the elements for sustaining life and developing the organization of the young subject, and place it upon a diet of hay or like innutritious trash, a whole truss of which would not afford one-half the quantity of nutriment contained in a quart of its mother's milk. However profitable and well-adapted hay may be for stock of mature growth and powerful digestive organs, it is a sad mistake to suppose that it will do for the young. A case of this kind came under my observation last year. The subject, aged two and a half years, died in a state of marasmus (a gradual wasting of the system without any apparent disease). A *post mortem* examination failed to detect any disease other than general emaciation. On making inquiries concerning the food, the owner replied, "The colt has been fed on good sweet hay and corn-stalks ever since it was weaned." This was capital food for the poor thing, provided, however, its stomach had a grist-mill within it; otherwise, it was hard fare, and must derange the digestive function, and, ere the colt has attained maturity, dyspepsia, in either a mild or aggravated form, has secured a victim.

Mr. PERCIVALL has defined strangles to be "a diffusible swelling under the jaw. The tumor consists in a circumscribed inflammation, having all the characteristics of simple phlegmon, attacking the subcutaneous cellular substance included between the branches of the jaw, which, in consequence, become gradually filled and distended with effusions of lymphy and serous matters, acquires a firm and solid feel, tenderness on pressure, and a sense of unnatural heat. This commonly proceeds to suppuration, ending, to all appearances, in a common submaxillary abscess. Now, this, and this alone, constitutes strangles. There are in the books a variety of other symptoms described, but they are all concomitant or accidental, none but these being, properly speaking, essential to its existence. The usual concomitants are, membranous inflammation, giving rise to soreness about the throat; reddening and discharge at the nose, and perhaps cough; tumefaction of the salivary glands, producing pain and difficulty of deglutition; and, lastly, some slight febrile commotion of the system."

Contagiousness of Strangles.—We learn from the "Veterinarian" that M. REYNAL, clinical professor at the Alfort School, submits a number of observations corroborative of the contagious character of strangles. He states that "young horses having strangles, and put into stables with horses of adult age, doing their duty, have communicated the disease to those of the latter, who have stood in adjoining stalls, though some few have only exhibited the disease in a catarrhal form. Even the foal has been known to suck the disease from its dam. Moreover, experiment has been had recourse to to inoculate for strangles. M. Damalix smeared with a sponge, impregnated with matter taken from the abscess of strangles, twice daily, both sides of the pituitary membrane and the internal surfaces of the linings of the eyelids, in a sound horse, about to be cast for spavin. This was continued for seven days. On the eighth, he remarked that the horse had lost his appetite, had commenced running from both nostrils, coughed softly and loosely, and had swelling under the jaw, which ended in resolution, all the symptoms terminating eight days from their commencement.

It has been remarked that strangles is more surely communicated at an early than a late stage, and in a certain form more readily than in others. Strangles will assume the herpetic character, simulate farcy and glanders, settle in the mesenteric glands, or may follow castration. In regard to contagion, may be mentioned, as most readily communicable, that form of strangles which assumes the character of eruptions on the lips, nose, and pituitary membrane."

The Tumor of Strangles is not always confined to the submaxillary space, as will be observed from the following case, which occurred in the author's practice, a short time ago, in a gray mare, and which speedily ran to a fatal termination. The subject had previously suffered from some internal disorder, and, on recovery from which, did not appear to regain its usual health, but remained in an unthrifty condition. A few days before our attention was called to the case, a tumor made its appearance on the near hip, which suppurated and discharged. On the morning of our visit, the animal had been attacked with abdominal pain, for which remedies had been prescribed. They not appearing to afford relief, the owner called in the aid of some person, who inserted a knife into the mouth, and wounded the palatine artery, from which the blood had flowed for several hours, but was now arrested. We found the extremities icy cold, and the pulse very feeble. The eye appeared glassy, mucous surfaces pale, and the lips hanging pendulous. There was a painful tumor on the near hind leg, close to the stifle, and another on the flank of the same side. The animal was exceedingly stiff and lame in the hind extremities, and appeared to be failing very fast. Considering the case a hopeless one, we merely recommended some restorative cordial and a few quarts of gruel. The animal died in the course of a few hours, and its death, no doubt, was hastened by the loss of blood; for, as the owner expressed himself, "the horse had some life in him before being bled," thereby intimating that the abstraction of blood deprived him of that life.

Treatment.—So soon as matter can be detected in the tumor or abscess beneath the jaw it should be laid open. This will liberate a quantity of pus. The orifice thus made must not be allowed to close, but must be kept open for a few days, or until matter ceases to run. During this period, a liniment, composed of cod-liver oil, half a pint, and powdered camphor, two ounces, must occasionally be rubbed about the region of the jaws. Half an ounce of powdered chlorate of potass should be mixed in the drink, or with some flaxseed tea, if the patient will drink it. A souple of drachms of fluid extract of pleurisy-root may be given, night and morning, for a short time. Active medicine is not indicated; good nursing, aided by pure air and voluntary exercise in an inclosure, will soon restore the patient to health.

We close our labors in the EQUINE DEPARTMENT by entering our solemn protest against the unwarrantable use of the lancet or knife in Veterinary Practice. It is a disgrace to humanity, an outrage on the rights and privileges of that noble animal, the horse, over whom God has given us dominion and power—not to abuse and render their life a weary toil, and extort from them all the labor we can, but to protect them, to administer to their wants and necessities, and render their condition as happy as possible. A moral responsibility rests on every man to see that the claims which these poor creatures have upon our race are respected; that their rights are not violated; that, when sick and in a dying con-dition, they shall be treated with mercy, and be protected from their tormentors, the unmerciful phlebotomists. Any misguided man who can draw from the veins of an emaciated or dying horse the little blood they contain should be sent to the insane asylum, and never again be permitted to disgrace the noble nature of man. We can respect the man who conscientiously practices blood-let-ting; but where is the medical man who can conscientiously bleed a dying brute with a view of saving life?

END OF EQUINE DEPARTMENT.

TABLES OF WEIGHTS AND MEASURES.

APOTHECARIES' WEIGHTS.

20 grains make one scruple, marked................. ℈j.
 3 scruples do.   drachm, do.   .... ʒj.
 8 drachms do.   ounce,   do.   .... ℥j.
12 ounces   do.   pound,   do.   .... ℔j.

LIQUID MEASURES.

60 drops make one drachm, marked.... .... ʒj.
 8 drachms do.   ounce,   do.   .... ℥j.
20 ounces   do.   pint,   do.   .... Oj.
 8 pints   do.   gallon, do.   .... Cong.

ORDINARY MODES OF MEASURING LIQUIDS.

A *tea-spoonful* of liquid equals 60 *drops*, or *one drachm.*
A *table-spoonful* of   do $\frac{1}{2}$ *ounce*, or *four drachms.*
A *dessert-spoonful* of do.   180 *drops*, or *three drachms.*
A *wine-glassful* of   do. $1\frac{1}{2}$ *ounces.*

The doses vary in quantity in different animals. The amount to be given in any case will be found clearly stated in the *treatment* of each disease, whether of horses, cattle, sheep, or hogs.

416

A LIST OF FLUID EXTRACTS,

RECENTLY INTRODUCED INTO VETERINARY PRACTICE

By G. H. DADD, V. S.,

WITH INSTRUCTION REGARDING THEIR ACTION, AND
THE QUANTITY TO BE ADMINISTERED.

THE introduction of medicines, in the form of FLUID EX-
TRACTS, for the treatment of diseases incidental to domestic
animals is considered a very great improvement over the old
method of drenching by the pint or quart, to the great disgust of
the patient as well as the practitioner, and the great danger attending the administration of a large quanty of fluid which was necessary when crude materials were used. In the use of fluid extracts
all danger is obviated and much labor saved, for the doses are
quite small, rarely exceeding one fluid ounce. This can be merely
placed on the tongue, the animal's head being slightly elevated,
and he swallows it without any difficulty or resistance.

The fluid extracts bear an exact relation to the crude materials—
that is, ounce for ounce—yet, from a variety of circumstances, they
may vary slightly from this standard. It is, however, the intention of the manufacturer to completely exhaust the active principle
of the crude material and render the medicine uniform in strength.
The fluid extracts used by the author are manufactured by Messrs.
GARRISON & Co., manufacturing chemists, of Chicago. They conduct their evaporations at a very low temperature, by means of an
improved steam bath, and use only select drugs, thereby preventing the possibility of adulteration. I have used large quantities
of medicine manufactured by the above-named firm, and find them

27   417

equal if not superior to any in the market; therefore I do not hesitate to recommend them as efficient and reliable medicines. *The doses here recommended apply to both horses and cattle.*

Fluid Extract of Chamomile.

(*Anthemis Nobilis.*)

This is a valuable tonic, and is used in cases of derangement of the digestive organs.   Dose, from two to four drachms.

Fluid Extract of Arnica.

(*Leopard's-bane.*)

Arnica is chiefly used in veterinary practice as an external application in the treatment of wounds, sprains, and bruises.   It is prepared as follows:

Fluid extract of arnica....................... **4 oz.**
Proof spirits (equal parts of water and alcohol)... **1 pint.**
Mix.

Apply a portion to the affected part two or three times daily.

Fluid Extract of Wormwood.

(*Artemisia Absinthium.*)

This is used as an anthelmintic for the expulsion of worms. The dose is one ounce (fluid), to be given every morning for a week, just before feeding-time.   It not only has a tendency to expel worms, but gives tone to the digestive organs, and corrects morbid appetite

Fluid Extract of Pleurisy Root.

(*Asclepias Tuberosa.*)

This is used in pleurisy and irritable sore throat.   Dose, four drachms, night and morning.   The powdered root enters into the composition of some of the most celebrated cough powders.

Fluid Extract of Male Fern.

(*Aspidium Filix Mas.*)

This is the celebrated remedy for the treatment of tape-worm, which is sometimes to be found in the intestines of young slts.

Dose, four drachms, to be given every morning, on an empty stomach, for six or eight successive days, and longer if necessary.

FLUID EXTRACT OF BUCHU.

(*Barosma Crenata.*)

This is used exclusively for the treatment of diseases of the urinary organs, retention of the urine, and for the treatment of the various forms of dropsy. Its direct action is diuretic, yet it also gives tone to the parts on which it acts. Dose, from four to eight drachms. Should an animal be suffering from retention of the urine, the diuretic action of the buchu may be augmented by adding to it a small quantity (equal parts) of sweet spirits of niter. For the treatment of spasm at the neck of the bladder, add to the dose of buchu an equal quantity of tincture of assafetida.

FLUID EXTRACT OF INDIAN HEMP (FOREIGN).

(*Cannabis Indicus Sativus.*)

This medicine acts as a narcotic and antispasmodic. Its use is indicated in all diseases attended with pain and restlessness, but is chiefly used for the treatment of tetanus or locked-jaw. Dose, two drachms, to be repeated as occasion may require.

FLUID EXTRACT OF PRINCE'S PINE.

(*Pipsissewa.*)

Prince's pine is a very excellent tonic and diuretic in chronic diseases of the urinary organs, and has been used with much success in cases of local dropsy of the limbs, known as swelled legs. Dose, from four to six drachms.

FLUID EXTRACT OF BLACK SNAKEROOT.

(*Cimicifuga Racemosa.*)

This extract is considered a valuable agent for the treatment of spasmodic affections, which arise in consequence of derangement of some portion of the nervous system. It is an efficient remedy in hysteria.

FLUID EXTRACT OF COLCHICUM.

(Colchicum Autumnale.)

This extract is prepared from the roots and seeds of the plant. It is celebrated for the treatment of rheumatism, inflammatory diseases of the joints, and inter-articular lameness. Dose, one drachm, two or three times per day.

FLUID EXTRACT OF YELLOW JESSAMINE.

(Gelseminum Sempervirens.)

This agent is used as a sedative in the treatment of pleurisy and pneumonia in the early or acute stage. It is also a powerful febrifuge, hence it is indicated in all acute fevers. It entirely dispenses with use of the fleam. Dose, from one to three drachms.

FLUID EXTRACT OF WITCH HAZEL.

(Hamamelis Virginica.)

Witch-hazel is used as a local astringent in the treatment of "bog and blood spavin." It is prepared for use as follows:

Fluid extract of witch-hazel....... .... 6 oz.
Proof spirits..................... .... 1 pint.
Mix.

Apply a portion twice daily.

FLUID EXTRACT OF HOPS.

(Humulus Lupulus.)

This preparation of hops is used as a tonic in general debility, and as a narcotic in diseases of the nervous system. It is a reliable remedy to induce sleep and relieve pain. Dose, from two to four drachms, to be repeated as occasion seems to require.

FLUID EXTRACT OF GOLDEN SEAL.

(Hydrastis Canadensis.)

The golden seal is a pure tonic, and its use is indicated in all cases of debility, loss of appetite, derangement of the digestive organs, and torpidity of the liver. Dose, from two to four drachms.

Fluid Extract of Sassafras.

All the preparations of sassafras are more or less alterative. The fluid extract is used in view of purifying the blood in diseases of the skin, and in cases of chronic rheumatism. Dose, from four to six drachms.

Fluid Extract of Culver's Root.
(*Leptandra Virginica.*)

This is employed, in the author's practice, as a substitute for calomel, in the treatment of acute and chronic diseases of the liver. Dose, from two to four drachms.

Fluid Extract of Lobelia.

Lobelia is a very valuable antispasmodic, and is useful in diseases of a spasmodic character. The author has used it, with great success, in obstinate cases of spasmodic colic. Dose, two drachms.

Fluid Extract of Bayberry.
(*Myrica Cerifera.*)

Bayberry is successfully used as a stimulant and astringent in scours, diarrhea, and super-purgation. Dose, two drachms, to be administered in a small quantity of milk porridge.

Fluid Extract of Poppy-heads.
(*Papaver Somniferium.*)

This is used as a narcotic, in all cases that seem to require the exhibition of opium. It mitigates pain, induces sleep and rest, and is much safer than crude opium. Dose, from two to six drachms.

Fluid Extract of Poke-root.
(*Phytolacca Decandra.*)

This is used as an alterative in a disease occurring among cows, known as "garget," or "caked udder." Dose, from two to four drachms.

FLUID EXTRACT OF MATICO.

(Piper Angustifolium.)

This is used as a styptic to bleeding wounds, internal hemorrhages, and chronic diarrhea.   Dose, from four to eight drachms.

FLUID EXTRACT OF CUBEBS.

This is a very valuable agent for the treatment of leucorrhea, gleet, nasal gleet, or discharges from mucous surfaces.   Dose, from three to six drachms.

FLUID EXTRACT OF BLACK PEPPER.

(Piper Nigrum.)

This is a carminative and stimulant to the digestive organs, and is often used successfully in the treatment of stomach staggers.   Dose, from two to four drachms

FLUID EXTRACT OF MANDRAKE.

(Podophyllum Peltatum.)

Mandrake is now used as a laxative and cathartic, in lieu of aloes.   About one ounce will usually purge a horse, if followed up by sloppy bran-mashes.

FLUID EXTRACT OF BLOODROOT.

(Sanguinaria Canadensis.)

This is used as an escharotic in the treatment of foul ulcers, and fungus growths.   It is used, also, as a gargle in suppurative laryngitis, croup, putrid sore throat, etc.   The gargle is prepared as follows:

Fluid extract of bloodroot....................... 4 oz.
Vinegar... 1 pt.
Mix.

It is applied to the mouth and throat by means of a sponge, affixed to a piece of rattan or whalebone.

Fluid Extract of Valerian.

(*Valerian Officinalis.*)

This is a very useful antispasmodic, and its use is indicated in all cases marked by irregular nervous action and hysterical affections. Dose, four drachms.

Fluid Extract of Ginger.

(*Zinziberis Officinalis.*)

Fluid extract of ginger is one of the most valuable diffusible stimulants to be found in the whole materia medica. The Jamaica ginger is preferable to any other; it is used in cases of indigestion, flatulency, etc. Dose, from three to six drachms.

Fluid Extract of Ginger enters into the composition of the celebrated colic drench, used by the author of this work and his students, during a period of twenty-five years, with marked success. The following is the formula:

Fluid extract of ginger..... } each......... 1 lb. fluid.
Fluid extract of golden seal.. }
Hyposulphite of soda..................... 6 oz.

Dose, four ounces (fluid) every four hours, until relief is obtained.

Remedy for Wounds.

Apply a compress of several folds of cotton cloth, soaked in a portion of the following:

Balsam fir................................ 1 lb.
Hyposulphite of soda...................... 2 oz.
Proof spirit.............................. 1 pt.
Mix.

Remedy for Tedious Labor.

The natural labor-pains may be excited by administering: fluid extract of ergot (ergota), two drachms, every half hour, until delivery is accomplished. The ergot operates with great energy upon the contractile function of the uterus, of both mares and cows, and to a certain extent prevents inordinate hemorrhage after parturition.

REMEDY FOR HEMORRHAGE IN PERFORMING OPERATIONS.

Apply, by means of a compress of linen or sponge, a small quantity of persulphite of iron. This is the most valuable styptic known to veterinary science.

FLUID EXTRACT OF ACONITE.

(Aconitum Napellus—Monk's-hood.)

This medicine is used as a very powerful sedative and narcotic, in all cases of active or inflammatory disease of the lungs or brain It is highly recommended by the professors of our art as a reliable antiphlogistic in the treatment of acute affections. The dose is from ten to twenty drops. This medicine is said to be very efficacious in inflammatory rheumatism, and as a fine topical remedy in localized painful affections, such as laminitis, myalgia, etc.

HYPOSULPHITE OF SODA.

The hyposulphite of soda is useful in all kinds of diseases known as rinderpest, pleuro-pneumonia, typhoid affections, and in enzoötic affections, and in all diseases caused by fermentation in the blood, such as Texas fever, malignant scarlet fever, typhus, etc. The hyposulphite of soda can be used with great success in the treatment of the above and other affections of an enzoötic origin, yet I have found it more efficacious and reliable when prepared as follows:

> Hyposulphite of soda.....)
> Hypophosphate of lime... } each............ 1 oz.
> Hypophosphate of potass..)
> Fluid extract of ginger..................... 2 oz.
> Sugar of milk............................. 1 oz.
> Proof spirit............................... 1 pt.

Dose, two ounces, two or three times daily.

A GUIDE TO DIAGNOSIS;

COMPLETE INDEX OF SYMPTOMS

THE following Index of Symptoms is arranged under four heads, viz.: DISEASES OF HORSES; DISEASES OF CATTLE; DISEASES OF SHEEP; DISEASES OF SWINE. Each part is arranged alphabetically, so that ready reference can be made to any symptom.

HOW TO USE THIS INDEX.

Animals when taken sick, not having the power of speech, can not make known the seat of their disease, except by certain signs or invariable symptoms. It therefore becomes of the highest importance to be able to understand the mute language of the poor dumb brute, and thus to find out what ails him, in order that the proper means of cure may be speedily made use of.

An instance of the use to be made of this Index may be found in the following case: "A horse is noticed to breathe faster than usual, with evident pain; back, arched; legs, straddling; head often turned toward the loins; urine, reddened and scanty." Now, a reference to these symptoms, as arranged alphabetically, enables us to turn directly to page 223, where "Inflammation of the Kidneys" is seen to be the disease we are looking for.

The same thing can be done in case a cow, sheep, or hog be the victims of disease.

In observing symptoms, attention should be directed to the

425

various parts of the body of the animal which give the earliest
warnings of disease. These are, usually,

THE EYES, TONGUE, MUZZLE, THROAT, PULSE, CHEST, HEART,
BREATH, BREATHING; DISCHARGES FROM THE MOUTH; DIS-
CHARGES FROM THE NOSTRILS; THE SKIN, THE COAT, THE
BACK, THE ATTITUDE, AND GENERAL APPEARANCE OF THE
WHOLE BODY.

EQUINE DEPARTMENT.

GLOSSARY

MEDICAL AND SCIENTIFIC TERMS.

ABNORMAL—Irregular.

ABSORBENTS—Medicines used for absorbing; also the vessels of the
body which suck up.

ACUTE—Sharp, severe.

ADIPOSE—Fatty.

ADYNAMIC—Debilitated.

ALAE—Wings.

ALTERATIVES—Medicines which change a disease for the better.

ANASARCA—Dropsy of cellular membrane.

ANASARCOUS—Dropsical.

ANÆMIA—Bloodlessness.

ANTISEPTICS—Medicines opposed to putrefaction.

ANTISPASMODICS—Remedies opposed to spasms or convulsions.

ANTIPHLOGISTIC—Opposed to inflammation.

APERIENTS—Medicines which open the bowels gently.

AQUEOUS—Watery.

ASCITES—Dropsy of the belly.

ATAXIC—Disordered.

AUSCULTATION—Examination by sounding and listening.

AUTOPSY—Post-mortem examination.

BIFURCATION—Division into two branches.

BOLUS—A large pill.

BUCCAL MEMBRANE—The lining of the mouth.

CANTHUS—Corner of the eye.

CAPSULE—Shell or case.

28

CARBONACEOUS—Containing carbon.

CARMINATIVES—Medicines which relieve pain by expelling wind from the bowels.

CARTILAGINOUS—Composed of cartilage.

CATHARTIC—Loosening.

CELLULAR—Cell-like.

CEREBELLUM—The brain.

CEREBRUM—The brain.

CHANCROUS—Cancerous.

CLINICAL—Relating to individual practice.

COMA—Stupor.

COMATOSE—Stupefied

CONJUNCTIVIAL MEMBRANE—The membrane which lines the eyelids and covers the eyeball.

CORNEA—Transparent coat of the eye.

CRANIAL—Pertaining to the skull.

CRANIUM—Skull.

CRUCIAL—Shaped like a cross.

DECARBONIZE—To purify by air.

DIAGNOSIS—The art of telling the nature of diseases.

DIAPHORETICS—Medicines which promote perspiration.

DIATHESIS—Predisposition to certain diseases.

DIETETICS—Regulation of diet.

DIURETICS—Medicines which increase the flow of urine.

DUCT—Canal.

DYNAMIC—Relating to the vital forces.

EMOLLIENTS—Substances used to reduce inflammations.

EMPHYSEMA—Distention by gas or wind of certain portions of the body.

EMUNCTORIES—Organs which carry off waste matters.

ENCEPHALON—The brain.

ENEMA—Injection.

ENZOOTIC—Endemic diseases among animals.

EPIZOOTIC—Epidemic among animals.

EQUILIBRIUM—Balance.

EQUINE—Relating to the horse.

ETIOLOGY—The doctrine of the causes of disorders.

EXCREMENTITIOUS—Useless.

EXCRETORY—Relating to vessels which throw off useless matter.

EXTRAVASATION—Escape of a fluid of the body from its vessel into surrounding parts.

EXUDATION—Oozing through a membrane.

FAUCES—The throat.
FLEAM—Used in bleeding.

GRAMINIVOROUS—Feeding on grass.

HEMATOSINE—The red coloring matter of the blood
HEMORRHAGE—Bleeding.
HIPPIATRIC—Relating to diseases of the horse.
HISTOLOGY—General anatomy.
HYDROCEPHALUS—Water in the head.
HYGIENE—Preservation of health.

ICHOROUS—Humory.
IDIOPATHIC—Primary affections.
IDIOSYNCRASY—Peculiarity of constitution.
INDURATED—Hardened.
INGUINAL—Belonging to the groin.
INTERCOSTAL—Between the ribs.
INUNCTION—The act of rubbing in.

LACHRYMAL GLANDS—Those which secrete tears.
LANCINATING—" Shooting."
LAXATIVES—Loosening medicines.
LESION—Disorder.
LIGAMENT--The substance which joins bones together.

MAMMÆ—Breasts.
MASSETERS—Muscles of the jaws.
MORBID—Diseased.
MORBIFIC—Producing disease.

NAVICULAR—One of the bones of the foot.
NEURO-PATHOLOGY—The nervous system in disease.
NODULOUS—Like a knot.
NOSOLOGY—Classification of diseases.

ŒDEMATOUS—Swollen.
OPAQUE—Not transparent.
OS CALCIS—Bone of the heel.
OSSEOUS—Bony.
OVOID—In form of an egg.

PALATINE--Relating to the palate.
PANZOOTIC—An epidemic affecting animals generally.
PARASITE—An animal which lives on another.
PAROTID—Largest salivary gland.
PATHOLOGY—The study of the body in disease.
PEDICLE--Narrow part of a tumor.
PETECHIAL—Resembling flea-bites.
PHTHISIS—Wasting away.
PITUITARY MEMBRANE—Lining of the nostrils.
PSEUDO-MEMBRANOUS—Relating to false membranes.
PUS—Matter.
PYLORUS—Entrance into intestines.

RALE—A watery sound heard in sounding the chest in some diseases.

SANATIVE—Health-giving.
SCHNEIDERIAN MEMBRANE—The lining of the nostrils.
SEBACEOUS—Of the nature of suet.
SEDATIVES—Medicines which produce sleep.
SEROUS—Watery.
SERUM—Watery part of the blood.
SOLVENT—That which dissolves.
SPORADIC—Scattered.
SUBMAXILLARY—Beneath the jaw.
SUDAMINA—Small eruptions.
SUPRA-RENAL--Above the kidney.

THORACIC—Relating to the chest.
THYROID—Shaped like a folding door.
TONICS—Medicines which give tone and strength to the body.
TUBERCULAR--Relating to tumors in the structure of an organ.
TUMEFACTION--Swelling.
TURBINATED BONES—Bones of the nose shaped like a top.
TURGESCENCE—Great amount of humors in any part.

VASCULAR--Full of blood-vessels
VENTRICLE—Cavity.
VIRUS—Poison.
VIS A FRONTE—Force from the front
VIS A TERGO—Force of propulsion.
VISCOUS—Sticky.
VOICE-BOX—Larynx.

INDEX.

437

www.ingramcontent.com/pod-product-compliance
Lightning Source LLC
Chambersburg PA
CBHW021345210326
41599CB00011B/756